Olov Sterner

Chemistry, Health
and Environment

 **WILEY-VCH**

Olov Sterner

Chemistry, Health and Environment

WILEY-VCH

Weinheim · New York · Chichester · Brisbane · Singapore · Toronto

Prof. Dr. Olov Sterner
Lund University
Department of Organic Chemistry 2
P. O. Box 124
S-22100 Lund
Sweden

Library of Congress Card No.: applied for

British Library Cataloguing-in-Publication Data:
A catalogue record for this book
is available from the British Library

Die Deutsche Bibliothek – CIP-Einheitsaufnahme
Chemistry, Health and Environment / O. Lund
Weinheim ; New York ; Chichester ; Brisbane ;
Singapore ; Toronto: Wiley-VCH, 1999
ISBN 3-527-30087-2

© WILEY-VCH Verlag GmbH, D-69469 Weinheim (Federal Republic of Germany), 1999

Printed on acid-free and chlorine-free paper

Composition: Text- und Software-Service Manuela Treindl, D-93059 Regensburg
Printing: strauss offsetdruck, D-69509 Mörlenbach
Bookbinding: Großbuchbinderei J. Schäffer, D-67269 Grünstadt
Printed in the Federal Republic of Germany.

To   Fabienne,
     Adeline and
     Marielle

Preface

Modern society makes use of a very large number of chemicals, and although we enjoy and prosper from the various benefits, e.g. the development of safe and efficient medicines, most of us worry about the possibility that all these chemicals somehow are hazardous to our health and to the environment. Our worries are well-grounded. During the last century numerous incidents in which the improper use of chemicals has led to severe damages to both people and environment have occurred. Today we are well aware of that. For example, most of the tumours that afflict humans are caused by chemicals, and, as a result of human activities, the release of enormous amounts of carbon dioxide and other greenhouse gases into the atmosphere pose a serious threat to the climate.

The hazards associated with a chemical do not arise by pure chance, instead they depend on a number of properties of that chemical, for example its solubility and reactivity. These will determine how a chemical may interact with the biochemical components of organisms, e.g. man, and thereby how toxic it is. Although it is difficult to assess chemical hazards theoretically, surprisingly many of the properties that determine the behavior of a chemical can be deduced from a simple visual inspection of the chemical structure. In essence, this is what this book is about. By identifying substructures or functional groups, that we know are associated with toxicity or with environmental effects, we can pay proper attention to a potentially hazardous chemical before it appears as a newspaper headline. In this process, we should be strongly aware of the ways that chemicals are modified by metabolic conversion in organisms and by chemical transformations in the environment, because the products formed will of course have significantly different chemical properties and thereby pose very different hazards.

The intention of the book is to make chemists, chemistry students and anybody working with chemicals, who maybe is or will be responsible for co-workers that handle chemicals, aware of the general relationships between the chemical structure of organic compounds and the effects on health and the environment. Although the basis for the discussion is chemical, the intention is that also those with relatively limited knowledge in chemistry should find the text informative, and the first chapters consequently sum up the basic chemistry needed for the following discussions.

As the author, I sincerely hope that the readers of this book will find it interesting and inspiring. In the event that somebody can make use the knowledge presented here to prevent a chemical accident from taking place, I would be a happy man.

Olov Sterner

Contents

1 Chemicals and society

Life on the planet earth can scientifically be described on a number of different levels. A lake, or a forest, can be defined as an ecosystem which consists of many different species and many individuals of each species. Some organisms, for example mammals, are composed of organ systems, such as the respiratory system, which are made up of organs, composed of tissues and cells. A cell is normally considered to be the smallest unit of life, but cells in turn contain various cell components which take care of important cell functions such as the generation of energy and the construction of proteins. The composition of such cell organelles may appear complex, but today we know in principle how they are made up by biomolecules such as amphipathic compounds, proteins and nucleic acids. The various processes that take place in a cell can in most cases be described in detail on a molecular level, and chemical structures and chemical reactions are eventually responsible for all cellular functions. It has been calculated that the number of chemical reactions that keep a relatively simple unicellular organism like a bacterium alive is approximately a few thousand. The cell of a mammal is of course much larger and more complex, and still not understood in all details, but it is reasonable to assume that the lives of our cells are also the result of a certain number (more than a few thousands but still finite) of chemical reactions. Such chemical reactions comprising life can be affected by many things, and this book will look closer on how various chemicals can disturb them and what consequences that may lead to.

The first Chapter will simply introduce the reader to the background and some general aspects on hazardous chemicals, and how the society responds to them. The key conclusion of Chapter 1 is that the lack of knowledge concerning hazardous chemicals is alarming, but that fundamental understanding of the relationships between chemical hazards and chemical structure/properties will help anybody handling chemicals to protect him or herself as well as the environment.

1.1 Basic problems

There are a number of basic problems that are important to consider when the effects of chemicals on health and environment are discussed, some are evident, e.g. that the number of people on earth is increasing rapidly and that more and more chemicals are used for various purposes, while others are more difficult to define. Consider, for example, the fictitious newspaper article shown in Figure 1.1.

At a first glance one may be concerned by this information, but a chemist would rapidly realize that the chemical DHMO is nothing but ordinary water. However, there are no lies in the article; everything said is in principle true. Thousands of people are killed by water yearly, in

BEWARE OF DIHYDRO –
MONOXIDE   (DHMO)

Recent reports have drawn the attention to a chemical that should be handled with the outmost care. The name of the chemical is dihydromonoxide, abbreviated DHMO, a deceptively tasteless and odourless chemical that yearly kills thousands of men, women and especially children. DHMO is used in enormous quantities by the chemical industry, for instance in nuclear power plants, and its use as an additive during the preparation of so called junk food has been demonstrated. The presence of high concentrations of DHMO in tumours has recently been reported. In addition, DHMO is a main component of acid rain and the large amounts of DHMO in our atmosphere makes a significant contribution to the greenhouse effect. Military sources have revealed that thousands of metric tons of DHMO are distributed through special underground tube systems, to, among others, secret weapon research plants.

Figure 1.1. A newspaper article that so far has not appeared.

drowning accidents, it is certainly used in large amounts by nuclear power plants, it is an ingredient in most "junk foods" and it is a major constituent in tumour cells just as in any other cell. Rain is water and acid rain is mostly water, gaseous water in the atmosphere will reflect heat radiation from the earth just as the more famous greenhouse gas carbon dioxide, and tap water is distributed to most industries, including secret weapon research plants (this fact could have been revealed by any source, including a military source).

Chemicals – compounds.

Chemicals is the general term for all substances that are used in or result from reactions involving changes to atoms and molecules, and comprise anything from ions and atoms to complex molecules. Compounds are molecules composed of two or more atoms linked together by chemical bonds, and all organic chemicals are compounds.

However, what is a joke for chemists that understand what the name DHMO stands for and what properties this chemical has, may cause severe anxiety in a person that has no background in chemistry. If we instead of water chose a less well-known chemical, that even chemists are unfamiliar with, we may imagine a situation when a chemical is described in different, and completely contrary ways, for example as a relative non-hazardous chemical that can be handled safely (by a manufacturer or a person that has worked with the chemical for a longer period) or as a dangerous and hazardous chemical with which all contact should be avoided (by somebody suffering from chemophobia). Few are able to assess information about chemicals critically and react to it in a rational way.

The example in Figure 1.1 puts the finger on several points:

1. That the hazards of a chemical can be described, formally correctly, in several different ways.
2. That chemical hazards cannot be defined in absolute terms.
3. That different hazards cannot be compared.

People perceive hazards in very different ways. Some have chosen to be smokers and/or to consume alcoholic beverages in spite of the fact that they are well aware that such habits will expose them to chemicals that in the long run will increase the risk of acquiring lung cancer and/or damaging the liver. The reason may be that they underestimate the risks, or they think the benefits of smoking and drinking outweigh the risks for damage. While such decisions perhaps can be left to the individual, others are more complicated, difficult, and affect many people and should consequently be based on very solid knowledge. For example: Should we produce electricity by nuclear power plants or by burning fossil fuels? We know that accidents in nuclear power plants will leak radioactivity to the environment, and that is certainly a severe hazard, but on the other hand we strongly suspect that the carbon dioxide added to the atmosphere by burning fossil fuels will increase global warming. Which hazard is more dangerous? Ask two persons and you are likely to get two answers, and there is simply not a straight yes-or-no answer to a question like: Is chemical X dangerous?

We will get back to this question several times in the book, simply because it is so fundamental. Not only do different persons perceive hazards differently, but commercial enterprises (manufacturing and selling chemicals) and government authorities (regulating their use and depending on the taxes they generate) will have their views on chemical hazards that not necessarily are the same as yours or mine. One should also be aware that in some situations there may exist reasons, for example economical or political, to twist the truth about chemical hazards in order to gain some kind of advantage.

Hazards – risks.

A chemical hazard can be defined as the intrinsic properties of a chemical or a situation when chemicals are handled that in particular circumstances could harm humans and/or the environment, and/or damage property. Highly toxic compounds may be considerably less hazardous than a relatively non-toxic chemical, depending on the conditions under which they are used. A risk differs slightly form a hazard in that it also considers the probability of a hazardous chemical to cause the harm or damage that is has the potential to do. However, the two terms are is often used alternately.

1.2 Definition of the sciences involved

The issues that will be discussed in this book are interdisciplinary, even if they mainly will be described from a chemical viewpoint, and we can note that the following major scientific disciplines are involved:

Organic chemistry describes the properties and reactions of organic compounds, which in principle are all compounds that contain carbon, while *inorganic chemistry* deals with chemicals that do not contain carbon (exceptions are for example metal carbonates, carbon dioxide and carbon monoxide, which are considered to be inorganic). Besides carbon, organic compounds can contain virtually any other element, but the absolute majority of them are only composed of carbon, hydrogen, oxygen and nitrogen. Almost all chemicals that take part in the chemical processes that keep organisms alive are organic, and the study of such processes is carried out within the discipline *biochemistry*. Most compounds that are known to have toxic effects are organic (among the toxic inorganic chemicals we will mainly get in contact with toxic metals in this book), and it is obvious that the carbon atoms of organic compounds play a central role for their structure and properties. As an atom with four valence electrons, the carbon atom has an urge to make chemical bonds to four other atoms in order to achieve the stabilizing electronic configuration of the noble gases. Carbon readily make chemical bonds to a range of other atoms, including carbon itself, and compounds containing only a few carbon atoms can form a large number of different molecules. Figure 1.2 indicate how almost infinitely the structures of organic compounds composed of a certain number of atoms can be varied, simply by changing the place of the bonds between the atoms. Even without considering the stereochemistry of the substituents on the rings and the double bonds, there are 44 compounds in Figure 2.1 all having the elemental composition $C_5H_{10}O$ and all having different chemical properties.

The carbon atoms will normally provide the backbone of an organic molecule, while the heteroatoms will constitute functional groups that give the compound its characteristic properties (discussed in Chapters 2 and 3). Adding to the usefulness of organic compounds, the energy of carbon–carbon and carbon–hydrogen bonds is high which makes organic compounds suitable as energy sources for, e.g. the biochemical reactions (discussed in Chapters 3, 4 and 10).

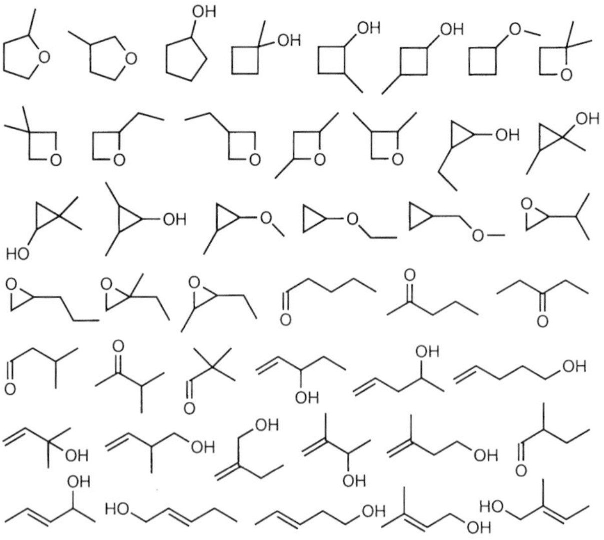

Figure 1.2. Examples of organic compounds having the composition $C_5H_{10}O$.

Although we will not get involved in any intricate biological problems, we will discuss some aspects of *toxicology* and *ecotoxicology*. Toxicology in general deals with the study of poisons and their effects on single organisms, especially on humans, while ecotoxicology is concerned with the effects on ecosystems. The difference should be noted; while an effect of one chemical for one or a few organisms may be dramatic (e.g. death) the effect on the ecosystem could well be negligible, and vice versa. The term *environmental toxicology* usually indicates that the interest is focused on what effects chemicals have on the environment and how these effects affect humans, for example the contamination by pesticides of species used by humans as food; and environmental toxicologists are not primarily interested in the effects on ecosystems. The main focus in this text will be on the chemical properties that are associated with toxic effects, directly or indirectly, on humans, for which the term *chemical toxicology* may be the most appropriate.

1.3 Trends and developments over the years

Anybody that has been working where chemicals are handled for a long time, has noticed that chemicals are treated differently today compared to, say 10 or 20 years ago. Attitudes have changed, procedures have been improved and regulations are more strict. Some decades ago, chemicals were handled in ways that could lead to health problems, but this is no longer considered acceptable. While chemical waste in the older days was simply discarded into the nearest river or buried, it is today taken care of and destroyed by specially developed processes. New bioactive chemicals were invented that, for example, could be used to treat illnesses and control pests, and that initially were considered to be miraculously suitable for their task, but later turned out to be hazardous. Examples are the insecticide DDT (DichloroDiphenylTrichloroethane) and the biocide 2,4-dinitrophenol that later was used in slimming cures (see Figure 1.3). DDT, an old compound that was rediscovered in the 1930s when it was found to be a very efficient insecticide, was used extensively during World War II and the following decade. It was, in those days, important for the successful control of illnesses (typhus and malaria in particular) that are spread by insects, and saved the lives of a great number of people. The effects of DDT were truly remarkable, as the incidence of malaria in a small country like Sri Lanka shows: 2,800,000 cases of malaria were reported in 1946, before the days of DDT, but only 110 cases in 1961 after DDT had been

DDT.

The compound DDT was first reported in 1874, but soon forgotten because it apparently had no interesting properties. The Swiss chemist Paul Müller discovered the insecticidal activity of DDT, and was awarded the Nobel prize in 1948 for his important and revolutionising finding. DDT is a nerve poison that affect the coordination of movements of insects and eventually paralyses them, but resistance against the compound in the form of increasing concentrations of DDT-degrading enzymes decreased the efficiency of the insecticide. Due to the amounts used especially in the 1950s and 1960s and to its chemical stability, DDT is present all over the globe.

used for several years, and then again 2,500,000 in 1969 when DDT had been banned for 4 years. However, something that was not considered was that DDT is an extremely lipophilic and stable compound, which is not degraded in our environment but instead efficiently extracted by organisms. Some species (e.g. birds of prey) in the top of food chains eventually accumulated so much DDT that their reproduction was threatened, and had the use of DDT continued at the same scale other species including man were at danger. The "magic bullet" against insects became an environmental nightmare, and we shall return to DDT and similar compounds in Chapter 12.

2,4-Dinitrophenol.

The dinitrophenols were among the first synthetic pesticides, manufactured already in the later part of the 19th century. Besides the parent compound, 6-methyl-2,4-dinitrophenol, originally patented 1892 by the German company Bayer AG under the name Antinonnin (in German "nonne" is an insect), and 6-butyl-2,4-dinitrophenol (sold under the name dinoseb) were important contact insecticides. Their toxic action is specific, but they are unselective as most pesticides of the early days and kill pests as well as man by blocking the cell respiration.

2,4-Dinitrophenol, an old pesticide, was launched in the 1930s as an efficient slimming agent for people that wanted to lose weight. It is indeed efficient, it blocks the production of energy which normally is associated with the degradation of nutrients and stimulates the body to consume more nutrient (e.g. fat) than it needs. The only problem was that a slight overdosing could be fatal, and many accidents took place. It was therefore eventually banned as a slimming agent, and in the 1970s it was even considered to be too toxic even for use as a pesticide and banned for this purpose in many countries.

The knowledge about the hazards associated with the use of chemicals has consequently increased over the years, often what we may call "the hard way", by making mistakes and learning from them. This has led to an extensive regulation of the use of chemicals, which will be the subject of the following Section, and the purpose of such regulations can be said to protect man and environment from the hazards of chemicals. Today we may consider the misjudgements concerning hazardous chemicals made in the 20th century as somewhat foolish, caused by a lack of knowledge, but it would be very stupid indeed to consider the level of knowledge we have today to be complete. Nor should we count on the legislation to protect us from chemical hazards completely, even if everyone follows the rules. Instead we can be convinced that 20–30 years from now experts will look back on our days and comment on the senseless use of chemicals X and Y and how this led to serious problems that easily could have been avoided "if only they knew what we know now". And they in turn will be the victims of the same afterwisdom some additional 20–30 years later ...

The conclusion, which is valid most of the time, is: Things have improved, but they are not perfect.

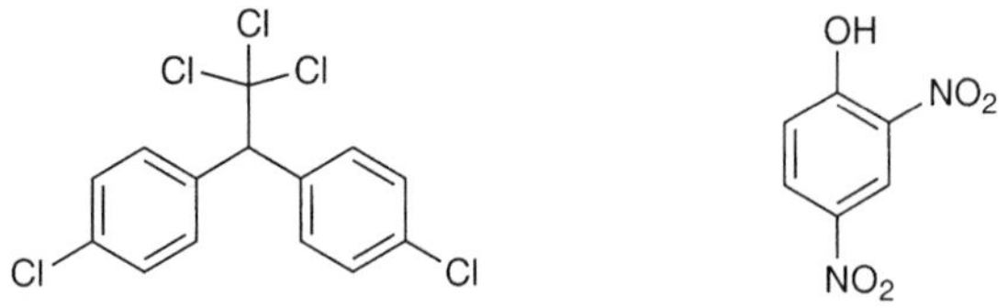

Figure 1.3. The structures of DDT and 2,4-dinitrophenol.

1.4 Legislation

The legislation that regulates the use of chemicals aims to avoid all avoidable risks to both man and environment associated with chemicals, just as traffic rules are intended to protect people from traffic accidents. However, as we all are strongly aware of, society as we know it will not function without chemicals or without traffic, and even if the intentions with the legislation are the best it is simply not possible to decrease the risks to zero, not in traffic and not when it comes to chemicals. So, at the same time as the legislation protects against chemical hazards it must also provide the means to use economically important chemicals, of which some are known to be for example carcinogenic but in spite of their toxicity are allowed to be used in workplaces. Nevertheless, modern legislation has improved the working conditions in the industry immensely, and in some cases also managed to decrease the pollution of the environment a lot. Any activity with chemicals, for example a business that involves the handling of chemicals, will be regulated by a number of laws and rules, and anybody that intends to get into such an activity has to get aquatinted with quite a lot of law-text. This makes knowledge about the legislation almost as important as knowledge about the hazards, and it should be a major issue to discuss in a text like this, but for two reasons this will not be the case. Firstly, the legislation differs from country to country, even if the chemical hazards and the people that should be protected are identical, and it is not possible to discuss "average" laws. Secondly, in contrast to the chemical hazards that the legislation should protect us against, the laws change frequently, due to new discoveries, to an improved or deteriorating financial situation, etc. and texts about laws and regulations therefore grow old quite quickly.

The legislation may in detail regulate which chemicals that are permitted to use, making sure that only chemicals that have been proven to be sufficiently safe are approved. However, that kind of legislation is unusual, and normally only applied with certain compound classes (e.g. pharmaceutical, pesticides and food additives). In most countries it regulates the concentrations of the most frequently used chemicals that are allowed in the respiration air in workplaces, and how chemical waste from different kinds of activities should be treated. In many countries there are also rule that make sure that the labels on containers of chemicals has information about any hazard associated with it, that written and more comprehensive information is available to those that purchase and use a chemical, and that workplaces where chemicals are handled are equipped with safety devices such as gas-masks and fire-extinguishers. The competence (education and training) of persons handling chemicals is not regulated in any detail, but this is of course a critical point as anybody that has an understanding and knowledge about a hazard will be able react in a adequate way. It may be argued that competence is even better than regulations in some instances, although a combination may prove best in the long run. The example shown in Figure 1.4 illustrates this.

Unlike the example in Figure 1.1, this has been taken from real life and is not made up. Anybody that gets a bottle of chemical X in his or her hand will be concerned when reading the warnings, obviously a chemical to treat with caution and respect. However, chemical X is seasand, a product that is used in the chemical laboratory for various purposes but not associated with any dramatic hazards. Instead, many of us dream about spending time on a sunny beach full of seasand, and the idea that the sand should pose a hazard is for most people far-fetched. It is true

CAUTION
Chemical X may be harmful if inhaled. Chemical X may cause
irritation. Chemical X may cause a rapidly-developing pulmonary
insufficiency, laboured breathing and cyanosis followed by cor
pulmonale and short survival time. Death may result from cardiac
failure or destruction of lung tissue with resulting anoxia. Skin
contact may cause irritation and dermatitis. Eye contact may cause
redness, irritation and conjunctivis.

TARGET ORGANS AFFECTED
Eyes, skin and mucous membranes.

FIRST AID — INHALATION
Remove from exposure area to fresh air immediately. Keep person
warm and at rest. Get medical attention immediately.

FIRST AID — SKIN
Remove contaminated clothing and shoes immediately. Wash with
soap and large amounts of water. Get medical attention.

FIRST AID — EYES
Wash eyes immediately with large amounts of water 15–20
minutes. Get medical attention.

Figure 1.4. Information about health hazards on the container of chemical X. (*cor pulmonale* is a heart disease resulting from disease of the lungs or pulmonary circulation.)

that inhaling seasand would give problems with the respiration, and that sand in the eyes can be rather painful, but this really comes as no surprise. This is instead an adoption of the producer, the chemical company, to a situation where they may be accused of not warning the consumer of any imaginable hazard, never mind how unlikely it is. For many, the initial feeling of respect for chemical X will be exchanged for confusion, which may lead to the general opinion that information about health hazards exaggerate the risks. In such case, the law that decrees this information is of little value. Such persons may instead chose to rely on their own judgement, based on knowledge about chemical hazards and experience of handling chemicals.

1.5 Knowledge about chemical hazards

Knowledge about the effects of chemicals on man and environment is crucial, as is a good sense to use that knowledge in a sensible way. Knowledge can be acquired in many ways, but if the effects of a chemical are described in the literature this is the first place to search. Today, huge databases where information about chemicals can be sought are available not only in chemical libraries but also via the internet. However, surprisingly few of the approximately 50–100,000 chemicals used today have been studied from a toxicological and ecotoxicological viewpoint, and it is only for a few percent that we can say that we complete knowledge (which includes long-term toxicity, e.g. carcinogenicity) about their effects on man. In general, the chemicals that have been most thoroughly studied are those that the legislation requires to be tested before they are approved. A pharmaceutical agent for example must be shown to be relatively safe before it may be used. Bulk chemicals on the other hand, although used in large quantities in the chemical industry, are not subject to the such regulations and little is known about the toxicity of most of them. In addition, they may contain considerable amounts of impurities (several %) which are not even declared on the label of contents. Many bulk chemicals have been used for decades and over the years we have had some experiences with them that indicate to us in what situations they may be hazardous, but even if a chemical from experience appears to be safe this is no proof that it really is so. Take carcinogenic chemicals as an example, all chemicals that have been proven to be carcinogenic to man have been disclosed in epidemiological investigations where the exposure to a chemical (at a workplace for example) is correlated with the incidence of cancer 20–30 years later, compared to a non-exposed control group. However, it is quite difficult to demonstrate such correlations, it has only been made with 30–40 chemicals, because it requires that the chemical is a potent carcinogen (e.g. bischloromethyl ether, used previously for the manufacture of ion exchange polymers), that the group exposed is huge (e.g. cigarette smokers vs non-smokers) or that the tumour formed is extremely unusual (e.g. vinyl chloride, which is a weak carcinogen but causes a very unusual form of liver cancer). Carcinogenic chemicals that are not potent, are not used by very large or isolated groups and do not cause unusual cancers will not be detected by epidemiological investigations, because their effects will not be noticeable in the number of tumours formed spontaneously.

An interesting example of a chemical that have been used for a long time (more than 100 years) and that we have a lot of experience of is acetylsalicylic acid (see Figure 1.5). It is used as an anti-inflammatory agent, as an antipyretic and an analgesic, and recently its ability to inhibit the ability of the blood to coagulate has been taken advantage of for the treatment of thrombosis.

It is not a non-toxic compound, 10–15 g may damage the kidneys of an adult, but the long-term effects (e.g. carcinogenicity and teratogenicity) of low doses of acetylsalicylic acid are based on our long experience with human exposure and known to be negligible. However, in animal experiments it has been noted that acetylsalicylic acid has a weak teratogenic effect, and it is doubtful if it would be approved if it was invented today.

The fact that we use so many chemicals without really knowing much about their effects on man and environment may come as a

Acetylsalicylic acid.

Acetylsalicylic acid is a derivative of salicylic acid, and so is salicin which is the active constituent of the bark of willow tree (*Salix alba*). In the later part of the 18th century reverend Edward Stone, according to the tale inspired by a local woman and by the bitterness of the bark that made Stone believe that it would have the same effects as Peruvian bark (containing quinine), treated a number of persons suffering from fevers with the bark. The treatment was successful, and the salicylic acid derivatives obtained from willow bark as well as from the plants meadowsweet and wintergreen became important to reduce fever and pain and to relieve gout and arthritis. The acetylated derivative was prepared in the end of the 19th century by the German company Bayer AG and sold under the name Aspirin.

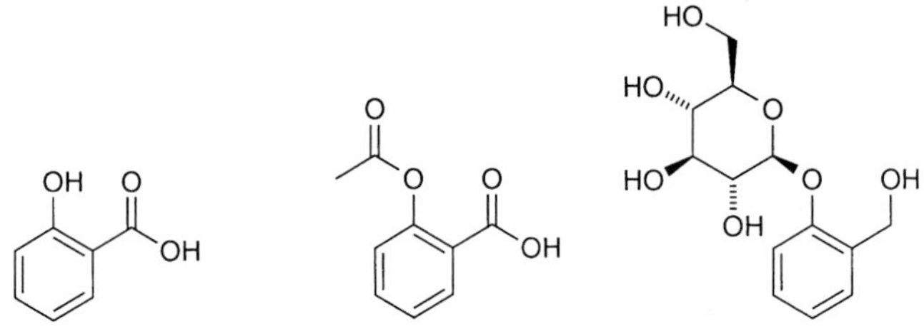

Salicylic acid   Acetylsalicylic acid   Salicin

Figure 1.5. The structures of salicylic acid and two of its derivatives.

shock. The reason is of course economic, new knowledge is unfortunately not free, and the costs to assay the toxicity and ecotoxicity of one chemical can climb to 1 million US$. Even if the ambition today is to raise the standards and demand more knowledge about all new chemicals that are added to the market, there are no possibilities to thoroughly investigate the toxicity of all the old chemicals already in use. It may sound reasonable to ban all chemicals until we know about their effects and can regulate their use, but such an action would overturn society. Instead, we will have to learn to live with this situation for many years to come. If knowledge about the hazards of a chemical cannot be found in the literature, and if it is considered to be too expensive to carry out the necessary biological tests by oneself, the remaining alternative is to make intelligent guesses based on known relationships between chemical structure/chemical properties and toxicity/ecotoxicity for similar compounds. This is essentially what this text aims to do, to show the reader that the effects on man and environment that we want to avoid are caused by chemical properties that we to a large extent can understand by analysing the chemical structure. In the future it is believed that such intelligent guesses, made by computers that have been fed all available data and taught to analyse the data in a relevant way, will be used routinely to sort out old chemicals that we know too little about but which have chemical properties that the computer has associated with toxicity, as well as chemicals in the pipe-lines of development but classified as potentially hazardous. However, until the day when computers take over the decision-making, it is good advice to keep an ability to make judgements by oneself.

1.6 Acceptable chemical risks

At the bottom line we will find that all chemicals have some hazards associated with them, even if the differences between those regarded as "safe" and "hazardous" are enormous. Toxicity and ecotoxicity do not come in black and white but as a grey scale, and apart from deciding if and how a chemical is hazardous we need to know something about its potency. This can be done by relating the toxic response of a chemical to the dose, and an example is shown in Figure 1.6 where the carcinogenicity of two chemicals is plotted against the dose. There is an obvious relation between dose and response, which can be approximated with a straight line (obtained by statistical analysis of the data or simply by putting a ruler along the xs and os). The slope of that line would be a measure for the potency of the compound. The slope of the line corresponding to the xs is greater than that corresponding to the os and it is possible to compare the carcinogenicity of the two chemicals.

in vitro
In the glass, experiments carried out with cells or tissues in test tubes or equivalents.
in vivo
In life, experiments carried out with living organisms.
vide supra – see above
vide infra – see below

The next difficulty is to translate our knowledge about the toxicity of a chemical, which in almost all cases refers to *in vitro* assays and animal experiments, to the species that we are interested in: Man. It is also not evident if it is relevant to use data from animal experiments obtained with very high doses, compared with those that humans would be exposed to. These problems will be discussed in more detail in Chapter 5, but in general the knowledge that we have acquired is not enough to make an absolutely certain assessment of the toxicity to man and approximations have to be made. Finally, knowing about the ifs, the hows and the how muchs, we have to decide which risks are acceptable in different activities. Obviously we have to accept all kinds of chemical hazards, even genotoxicity, but we can still regulate for example which concentrations of a carcinogenic chemical in the respiration air at a place of work that we consider acceptable. Some chemicals, e.g. pesticides and preservatives, are considered by many to be superfluous as they in general are toxic to at least some organisms, but they also save human lives by improving crops, killing disease carriers (e.g. DDT, *vide supra*), and avoiding food poisoning by bacteria. Such decisions, to say what risks we are ready to accept in order to benefit from the various uses of chemicals are as already noted extremely difficult, as individuals as well as authorities and companies value risks and benefits associated with chemicals very differently. In addition to the

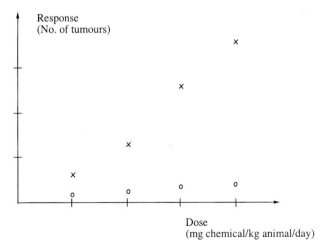

Figure 1.6. Dose-response relationships for two carcinogenic chemicals.

scientific and humanistic dimensions, it also has a political aspect, as the safety of people and the protection of the environment have to be considered together with economic factors. Looking back, there are examples of chemicals that have received exaggerated attention (because of a lack of knowledge, suspicion, fear, and/or political opportunism) while others have not received the attention they merit. One of the major threats today is that the rapidly increasing concentrations of carbon dioxide in the atmosphere, formed by combustion of mineral fuels, will affect the climate and thereby the weather systems of the earth. Little is today known about this so called greenhouse effect (it will be further discussed in Chapter 12), but if it exist and the additional carbon dioxide does change the climate in a dramatic way, we will be very sorry that we did not take action sooner.

2 Organic compounds

Everything in this world, from rocks and rivers to plants and animals, is composed of atoms and molecules, and in principle it is even possible to describe organisms at a molecular level. We are not there yet, but the rapidly increasing knowledge of how the processes of life are regulated by complex molecular systems conveying chemical signals inside a cell and between cells, has revolutionized not only the biosciences but also our attitude to life on earth. As will be discussed in Chapter 4, the biochemical processes that constitute life are composed of a large but nevertheless finite number of chemical reactions. The well-being of any organism depends on how these chemical reactions can be performed, i.e. that the environment where they take place has a suitable temperature and contains the chemicals needed in the right amounts and forms. It also depends on the absence (or the presence in insignificant concentrations) of chemicals that may disturb the biochemical processes, which can be said to be potentially harmful and able to give rise to toxic effects. If we want to understand why most chemicals actually are toxic and harmful, we need, at least to some extent, to understand how the "chemical machinery" of organisms works and how toxic/harmful chemicals interact on a molecular level. Obviously we need to employ chemistry for our discussions. For the reader who has forgotten most of the chemistry courses taken years ago, or wants to rehearse the most important points in this context, Chapter 2 may be useful. Chemists may find this Chapter elementary, and are recommended to proceed to the following Chapters.

2.1 From atoms to molecules

2.1.1 Atoms

Although the periodic system lists over 100 atoms, only a few are present in any significant amounts in the organic chemicals most frequently encountered. As a matter of definition all organic compounds contain carbon (C), and hydrogen (H) and oxygen (O) are also present in most organic compounds. In addition, nitrogen (N), phosphorus (P) and sulphur (S) are frequently present in organic compounds, especially in biomolecules. The halogens, fluorine (F), chlorine (Cl), bromine (Br) and iodine (I), are often associated with compounds hazardous to the environment (e.g. DDT), but are actually not unusual in natural products produced for example by marine organisms. Many halogenated compounds are indeed toxic and/or difficult to degrade in nature, and their use in industry has decreased lately, but they still have important uses and are consumed in large amounts and we will return to this class of compound. Apart from chlorine, mainly as chloride ions, our biochemistry does not contain halogenated compounds. A number of other elements are also present in organisms, e.g. sodium (Na), magnesium (Mg), potassium (K) and calcium (Ca), without which any organism will die, and this will be discussed in detail in Chapter 4. Among the metals, some (e.g. iron [Fe]) are needed while others (e.g. mercury [Hg]) are strongly toxic, and the effects of metals, as free ions as well as organic forms, will be discussed in Chapter 12.

2.1.2 Covalent bonds

Atoms form chemical bonds to other atoms when this will lower their energy and thereby increase the stability of the system. For example, two hydrogen atoms and one oxygen atom will readily form a water molecule (H_2O) with two hydrogen–oxygen bonds, because the energy of this three-atom system is lower in the form of water than as three separate atoms. The basic principle for bonding between atoms is that the energy is lowest (and the system most stable) when the electrons are together in pairs and when the atoms involved have the same electron configuration as the stable and unreactive noble gases to the far right in the periodic system: Helium (He) with 2 electrons, neon (Ne) with 8 electrons and argon (Ar) also with 8 electrons in the outermost electron shell (the valence electrons). The second row elements (e.g. sodium, carbon, nitrogen, oxygen, fluorine and neon) all have a filled inner (first) electron shell (with 2 electrons, as helium) that does not participate in their chemical bonds, but in addition they have a second shell (which is their outermost shell, or valence shell). The second shell is filled when it contains 8 electrons, as in neon. In this shell, as shown in Figure 2.1, sodium has 1 electron, carbon 4, nitrogen 5, oxygen 6, fluorine 7 and neon 8 electrons (filled). Oxygen needs 2 electrons to fill its valence shell and become like neon, and by sharing electrons with for example two hydrogens this will be accomplished. Third row elements have two filled inner shells (containing 2 and 8 electrons), and make bonds with their third shell (the valence shell) which is partly filled when it contains 8 electrons, as argon).

Shell 1	**H** (1)							**He** (2)
Shell 2	**Li** (1)	**Be** (2)	**B** (3)	**C** (4)	**N** (5)	**O** (6)	**F** (7)	**Ne** (8)
Shell 3	**Na** (1)	**Mg** (2)	-	-	**P** (5)	**S** (6)	**Cl** (7)	**Ar** (8)

Figure 2.1. Some selected elements and the number of electrons in their valence shells.

Hydrogen and the halogens can obtain the electronic configuration of the noble gases by adding 1 electron, oxygen and sulphur need 2 electrons, nitrogen and phosphorus 3 while carbon needs 4 electrons. On the other hand, sodium and potassium need to lose 1 electron while magnesium and calcium have to lose 2. One way to accomplish this is simply to accept or donate electrons, and sodium, potassium, magnesium, calcium and the halogens have a strong tendency to form the corresponding ions (e.g. Na^+, Mg^{2+}, Cl^- and Br^-). Ions with opposite charges form ionic bonds, which will be discussed in Chapter 3. Another way is by sharing electrons with other atoms, as the two hydrogen atoms and the oxygen atom in H_2O. By combining the electrons of the two hydrogen atoms with 2 electrons from the oxygen atom, H_2O with two hydrogen–oxygen bonds (each consisting of two electrons) is formed (see Figure 2.2). The 2 electrons in the bonds are shared by both atoms, and consequently both hydrogens in water have two electrons (as He!) in their valence shells while oxygen has 8 (as Ne!). This bond between atoms is called a covalent bond.

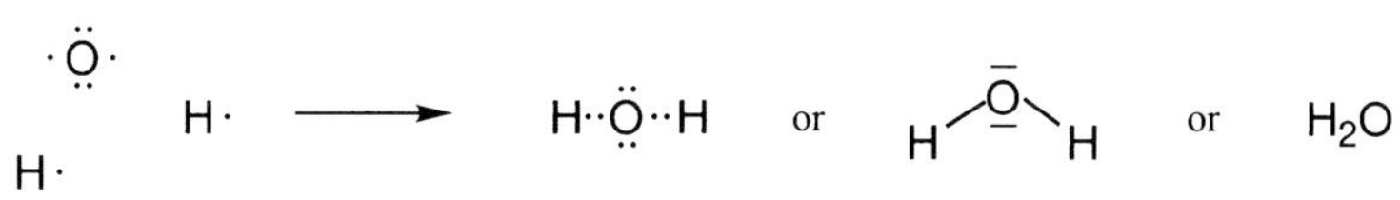

Figure 2.2. Water is more stable than two hydrogen and one oxygen atoms by themselves.

Covalent bonds are normally drawn as a line, which represents the 2 electrons shared by the atoms bonded. In organic molecules, hydrogen is able to form one covalent bond, and so are in principle also the halogens. Oxygen forms two covalent bonds, nitrogen three and carbon four. Third row elements behave similar to the corresponding second row elements, sulphur often forms two covalent while phosphorus may form three, but the valence shell of these elements is bigger and they may also form additional bonds. The remaining electrons in the valence shell of for instance oxygen and nitrogen, that do not participate in covalent bonds, are important and should not be forgotten. They may be indicated by dots or lines (a dot for an electron and a line for a pair of electrons) close to the atom in question, as in Figure 2.2, although as their presence is obvious they are most often omitted. These unshared electron pairs can be seen as local sources of electrons, they take part in chemical reactions and they are important for intermolecular bonds (attractive forces between molecules), as will be discussed in the following Chapters.

The 4 electrons in the valence shell of a carbon atom and that may participate in four covalent bonds to that carbon, are not formally equivalent but originate from "*s*" and "*p*" orbitals. Orbitals are subshells that can hold 2 electrons each, they can be described as regions of space where electrons spend most of their time, and "*s*" and "*p*" orbitals have different shape (see Figure 2.3). The first shell contains one "*s*" orbital (filled with 2 electrons), the second one "*s*" and three "*p*" orbitals (filled with 8 electrons), while the third contains one "*s*", three "*p*" and in addition also "*d*" orbitals (a more comprehensive description of orbitals and the electronic configuration of atoms will be found in any textbook on general chemistry). However, in a molecule as methane (CH_4), all four carbon–hydrogen bonds are completely equivalent, and this is due to a phenomenon called hybridization. Thanks to hybridization, the carbon atom can mix its four orbitals (one *s* and three *p*) and create four new hybride orbitals (called *sp*³ orbitals) that are identical. The four *sp*³ orbitals are directed in space towards the corners of a tetraeder (with the carbon atom in the centre) as shown in Figure 2.3, whereby the orbitals containing electrons are the furthest possible apart. The *sp*³ orbitals are used for all bonds in saturated molecules, for example in methane as shown in Figure 2.3. However, by limiting the hybridization to one *s* and only two *p* orbitals, leaving the third *p* orbital as it is, three equivalent *sp*² orbitals are formed.

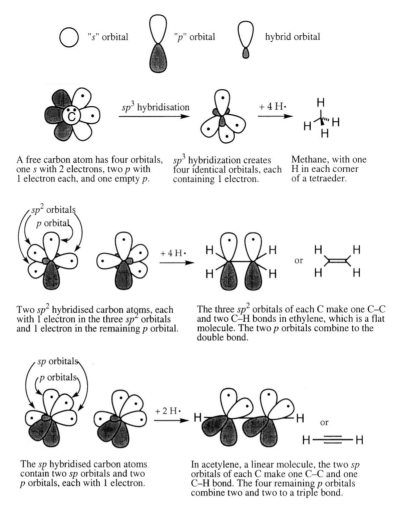

A free carbon atom has four orbitals, one *s* with 2 electrons, two *p* with 1 electron each, and one empty *p*.

sp^3 hybridization creates four identical orbitals, each containing 1 electron.

Methane, with one H in each corner of a tetraeder.

Two sp^2 hybridised carbon atoms, each with 1 electron in the three sp^2 orbitals and 1 electron in the remaining *p* orbital.

The three sp^2 orbitals of each C make one C–C and two C–H bonds in ethylene, which is a flat molecule. The two *p* orbitals combine to the double bond.

The *sp* hybridised carbon atoms contain two *sp* orbitals and two *p* orbitals, each with 1 electron.

In acetylene, a linear molecule, the two *sp* orbitals of each C make one C–C and one C–H bond. The four remaining *p* orbitals combine two and two to a triple bond.

Figure 2.3. The ways the binding orbitals of carbon can hybridize.

This takes place when carbon is part of a double bond, as in ethylene, and overlap between the remaining *p* orbitals (one on each carbon) forms the actual double bond (see Figure 2.3).

A third possibility for a carbon atom is to use one *s* and one *p* orbital to form two equivalent *sp* orbitals, while two *p* orbitals remain unchanged. This is used in triple bonds, e.g. in acetylene, and the second and third bond between the carbon atoms are formed by overlap between the *p* orbitals as shown in Figure 2.3. Other atoms may also form double (e.g. nitrogen and oxygen) and triple (e.g. nitrogen) bonds, and this follows the same principle. Nitrogen for example, may be sp^3 hydridized as shown in Figure 2.4, and have three sp^3 orbitals each contain-

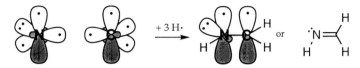

In a sp^3 hybridized N, the three orbitals containing a single electron can make N–H (or N–C, N–O, N–N, etc) bonds as in ammonia, while the fourth sp^3 orbital contains the unshared electron pair.

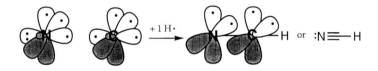

sp^2 hybridised N take part in double bonds, for example in imines containing a C=N bond. The sp^2 orbital containing 2 electrons remain as an unshared electron pair.

N may also take part in triple bonds, for example in hydrogen cyanide. The two sp orbitals contain 1 and 2 electrons, respectively.

Figure 2.4. Also other atoms, e.g. nitrogen, may profit from the advantages of hybridization. Ammonia and hydrogen cyanide are indeed not organic compounds, but serve as good examples nevertheless.

ing 1 electron available for bonding while the fourth is filled with the remaining 2 electrons and appears as an unshared electron pair. If nitrogen only uses one s and two p orbitals (to form three sp^2 orbitals), the remaining p orbital may be used in a double bond, with for example carbon to form an imine. A sp hybridized nitrogen, having one sp orbital with 1 electron that can be used for a chemical bond and another sp orbital with 2 electrons as an unshared electron pair, can use its two remaining p orbitals to make a triple bond with carbon (forming hydrogen cyanide or a nitrile).

The bonds made up of sp^3, sp^2 and sp orbitals are called sigma (σ) bonds, and are characterized by their relative strength and that the electrons are more localized to the molecular orbitals that result from the combination of the atomic orbitals. The part of double and triple bonds that the interaction of adjacent p orbitals is responsible for is called a pi (π) bond. π bonds are weaker than σ bonds and the electrons in the overlapping p orbitals are more mobile, especially in conjugated systems when there are several p orbitals in a row. To some extent they can be compared with unshared electron pairs, and they will influence the reactivity of organic compounds significantly. We shall return to this, and especially conjugated unsaturated systems in organic compounds have properties that make compounds potentially toxic.

2.2 Classes of organic compounds

The remainder of this Chapter will be spent on brief descriptions of the major classes of organic compounds and chemical functionalities. The main purpose is to make the reader familiar with the structures and some basic features, although most chemical properties that will determine the toxicity of chemicals will be discussed in Chapter 3.

2.2.1 Hydrocarbons

As the name implies, hydrocarbons are made up of carbon and hydrogen, and nothing more. Hydrocarbons form the base of all organic compounds, which can be said to be derivatives of hydrocarbons or combinations of hydrocarbons. The major source for hydrocarbons is of course petroleum, and many hydrocarbons are consequently readily available and relatively cheap. In general, they are volatile and insoluble in water, and primarily hazardous because they are flammable and their vapours form explosive mixtures with air. Crude oil is a serious threat to the environment mainly because it is handled in such enormous quantities, both during the drilling for oil and during its transportation (intentional as well as accidental discharges). Traditionally the hydrocarbons have been divided into aliphatic and aromatic hydrocarbons, but today it is more convenient to divide them into the subgroups saturated, unsaturated and aromatic.

Nomenclature of alkanes.	
No of C:	Name:
1	Methane
2	Ethane
3	Propane
4	Butane
5	Pentane
6	Hexane
7	Heptane
8	Octane
9	Nonane
10	Decane
11	Undecane
12	Dodecane
13	Tridecane
14	Tetradecane
15	Pentadecane
16	Hexadecane
17	Heptadecane
18	Octadecane
19	Nonadecane
20	Eicosane

2.2.1.1 Saturated hydrocarbons

The saturated hydrocarbons (alkanes and cycloalkanes) contain only single bonds between the carbon atoms, and the absence of functional groups and π-bonds make this group of organic compounds both unreactive and water insoluble. Due to the fact that they are readily available, very cheap and have useful properties, saturated hydrocarbons are used in extreme amounts as for example solvents and fuels. Most people come in contact with saturated hydrocarbons frequently, at the petrol station, at home in houses heated by oil or gas, at work, etc., and besides the risk for fires and explosions the contacts with hydrocarbons are normally undramatic. However, the exposure to larger amounts will be toxic, and one of the most sensitive organs is the central nervous system (as for example sniffers of butane gas will experience). Figure 2.5 shows some of the

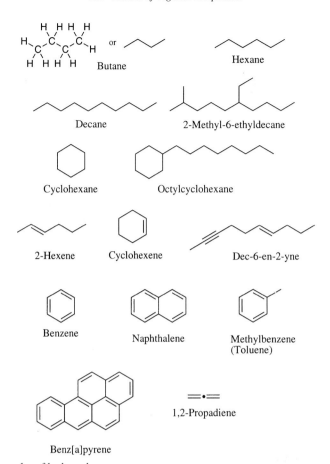

Figure 2.5. Examples of hydrocarbons.

most common saturated hydrocarbons, the smallest (C_1–C_4) are components of natural gas, the slightly bigger (C_6–C_{10}) of petrol, while diesel fuel and lubricating oils are even bigger (C_{12}–C_{14} and C_{18}–C_{20}, respectively). Cyclic saturated hydrocarbons are given the prefix *cyclo-*. Note that while the names of saturated hydrocarbons end with *-ane* (because they are alk*ane*s), saturated hydrocarbons as substituents in other molecules end with *-yl* (alk*yl* groups). To simplify the drawing of chemical structures, the hydrogens bound to carbons are often omitted as shown for butane.

2.2.1.2 Unsaturated hydrocarbons

Besides carbon–hydrogen and (possibly) carbon–carbon single bonds, unsaturated hydro-carbons contain one or several carbon–carbon double (alkenes) or triple (alkynes) bonds. They have similar chemical properties as the saturated hydrocarbons, but are slightly more prone to take part in chemical reactions due to the presence of π-bonds. The most simple unsaturated hydrocarbons, ethene (note the suffix *-ene* in alk*ene*s) and acetylene have already been discussed (see Figure 2.3). The name acetylene is what is called a trivial name, the systematic name of this compound is ethyne (with the suffix *-yne* for an alk*yne*) but both names can be and are used. Chemicals that have been used in significant quantities for a long time or for some reason are considered especially interesting (e.g. natural products) are often given trivial names, and commercially important chemicals may even have several names. The advantage is that, in general, trivial names are easier to use compared to systematic names (that can be extremely complicated but always are unambiguous for those that know the rules). (Note that the use of trivial names may pose a problem for anybody that wants to find information about a compound in the literature, if it is listed under another name!)

This fungal antibiotic has both a trivial name (isovelleral) and a systematic name [(+)-1*S*, 2*R*, 7*R*-2,9,9-tri-methyltricyclo[5.3.0.0^{2,4}]-dec-5-en-4,5-dicarbaldehyde].

If an unsaturated hydrocarbon contains more than one unsaturation and these are situated one after the other, it is called a conjugated unsaturated hydrocarbon which has some special properties. A special case is the allene functionality, having two adjacent C=C bonds emerging from the same carbon atom. Allenes, e.g. 1,2-propadiene, are exclusive and reactive compounds. Some examples of unsaturated hydrocarbons are shown in the lower part of Figure 2.5.

2.2.1.3 Aromatic hydrocarbons

The most important representative of the aromatic hydrocarbons is the simplest, benzene, which in principle is a cyclic and conjugated unsaturated hydrocarbon. However, in cyclic compounds consisting of atoms having an unused *p* orbital, unique properties are observed if the number of *p* electrons (electrons in the *p* orbitals) in the ring is 2, 6, 10, 14, etc., and such compounds are called aromatic. This is the case for benzene, with 6 *p* electrons, benzene is for example more stable and reacts differently compared to for example the non-cyclic but otherwise similar compound 1,3,5-hexatriene. When a benzene ring is a substituent it is called a phenyl group, and an aromatic substituent in general is called an aryl group. Although benzene is an excellent organic solvent and has been used in a number of chemical processes, it is carcinogenic and its use has been restricted. Other aromatic solvents that are less toxic, e.g. toluene, have replaced it. The polyaromatic hydrocarbons (PAHs, with three or more aromatic rings joined together) are present in high concentrations in rest products like coal-tar and mineral-tar, and are formed as byproducts during combustion of any organic matter. Several PAHs, e.g. benz[a]pyrene, are potent carcinogens, and their formation and concentration in for example city air is constantly monitored in many countries.

2.2.2 Chemical functionalities, first oxidation level

Saturated hydrocarbons are oxidized either by the introduction of unsaturations or the introduction of heteroatoms, e.g. nitrogen, oxygen, sulphur or a halogen. Carbon–carbon double and triple bonds can be viewed as oxidized functionalities (as in principle also the ring in a cycloalkane is), and it is obvious that an alkyne is more oxidized than an alkene. Consequently one can talk about different oxidation levels of an organic compound, and a C=C bond is oxidized to the 1st level while the triple bond of an alkyne is at the 2nd level. However, it is not practical to follow this division strictly, and all hydrocarbons were treated in Section 2.2.1 while all types of halogenated hydrocarbons will be mentioned in subsection 2.2.2.3. It should also be remembered that it is impossible to cover all functional groups in this summary, only those that are important for the following discussions in this text will be included.

2.2.2.1 Functionalities containing oxygen

The basic oxygenated organic compound contain a hydroxyl group and are called alcohols, their systematic names consequently have -ol as in ethanol as suffix. If the carbon to which the hydroxyl group is attached has one additional carbon atom bound to it the alcohol is called primary, if it is two or three carbons the alcohol is classified as a secondary or a tertiary alcohol.

Simple alcohols have both a hydrocarbon-like and a water-like character, and alcohols are miscible with most solvents. The hydroxyl group *per se* is not associated with any pronounced toxic effects, but in combinations with other functional groups it may become considerably more harmful. Compounds that have a hydroxyl group attached directly to an aromatic ring are called phenols, and they differ from alcohols in several respects. The phenolic hydroxyl group is more acidic (see subsection 2.2.2.2) compared to alcohols making phenols more irritating and toxic in general, and phenol itself was one of the first antiseptics used in hospitals. Chlorinated phenols have been

Primary carbons are bonded to one other carbon atom.

Secondary carbons are bonded to two other carbon atoms.

Tertiary carbons are bonded to three other carbon atoms.

Quaternary carbons are bonded to four other carbon atoms.

n– as a prefix indicates that an open-chained hydrocarbon is unbranched.

sec– as a prefix indicates that the functionality of interest is on a secondary carbon.

tert– as a prefix indicates that the functionality of interest is on a tertiary carbon.

iso– as a prefix in hydrocarbons indicates that an otherwise unbranched chain ends with $-CH(CH_3)_2$, although in general it stands for isomer.

neo– as a prefix indicates that an otherwise unbranched chain ends with $-C(CH_3)_3$.

used as biocides for many years and are part of the well-known environmental hazards such as 2,4,5-T [(2,4,5-trichlorophenoxy)acetic acid] and TCDD (dioxin). In ethers both hydrogens of water have been exchanged for alkyl (or aryl) groups. Ethers are in general volatile and flamma-

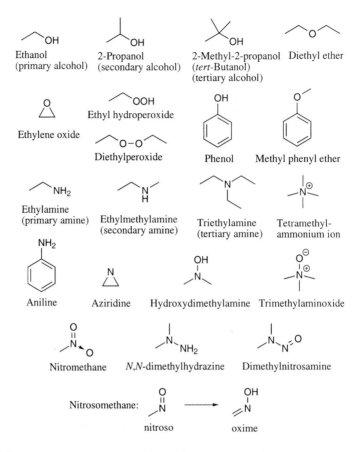

Figure 2.6. Examples of organic compounds containing oxygen and nitrogen.

ble, and the risk for fires and explosions is probably the biggest hazard when ethers are handled in workplaces. The ether functionality may also be oxidized by the oxygen in air, and the peroxides formed are extremely hazardous due to their tendency to explode when concentrated. The corresponding alcohols with an extra oxygen in the hydroxyl group are called hydroperoxides. Cyclic ethers, in which the ether oxygen is part of a ring, are used as for example reagents and solvents. Especially interesting are the three-membered cyclic ethers, the epoxides or the oxiranes, exemplified in Figure 2.6 by the most simple member, ethylene oxide, which is used in large quantities. Many of the epoxides, including ethylene oxide, are highly toxic and carcinogenic, and we shall come back to them.

2.2.2.2 Functionalities containing nitrogen

Amines contain a nitrogen atom, which is substituted with one (primary amines with a -NH$_2$ group, as for example ethylamine in Figure 2.6), two (secondary amines with a -NH- group) or three (tertiary amines with no hydrogens bound to the nitrogen) alkyl/aryl groups. The nitrogen may even bind four alkyl groups, in quaternary ammonium ions (e.g. the tetramethylammonium ion), then the unshared electron pair of nitrogen has been used for the forth bond and nitrogen has become electron-deficient (and consequently has a positive charge). In analogy with the hydroxyl group that gives alcohols some water-like properties, amines will to some extent resemble ammonia. This is especially pronounced in the primary amines, but also secondary and tertiary amines have basic properties (*vide infra*).

A base is the converse to an acid, in the respect that bases have a tendency to accept protons (hydrogen atoms without an electron, H$^+$) while acids donate protons. Water (and alcohols) can be both a (very weak) base and an (very weak) acid, and in pure water a few molecules (as acids) will donate a proton to another few (as bases) that will accept them according to Figure 2.7. (In Figure 2.7 the mechanism of the reaction is also shown, where the curved arrows indicate the "movement" of electron pairs, this will be explained in more detail in Chapter 3.) The numbers of oxonium and hydroxide ions formed in pure water are equal, and the pH of water (pH is actually a measure of the oxonium ion concentration) is 7. Acidic water solutions have a pH lower than 7 while basic have a pH over 7. Primary and secondary amines, and of course ammonia, can also be both a base and an acid (tertiary amines lack protons bound to the N and can only be basic), and compared to water and alcohols amines are stronger bases but even weaker acids. It is the unshared electron pair on the nitrogen that is responsible for the basicity of amines, as it is more readily protonated by an acid compared to the unshared electron pair of the oxygen atom in water or an alcohol. In a water solution of an amine, water is the stronger acid (strong compared to the amine, not for example to hydrochloric acid!) while the amine is the stronger base, and the reaction depicted in Figure 2.7 will take place. However, the ammonium ion formed is a stronger acid than water, and the hydroxide ion a stronger base than the amine, so an equilibrium will establish itself. As a consequence of their basic properties, amines are in general irritating and in some cases corrosive chemicals that should be handled with care. Aro-

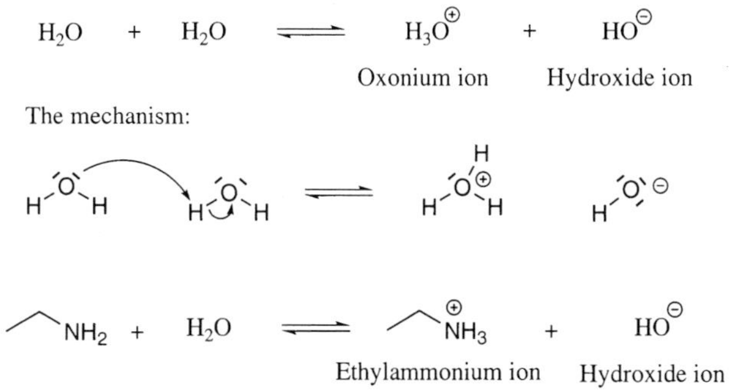

Figure 2.7. Acid–base reactions of water and ethylamine.

matic amines, e.g. aniline (Figure 2.6), are considerably more toxic and in some cases potent carcinogens. As with epoxides, cyclic amines where the nitrogen is part of an three-membered ring (aziridines) are potentially toxic and carcinogenic. The nitrogen atom in amines may be oxidized to a hydroxylamine (primary and secondary amines), an aminoxide (tertiary amines), to a nitroso or a nitro group (see Figure 2.6). The nitroso group is only stable if there are no α-hydrogens, if this is the case it will spontaneously be transformed to the corresponding oxime. Compounds with two adjacent amino groups are called hydrazines, and if the second amino group is oxidized it is a *N*-nitrosamines. Both hydrazines and *N*-nitrosamines are also strongly associated with carcinogenicity.

2.2.2.3 Functionalities containing halogens

Organic chemicals can contain any of the four halogens fluorine, chlorine, bromine and iodine, although the chlorinated are most common. The same nomenclature as for the alcohols is used to indicate whether the halogen is positioned on a carbon that has one or two other carbons bound to it. In this subgroup also halogenated hydrocarbons formally belonging to higher oxidation levels will be included, for simplicity. Halogenated hydrocarbons have found a large number of uses, as the exchange of a hydrogen for a halogen in some instances makes a compound more stable while the halogen in other cases renders the compound reactive. The ozone-degrading freons (e.g. CFC (freon) 12), DDT (see Figure 1.3) and the PCBs (polychlorinated biphenyls) are examples of extremely stable (even too stable!) halogenated compounds, while others, e.g. allylbromide, are toxic and carcinogenic due to their chemical reactivity. Large amounts of vinylchloride are used for the manufacture of PVC (poly-vinyl-chloride), and besides generating hazardous by-products during the production, PVC (and other halogenated organic materials) will be responsible for the formation of for example dioxins when burnt at refuse dumps. A number of halogenated solvents, which are very efficient but unfortunately toxic to man and environment in various ways, have been and are still used in large quantities, examples are carbon tetrachloride, chloroform, methylene chloride, 1,1,1-trichloroethane, trichlorethylene and tetrachlorethylene. In recent years, the use of halogenated hydrocarbons have decreased as we have become more and more aware about the hazards associated with them.

2.2.2.4 Functionalities containing sulphur

Sulphur is below oxygen in the periodic system, and can replace oxygen in alcohols and ethers resulting in thiols and sulphides. The very intensive odour of both thiols and sulphides explains at least to some extent the limited usefulness of such compounds, only few organic compounds that contain sulphur are used industrially and the exposure to humans is consequently relatively small. However, it should be noted that the sulphur analogue of water, hydrogen sulphide, is highly toxic. As the sulphur atom is bigger than oxygen it can form additional bonds, and is frequently oxidized to sulphoxides, sulphones, sulphonates and sulphates. The structures of some derivatives are shown in Figure 2.8.

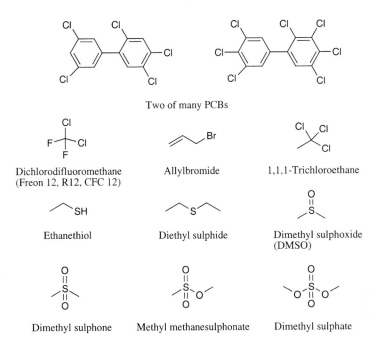

Two of many PCBs

Dichlorodifluoromethane
(Freon 12, R12, CFC 12)

Allylbromide

1,1,1-Trichloroethane

Ethanethiol

Diethyl sulphide

Dimethyl sulphoxide
(DMSO)

Dimethyl sulphone

Methyl methanesulphonate

Dimethyl sulphate

Figure 2.8. Examples of organic compounds containing sulphur and halogens.

2.2.3 Chemical functionalities, second oxidation level

2.2.3.1 Aldehydes and ketones

Both aldehydes and ketones are characterized by the presence of a carbonyl group, in which an oxygen is connected to a carbon by a double bond. Aldehydes may in systematic names have the suffix *-al* while ketones may have the suffix *-one*. In aldehydes the carbonyl group is attached to at least one hydrogen while it in ketones is attached to two alkyl/aryl groups. The most simple representatives are formaldehyde and acetone, two industrially very important chemicals. While formaldehyde is highly toxic, allergenic and suspected to be carcinogenic, acetone is a relatively harmless organic solvent. Also acetaldehyde is considerably more hazardous compared to acetone. Thus, in spite of the apparent similarities between the two types of compounds, there are differences in the way they react. As we will see, the toxic effects of aldehydes and ketones are especially interesting when they are combined with other functional groups, for example C=C bonds and halogens.

2.2.3.2 Hemiacetals and acetals

The carbonyl group of an aldehyde or a ketone may be transformed to a hydrate by the addition of water (formaldehyde forms formalin), or a hemiacetal or an acetal by the addition of an alcohol to the carbonyl group. Acetone and methanol will quickly react to form a hemiacetal, a reversible reaction, but the transformation of the hemiacetal to an acetal normally requires catalysis by an acid. The corresponding products may form if amines, thiols or halide ions add to the carbonyl. The chemical properties of any aldehyde or ketone will naturally change dramatically if the carbonyl group disappears, as will those of an acetal if it is transformed to an aldehyde or ketone. Hemiacetals (and corresponding compounds with two heteroatoms of which one has a hydrogen, bound to the former carbonyl carbon) are generally not stable in contact with water and will regenerate the aldehyde or ketone spontaneously. Acetals (and corresponding compounds with two heteroatoms with no hydrogen bound to the former carbonyl carbon) are considerably more stable, but will eventually be hydrolyzed in the presence of water. The interconversion of carbonyl compounds and hydrates as well as hemiacetals will be further discussed in Chapter 3.

2.2.3.3 Enols and similar compounds

An enol has as the name implies a C=C bond connected to a hydroxyl group. Most enols are not stable and will spontaneously be transformed to the corresponding carbonyl compound (generally a ketone) as indicated in Figure 2.9. Actually, all ketones possessing an α-hydrogen (α is the position next to the carbonyl group, β the next, etc.) are in equilibrium with the enol form (the equilibrium is called tautomerism), and for acetone for example the keto form is so much more stable that the enol form is not observed. However, the equilibrium still exists, and will influence the way acetone reacts. Some compounds are more stable in their enol form than in the keto form, and an example is phenol which is aromatic (remember that aromaticity is stabilizing!) as the enol and non-aromatic as the ketone. If trapped as for example the an enol ether (see Figure 2.9), the transformation to the keto form is impossible and the enol is stable. Compounds corresponding to enols but with other heteroatoms than oxygen, e.g. enamines and thioenols, behave in a similar way and may in contact with water be hydrolyzed to the carbonyl compound. The "keto" form of an enamine is called an imine, and imines are easily hydrolyzed to the corresponding carbonyl compound. Enol, enamine and imine functionalities are frequently encountered in so called heterocyclic compounds. These are often aromatic as the heteroatom either participates in a π-bond (as pyridine in Figure 2.9) or keeps an unshared electron pair in a p orbital which can participate in the aromatic system with 2 electrons (as furan).

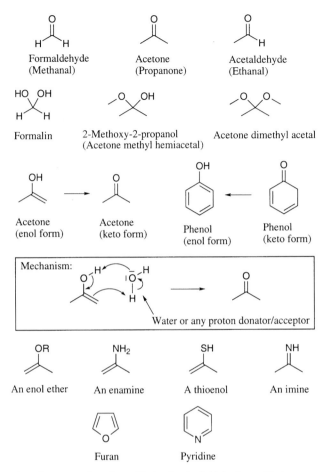

Figure 2.9. Examples of compounds at the oxidation level of aldehydes and ketones.

2.2.4 Chemical functionalities, third oxidation level

2.2.4.1 Carboxylic acids, anhydrides and halides

The carboxylic acids contain the –COOH functionality, which primarily is characterized by its relative acidity. The simplest and most well-known representatives for this class of compounds are formic acid and acetic acid, which both are hazardous because they are relatively strong organic acids. The acidity is the result of the fact that the anion formed when a carboxylic acid has donated its acidic proton to a base (e.g. water) is stabilized, because the negative charge is not positioned on only one atom but can be distributed on two oxygens (see Figure 2.10). This phenomenon is called resonance, the two different forms of the acetate ion are resonance struc-

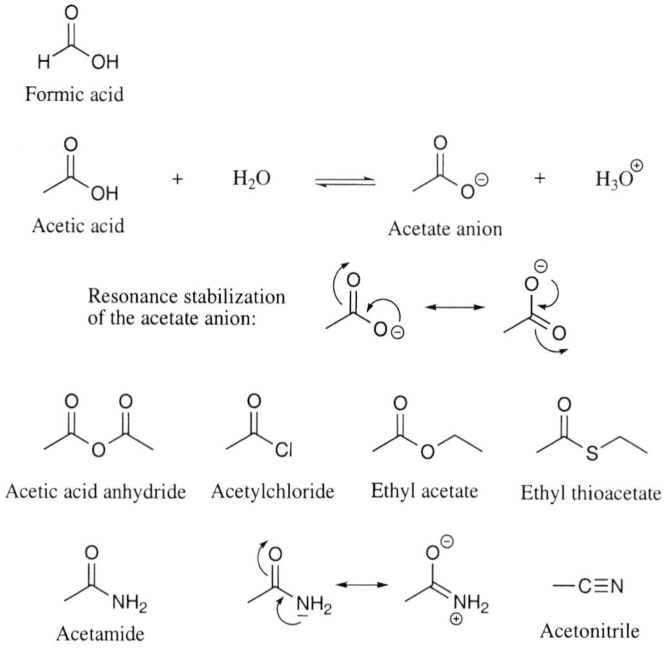

Figure 2.10. Examples of compounds at the oxidation level of carboxylic acids.

tures, and the stabilization is due to the delocalization of the negative charge to several electron-egative atoms. However, the acetate ion is not a 1:1 mixture of the two resonance structures, instead it is an average structure with some charge on both oxygens. Although many carboxylic acids are irritating and corrosive, it should be remembered that they in general are much weaker acids compared with the inorganic acids such as hydrochloric acid (HCl) and sulphuric acid (H_2SO_4), and acetic acid in water is actually only ionized to a small extent while HCl and H_2SO_4 are completely ionized. Carboxylic acid anhydrides (e.g. acetic acid anhydride in Figure 2.10) and halides (e.g. acetylchloride) are derivatives of carboxylic acids, prepared and used because they are reactive and can take part in chemical transformations.

2.2.4.2 Carboxylic acid esters, thioesters and amides

The transformation of a carboxylic acid to an ester instead decreases the reactivity and takes away the acidity, and esters are used in large amounts as for example organic solvents. An ester (e.g. ethyl acetate) is easily prepared from the corresponding carboxylic acid and the alcohol (acetic acid and ethanol form ethyl acetate), and esters can consequently be hydrolyzed to the carboxylic acid and the alcohol by water. The hydrolysis of esters is one of the most efficient enzymatic reactions that takes place in organisms, and we shall return to this issue in Chapter 7. Thioesters are the esters between a carboxylic acid and a thiol, ethyl thioacetate is consequently formed from acetic acid and ethanethiol. Amides are the condensation products of carboxylic acids and amines, and the properties of amides differ significantly from those of the esters and thioesters. Amides are also important chemicals, and are for example used as solvents. The amide bond between the nitrogen and the carbonyl carbon is extremely important for the biochemistry of all organisms, as it connects the amino acids in proteins (Section 4.4.2). It is relatively strong, because the nitrogen readily (more readily than oxygen) shares its unshared electron pair with the carbonyl group in the way indicated in Figure 2.10. This gives the N–CO bond a partial double bond character.

2.2.4.3 Nitriles

Nitriles, for example acetonitrile (see Figure 2.10), containing a carbon with a triple bond to nitrogen (a cyanide group), are not uncommon chemicals in industry as well as in nature. The cyanide group resembles to some extent the carbonyl group and may also activate other chemical functionalities to be more toxic. In addition, the cyanide ion can in some situations be lost from nitriles, generating the highly toxic and well-known chemical hydrogen cyanide (HCN).

2.2.5 Chemical functionalities, fourth oxidation level

One final oxidation level remains, at which a carbon atom is fully oxidized and only surrounded by heteroatoms. Isocyanates are reactive chemicals that are used in several circumstances, it is for example an important component in certain paints. Methylisocyanate is highly reactive and was responsible for the killing of 2500 persons in India at an accident in Bhopal in 1984 (where it was used for the synthesis of an insecticide). The end-product of the oxidative degradation of the nutrients in organisms is carbon dioxode, which can add water and form carbonic acid. Carbon dioxide, carbonic acid, bicarbonate and carbonate (anions of carbonic acid) are regarded as inorganic compounds, and a number of derivatives (also organic) of carbonic acid have found uses. Examples are diethyl carbonate (used as a solvent) and phosgene (a highly toxic reagent also used as a war gas).

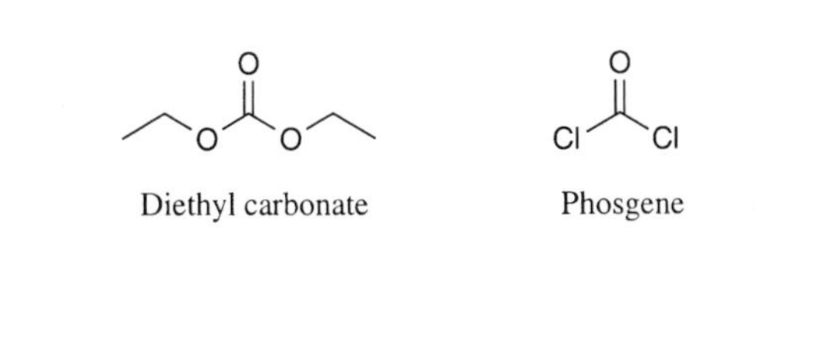

$$—N{=}C{=}O$$

Methylisocyanate

$$O{=}C{=}O \quad + \quad H_2O \quad \rightleftharpoons \quad \underset{\text{Carbonic acid}}{HO{-}\overset{\displaystyle O}{\overset{\|}{C}}{-}OH}$$

Carbon dioxide

Figure 2.11. Examples of compounds at the oxidation level of carbonic acid.

Diethyl carbonate

Phosgene

3 Chemical properties

Chemicals that are toxic to organisms may be described, for example, as cytotoxic, carcinogenic, allergenic, etc. The potential of a compound to produce toxic effects depends on its chemical properties, e.g. its volatility, solubility, reactivity, etc. which in turn depends on the molecular properties of a compound. Important molecular properties are, for example, the size and form of a molecule, the distribution of charge in different parts (i.e. its polarity), the polarizability (i.e. how easily the electrons can move in the molecule), and the strength of the bonds within the molecule. It is evident that one can discuss and describe hazardous chemicals from their effects on biological systems, from the chemical properties that make them hazardous, or from the molecular properties that determine the chemical properties. To some extent, this text attempts to mix all three points and explain how a toxic effect by a chemical (e.g. carcinogenicity) is caused by a chemical property (such as chemical reactivity) due to the presence of certain functional groups in the molecule (that for example enable the chemical to form new bonds with the genetic material). In this Chapter, a brief survey of the most important chemical properties involved in toxicity will be made, starting with a summary of the different intermolecular (i.e. between molecules) chemical forces and bonds.

3.1 Electronegativity, dipoles and ionic bonds

In principle, all forces between molecules are electrostatic, and depend on an attraction between different charges (plus and minus) and a repulsion between like charges (plus and plus, or minus and minus). In bonds between different atoms in a molecule there is an unequal distribution of electrons and thereby electric charge along the bond, creating what is called a dipole moment. This is caused by the fact that different atoms have different electronegativities. Electronegativity is a measure of the force by which the nucleus (of an atom) attracts the electrons in its vicinity, including those shared with other atoms in a covalent bond. Therefore, in a covalent bond between for example carbon and chlorine, chlorine is more electronegative (see Table 3.1) and the 2 electrons in the bond are, on average, closer to the chlorine than to the carbon. The carbon–chlorine bond is a dipole, and in chloromethane (or methyl chloride, see Figure 3.1) this gives the molecule a dipole moment directed along the carbon–chlorine bond and with the negative end in the direction of the chlorine. The carbon atom has a small deficit of electrons around it while the chlorine has a small surplus. We may indicate this by a $\delta+$ (a partial positive charge) and a $\delta-$ (a partial negative charge) close to the atoms, as shown in Figure 3.1.

If the molecule contains several dipoles, the net dipole moment is the sum of all individual dipoles. In a molecule that is symmetrical in space, like tetrachloromethane (or carbon tetrachloride), in which the dipoles are directed from the centre to the four corners of a tetraeder, and

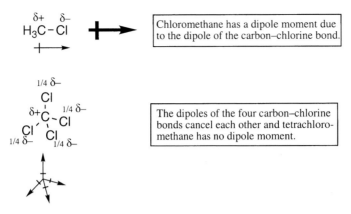

Figure 3.1. Electronegativity, bond dipoles and dipole moment.

carbon dioxide, which is a linear molecule with two equal dipoles directed in opposite directions, the dipoles will cancel each other and the total dipole moment is zero. (The magnitude of the partial negative charges on the chlorines in tetrachloromethane in Figure 3.1 is only 25 % of the partial positive charge of the carbon, because the net electrical charge is zero.) The approximate electronegativities of some elements are given in Table 3.1, and a useful rule of thumb places the electronegativities of the most common elements in the order F>O>Cl≈N>C≈H≈S. As a general rule, the electronegativity increases to the right and to the top of the periodic system, while it decreases to the left and to the bottom. Compare for example the elements carbon, nitrogen, oxygen and fluorine in the second row. When they participate in covalent bonds all are surrounded by an octet of electrons in their outer electron shell. However, the atoms to the right have more protons and thereby a larger positive charge in their nuclei. This will attract the electrons shared in a covalent bond and create the electrical dipole. Bond dipoles will normally give a dipole moment to the molecule in which they are part, making the molecule polar, but as mentioned above molecules having bond dipoles may be nonpolar if the dipoles cancel each other. A compound is of course also nonpolar if it consists of atoms with essentially the same electronegativity, as the hydrocarbons. It is important to realize that the electronegativities given in Table 3.1 are average values, and that the actual value for an atom depends on the atoms to which it is bound. Compare, for example, 1,1,1-trichloroethane and ethane. The carbon–carbon bond is polarized in the former but not in the latter because the three chlorines in 1,1,1-trichloroethane will make the carbon they are attached to more electron deficient than usual and thereby increase its electronegativity. This will give rise to the inductive effect discussed later in this Chapter. In addition, the hybridization (Section 2.1.2) of an atom will affect its electronegativity, *sp* hybridized carbon (in triple bonds) are considerably more electronegative (see Table 3.1) due to a larger proportion (50 %) of *s* orbital (which is closer to the nucleus than the *p* orbital) compared to *sp*^{2} carbons (33 % *s* character), which in turn are more electronegative than the saturated *sp*^{3} carbon (25 % *s* character). The hybridization of other atoms will influence their electronegativity correspondingly.

Table 3.1. The approximate electronegativities of selected atoms.

H	2.1											
Li	1.0			B	2.0	C	2.5	N	3.0	O	3.5	F	4.0
Na	0.9	Mg	1.2			Si	1.8	P	2.1	S	2.5	Cl	3.0
K	0.8									Se	2.4	Br	2.8
												I	2.5

sp^3 C 2.5	sp^2 C 2.8	sp C 3.3

In this scale of electronegativities (the Pauling scale) fluorine has been given the value 4.0 and the other elements are assigned values in relation to fluorine. When the difference in electronegativity between two atoms in a bond is small, less than 0.5, the bond is characterized as a nonpolar covalent bond, while the difference in electronegativity in polar covalent bonds is between 0.5 and 1.9. If there is a large difference, greater than 1.9, the bond is said to be ionic. Obviously, there is in fact a sliding transition between covalent and ionic bonds. In a ionic bond, the two electrons are completely drawn to the more electronegative end [i.e. the chlorine in sodium chloride (table salt)] and the bond is formed by the attractive force between ions with different charge. Ionic bonds are strong. If one wants to break the ionic bond between, for example, the sodium and chloride ions in a crystal of NaCl, it is necessary to put in a lot of energy. In fact, the strength of the electrostatic force between ions can be compared to covalent bonds, and for example it takes heating to 800 °C before the crystals of sodium chloride will melt and the ions can move freely.

3.2 Intermolecular forces

3.2.1 Covalent bonds

Covalent bonds are intramolecular by definition, not intermolecular, but they may be formed transiently between two molecules containing suitable functionalities. In such cases covalent bonds will act as if they were intermolecular, and keep certain molecules closer together than expected. Examples are aldehyde and keto functions, which react with alcohols (to form hemiacetals), thiols and amines. The products formed are normally not stable and the reaction will spontaneously reverse to the original aldehyde/ketone and alcohol/thiol/amine, but the effect is still that such functions will be a sort of "glue" that prolongs the time that an aldehyde or a ketone for example stays next to a protein containing alcohol and amine functions. If, and we will see examples later, another part of the molecule has properties that may destroy the protein, it is obvious that its potential to do so is greater if the molecule and the protein are kept together by the aldehyde/keto functionality.

Figure 3.2 shows the corresponding reaction between formaldehyde, acetaldehyde and acetone, yielding hydrates. The reaction is readily reversible and goes back and forth all the time, but the equilibria are as indicated by the arrows. Thus, formaldehyde in water is present entirely

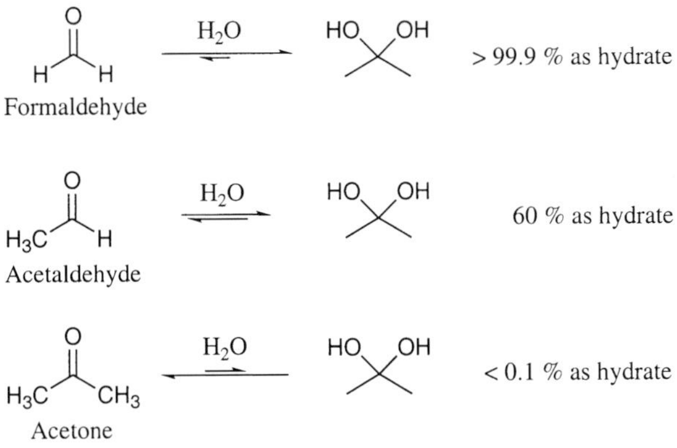

Formaldehyde

H_2O

> 99.9 % as hydrate

Acetaldehyde

H_2O

60 % as hydrate

Acetone

H_2O

< 0.1 % as hydrate

Figure 3.2. Formaldehyde, acetaldehyde and acetone react with water to form hydrates.

as the hydrate (37 % formaldehyde in water is called formalin), while acetone is essentially present as the ketone. The equilibrium depends on the groups attached to the carbonyl group, and for some compounds, e.g. acetaldehyde, both forms are approximately equally stable.

3.2.2 Forces between ions and dipoles

Although many ions, inorganic as well as organic, take part in the biochemical processes that we are interested in, the ionic bond is of little relevance in biological systems. The reason for this is that the ion–ion forces to a large extent are cancelled by weaker but numerous ion–dipole forces, especially between ions and water molecules.

As discussed above, molecules with a dipole moment will have an end with a partial negative charge and another with a partial positive charge. These will of course interact, and there will be attractive forces between the positive end in one molecule and the negative in another (see Figure 3.3). An acetone molecule, for example, has a dipole moment due to the presence of a

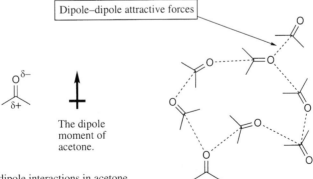

Dipole–dipole attractive forces

The dipole moment of acetone.

Figure 3.3. Dipole–dipole interactions in acetone.

carbon–oxygen double bond, and the oxygen (with a δ–) is attracted by the carbonyl carbon (with a δ+). This attraction holds the acetone molecules together and results in a considerably higher boiling point compared to, for example, 2-methyl propane which has approximately the same size.

The strength of the force between two dipoles can be anything from very weak to rather strong, although they are always considerably weaker than ionic and covalent bonds.

3.2.3 Hydrogen bonds

Hydrogens bonded to strongly electronegative atoms, e.g. the hydrogens of water, will be attracted to other strongly electronegative atoms, e.g. the oxygen of another water molecule, by an attractive force called the hydrogen bond. Although this in reality is a dipole–dipole interaction, its relative strength and great importance in all biological systems makes it special. Hydrogen bonds are stronger than most dipole–dipole interactions because when the electrons in a covalent bond, between for example hydrogen and oxygen in water, are drawn towards the oxygen, the hydrogen, lacking inner electron shells to conceal the positively charged nucleus, will exhibit an relatively naked nucleus. The importance in biological systems comes from the fact that water is the solvent for the biochemical processes. Hydrogen bonds have a direction in the sense that a certain molecule may in some situations take part in hydrogen bonds and in others not. The terms hydrogen bond donor and hydrogen bond acceptor are used to indicate the role of a molecule in a hydrogen bond. Water and ethanol are both a donors and acceptors, but acetone, with no hydrogen bonded to an electronegative atom but having an electronegative atom (the carbonyl oxygen), is only an acceptor. In pure acetone there will be no hydrogen bonds, but in acetone–water mixtures the water molecules will give hydrogen bonds to both other water molecules as well as to acetone molecules (see Figure 3.4).

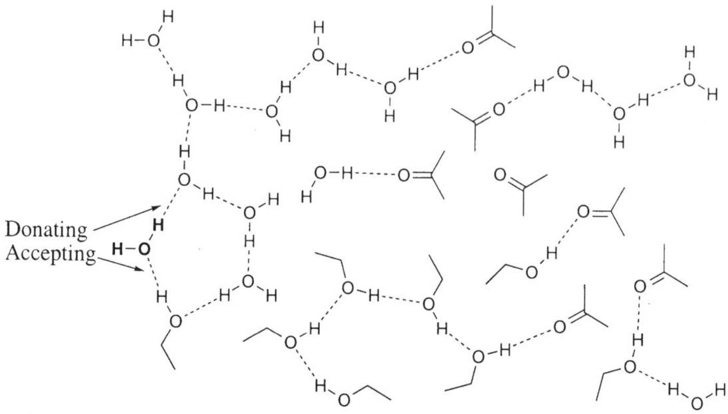

Figure 3.4. Hydrogen bonds (- - -) between water, acetone, ethanol.

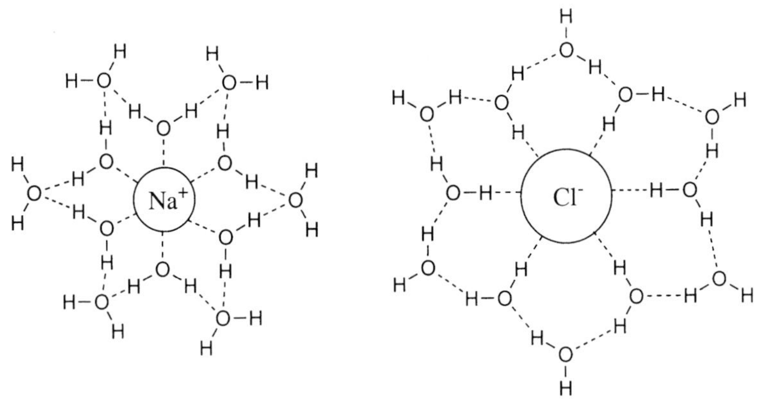

Figure 3.5. Interactions between water molecules and ions in a water solution.

Water is a good solvent for polar molecules because it readily participates in dipole–dipole and ion–dipole interactions with the solute. It even dissolves salts as sodium chloride, despite the strong ionic bonds, because each ion can bind many water molecules in several layers (see Figure 3.5) and the sum of attractive forces exceeds the energy of the ionic bond.

3.2.4 Charge transfer forces

Charge transfer forces can arise between molecules that have the ability to donate and accept electrons, respectively, for example between π-electron rich heterocycles (e.g. furan) and aromates with electron withdrawing groups (e.g. picric acid). The attraction, which still is not completely understood, is caused by the electrostatic force between the two species due to the induced changes in electron distribution. An example of a charge transfer force that actually can be observed is that between iodine, which is an acceptor of electrons, and cyclohexene, a donor with a double bond. In a nonpolar solvent such as carbon tetrachloride, iodine will have a violet colour, but in cyclohexene iodine has a brown colour. The charge transfer force will change the electronic configuration of iodine, and thereby its light absorption.

3.2.5 van der Waals forces

van der Waals forces (or dispersion forces) are generally weak, and although they are present between all kinds of molecules they only have a practical importance for molecules that do not participate in stronger electrostatic intermolecular forces. In nonpolar compounds such as hydrocarbons, there is no significant difference in the electronegativity of the atoms, and no bond dipoles to speak of, and the van der Waals forces become important. They are caused by the weak dipoles (called induced dipoles) formed by the influence of neighbouring molecules. The

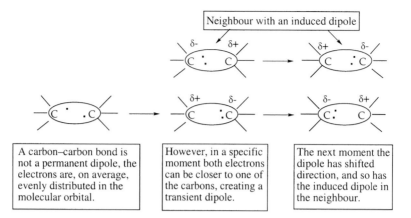

| A carbon–carbon bond is not a permanent dipole, the electrons are, on average, evenly distributed in the molecular orbital. | However, in a specific moment both electrons can be closer to one of the carbons, creating a transient dipole. | The next moment the dipole has shifted direction, and so has the induced dipole in the neighbour. |

Figure 3.6. van der Waals forces.

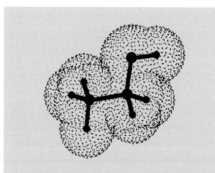

The van der Waals forces (J.D. van der Waals was a Dutch scientist that studied these phenomena in the 19th century) are attractive up to a point, when the molecules come so close that the electron clouds get in contact with and repel each other. The surface around an atom or a molecule where the attractive forces are the strongest is sometimes depicted as with ethanol above, because it is a convenient way to describe how much space a molecule occupies. The distance from an atom to the surrounding surface is called the van der Waals radius of that atom.

covalent bond between two saturated carbon atoms (σ bond) consists of a molecular orbital surrounding both atoms and containing two electrons. On average, the electrons are evenly distributed between the two carbons, i.e. in the molecular orbital, but at a certain moment they may well both be situated closer to one of the carbons. At that moment the bond is polarized, but the dipole will disappear and may even be inverted the next moment. However, the alternating dipole will influence neighbouring molecules and induce dipoles in them with the opposite dipole moment (as indicated in Figure 3.6). Such induced dipoles, that constantly change direction, create a weak electrostatic attractive force which is important for nonpolar compounds. Typical for van der Waals forces is that the strength of the interaction between two molecules depends on their surface area, and van der Waals forces can be quite strong for larger molecules. In addition, van der Waals forces are extremely sensitive to the distance between the interacting molecules and will only be significant when the molecules are close together.

3.3 Chemical properties

3.3.1 Volatility

As will be discussed in Chapter 6, the principal route for exposure to chemicals in the workplace is via the lungs, and the chemicals inhaled are present in the air either as a vapour or as an aerosol. A compound's volatility will therefore to a large extent determine the risk for a chemical to be absorbed by persons exposed to it at workplaces, as well as influence its distribution in the environment. Many commercially important chemicals, for example most organic solvents, are liquid at room temperature but evaporate readily. A liquid stays liquid because the intermolecular forces hold the molecules together. However, by increasing the kinetic energy of the molecules, by for example increasing the temperature, molecules may become energetic enough to break free from the intermolecular forces and pass into the gaseous phase. The energy of the molecules in a liquid is not uniform, but distributed according to Gauss. Even at temperatures much lower than the boiling point, a small proportion of the molecules will have sufficient kinetic energy to leave the liquid. Therefore, all liquids (and in principle also all solids) have a vapour pressure at all temperatures, and wet clothes will dry at room temperature (and eventually even at temperatures below 0 °C when the water is frozen). When the temperature increases the vapour pressure does the same, and at temperatures when the vapour pressure is the same as the air pressure the liquid will boil.

The volatility of a chemical is influenced by several other factors as well, besides the temperature and the strength of the intermolecular forces between its molecules. The air pressure, the vapour pressure of the evaporating compound in the gas phase above the liquid, the composition of the liquid, as well as the molecular weight of the compound in question may also be important. For example, at high altitudes where the air pressure is low compared to sea level, water will boil below 100 °C, and in technical systems it is common to decrease the pressure in order to make a solvent evaporate faster. If the pressure is increased the boiling point increases, as for instance in a pressure-cooker where water is liquid even at 120 °C. If the gas phase above the liquid contains a lot of the volatile compound this will slow down the evaporation, because condensation (from the gas to the liquid phase) will also take place. Wet laundry will dry faster on a dry day then on a humid day, even if the temperature is the same. When the compound evaporating is present in a mixed liquid consisting of several chemicals, the intermolecular forces experienced by one molecule depend on its neighbours. However, the volatility of the components of a mixture will normally not differ a lot from the volatility of the pure compounds. The molecular weight will influence the volatility because the kinetic energy of a molecule depends on its mass. If two molecules with different molecular weight experience the same intermolecular forces, the lighter will be more volatile (have lower boiling point) because the heavier requires more heat to obtain the same kinetic energy.

The volatility of a pure compound may be characterized in several ways, e.g. by the partial pressure of the compound over the liquid at a certain temperature, or the boiling point of the pure liquid at a certain air pressure. For practical reasons, it is easier to use the boiling point at normal sea level atmospheric pressure, as such values can easily be extracted from the chemical

Table 3.2. Comparison of the volatility of chemically different compounds.

Name	Structure	Molecular weight	Boiling point (°C)
Butane		58	0
2,2-Dimethylpropane		72	9
2-Methylbutane		72	30
1-Fluorobutane	F	76	32
Diethyl ether	O	74	35
Pentane		72	36
Methyl propyl ether	O	74	38
Propylchloride	Cl	78	47
Methylacetate	O O	74	57
1-Propanethiol	SH	76	68
Hexane		86	69
Butanal	O H	72	76
1-Aminobutane	NH$_2$	73	78
Butanone	O	72	80
Nitroethane	NO$_2$	75	115
1-Butanol	OH	74	117
Butyronitrile	CN	69	118
Propanoic acid	OH O	74	141
Propionamide	NH$_2$ O	73	213

literature. In Table 3.2 the boiling points of a number of compounds with comparable molecular weight but containing different chemical functionalities are listed.

While it is obvious that butane is more volatile than pentane which in turn is more volatile than hexane, the difference between the pentane isomers 2,2-dimethylpropane (neopentane), 2-methylpentane (isopentane) and pentane is not self-evident. However, the van der Waals forces are the only attractive forces for these hydrocarbons, and their strength is proportional to the surface area of the molecule. A straight-chained molecule like pentane is elongated and has a larger surface compared to 2,2-dimethylpropane, which is more shaped like a ball. This effect is general for all molecules in which van der Waals forces make a significant contribution to the intermolecular forces, and 2-chloro-2-methylpropane (*tert*-butyl chloride) for example boils at 51 °C compared to 77 °C for 1-chlorobutane. It is perhaps surprising to find that compounds containing strongly electronegative atoms such as oxygen and fluorine (1-fluorobutane, diethyl ether (ether) and methyl propyl ether in Table 3.2) do not have higher boiling points than the nonpolar hydrocarbon pentane. However, the dipole moment of a molecule does not only depend on the difference in electronegativity between two atoms, also the length of the bond plays a role. Fluorine and oxygen are relatively small atoms that are bound more closely to carbon (bond lengths 1,39 and 1.43 Å, respectively) than for example chlorine (bond length 1,78), and the dipole moment of chlorinated hydrocarbons is actually larger than the corresponding fluorinated hydrocarbons. The effect of the dipole created by the fluor–carbon and oxygen–carbon bond is therefore limited, and comparable with the van der Waals forces. Propyl chloride has a significantly higher boiling point (see Table 3.2), which is due to its larger dipole moment as well as its greater ability to be polarized (giving rise to stronger van der Waals forces), because the three unshared electron pairs of chlorine are less tightly held by the chlorine nucleus compared to situation in the smaller fluorine. The same effect is responsible for the increased boiling point of 1-propanethiol. Although sulphur has approximately the same electronegativity as carbon (see Table 3.1) it is more easily polarized. The thiol group may in some cases take part in weak hydrogen bonds, although this is not the case in 1-propanethiol. However, the effects are still small, and comparable with the addition of an methylene group to pentane (to give hexane). The carbonyl groups of aldehydes and ketones are considerably more polar, it is electron withdrawing both because oxygen is more electronegative than carbon and due to the possibility of resonance (discussed later in this Chapter). Butanal and butanone therefore have a similar boiling point as 1-aminobutane, which is our first example of a compound that forms hydrogen bonds to itself. The hydrogen bonds in 1-aminobutane are relatively weak compared to those in 1-butanol, because of the difference in electronegativity between nitrogen and oxygen, and the boiling points of the two compounds differ almost 40 °C. 1-Butanol has a similar boiling point as nitroethane and butyronitrile, indicating the polarity of the nitro and the cyano groups which both are unable to donate hydrogen bonds. Carboxylic acids and amides are even less volatile, due to the polarity and the ability of these functionalities to form hydrogen bonds. Two carboxylic acids (or amides) may form a dimer (by two hydrogen bonds to each other, see Figure 3.7), which formally may be regarded as a polar entity with the double molecular weight (and thereby considerably less volatile). The difference between carboxylic acids and amides (e.g. propanoic acid and propionamide in Table 3.2) is due to the greater tendency of nitrogen to donate its unshared electron pair to the carbonyl group, pushing electrons to the carbonyl oxygen which will have an even larger negative charge (see Figure 3.7).

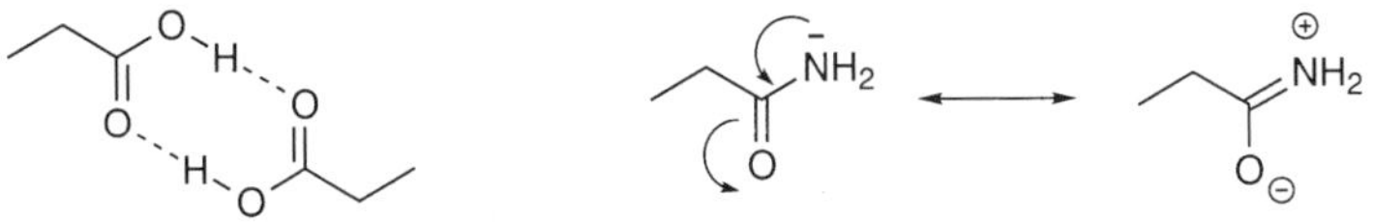

Figure 3.7. A hydrogen bond dimer of propanoic acid (left), and the electron flow from nitrogen to oxygen in propionamide resulting in an increased polarity (right).

3.3.2 Solubility

The solubility of a compound is an important property that strongly influences the mobility of a chemical in the environment and between different organisms, as well as its absorbtion into and distribution in organisms. As will be discussed in the next Chapter, cells are surrounded by membranes that are more easily penetrated by nonpolar compounds than polar or ionic compounds, and most toxic compounds need to get into cells in order to be toxic. In general, a compound (a solute) will be dissolved by another (a solvent) if the intermolecular forces present in the final solution are stronger than those present in the solute and solvent separately. Nonpolar solvents will, in general, dissolve nonpolar compounds easily, while polar solvents will dissolve polar compounds. Because fat is nonpolar while water is a polar solvent, the term lipophilic ("fat-loving") or hydrophobic ("water-fearing") is used to indicate that a compound is nonpolar, while hydrophilic and lipophobic indicate the reverse. Amphipathic compounds have both a lipophilic and a hydrophilic part, and prefer surfaces and phase borders (such as cell membranes).

To characterize nonpolar compounds as "hydrophobic" indicates that there is some kind of repulsion between water and nonpolar molecules that forces them to stay apart, and this is a general misunderstanding. There is no such repulsion, water and fat molecules will be attracted to each other by van der Waals forces. The reason that water and fat do not mix is instead that if fat molecules were to be inserted among the water molecules, strong, stabilizing hydrogen bonds between the water molecules would have to be broken and that would cost the system

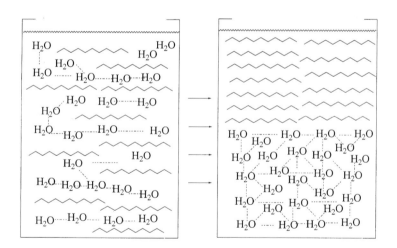

The system is more stable if the maximum number of strong hydrogen bonds between the water molecules can be formed, and will spontaneously go from left to right.

Figure 3.8. Nonpolar compounds will not dissolve in water.

energy as indicated in Figure 3.8. This effect is important and useful in biological systems, where it is called the hydrophobic effect or hydrophobic repulsion.

Although many organic compounds can be classified either as hydrophilic (e.g. carbohydrates) or lipophilic (e.g. fats and waxes), most fall between these two extremes. It has therefore been found suitable to describe the solubility of a certain compound in terms of its distribution between two solvents that are not soluble in each other. If for instance 1 mg of a chemical is mixed with 1 ml each of the two inmiscible solvents water and cyclohexane and the mixture is allowed to reach equilibrium, the relative amounts of the compound that dissolve in the two phases (water/cyclohexane) give an indication about its solubility. As octanol to some extent resembles the amphipathic compounds present in biological membranes (see Chapter 4), the two-phase system octanol/water is most frequently used today to characterize a compound's solubility. The proportion of the compound in the two phases is called P, and for practical reasons (for highly lipophilic compounds as DDT P is over 1,000,000) the logarithm, logP, is often used. In Table 3.3, the solubility in water and the P values of some common organic compounds are listed for comparison.

The smallest alcohols, methanol, ethanol and the two propanols, are all mixable with water in all concentrations at 20 °C, because of the strong hydrogen bonding to the hydroxyl groups and the relatively small effect that the alkyl groups have on the water–water interactions. However, for three of the butanols, 1-butanol, 2-methyl-1-propanol (isobutanol) and 2-butanol, the water solubility is limited and the P values over 4, as the larger alkyl group will interfere with more hydrogen bonds between water molecules and will increase the van der Waals forces between the butanol molecules. There is a trend in the water solubility in the butanol series, and the fourth member, 2-methyl-2-propanol (*tert*-butanol), is completely soluble in water. This is due to the different total size and surface area of the alkyl groups which to different degrees will disturb the hydrogen bonds between water molecules and give rise to attractive van der Waals forces between the organic molecules. For pentanol and higher alcohols the water solubility is low and the P values high. The two cyclic ethers dioxane and tetrahydrofuran (THF) are mixable with water, while diethyl ether, which can be seen as a open form of tetrahydrofuran, and diisopropyl ether have limited water solubility and higher P value. The comparison between tetrahydrofuran and diethyl ether is interesting, as tetrahydrofuran also has a considerably higher boiling point (66 °C) than diethyl ether. In diethyl ether there is free rotation around all bonds, resulting in a partial cancelling of the dipole resulting from the carbon–oxygen bonds by the smaller carbon–(carbon–with–oxygen) dipoles, while the dipoles in tetrahydrofuran are forced to work together, making the latter a more polar molecule. The carbonyl function is polar as it is, and may in addition take part in the formation of an hydrate (see Figure 3.2) or an enol (see Figure 3.14). Small aldehydes and ketones are mixable with water, but starting with propanal and butanone the solubility is limited (note the similarity in water solubility of 2-butanol and butanone, Table 3.3). Carboxylic acids are mixable with water up to pentanoic acid (and its isomers) while all esters (even methyl formate) have limited water solubility.

The lipophilicity of organic compounds is a central factor during bioaccumulation, the "extraction" of chemicals from the environment by organisms and the transportation and concentration from species to species in food chains, and this will be further discussed in Chapter 6.

Table 3.3. The relative solubility in octanol/water (P) compared with water solubility.

Name	Structure	Solubility in water (% w/w, 20 °C)	P (octanol/water)
Methanol		∞	0.2
Ethanol		∞	0.5
1-Propanol		∞	1.8
2-Propanol		∞	1.1
1-Butanol		7.8	7.6
2-Methyl-1-propanol		8.5	5.8
2-Butanol		12	4.1
2-Methyl-2-propanol		∞	2.2
1-Pentanol		2.5	36
Dioxane		∞	0.5
Tetrahydrofuran		∞	2.9
Diethyl ether		6.9	7.8
Diisopropyl ether		0.9	33
Acetone		∞	0.6
Butanone		26	2.0
Cyclohexanone		2.3	6.5
Acetic acid		∞	0.7
Methylacetate		24	1.5
Ethylacetate		7.9	5.4

3.3.3 Adsorption and absorption

Sorption is the process when a chemical (a sorbate) becomes associated with a solid material (a sorbent), adsorption refers to sorption on a surface (two dimensions) while absorption is sorption into a material (three dimensions). As most organic compounds emitted to the environment eventually contact solid materials, sorption to organic and inorganic particles in the soil for example will influence the fate of an organic pollutant. Adsorption depends on intermolecular forces between the sorbent and the sorbate, and results in a equilibrium between the adsorbed and free state. The free state is normally as a solution in the surrounding water, but it can of course also be the gas phase. Adsorption means that a compound is taken out of circulation, at least temporarily, and can be seen as something that protects organisms from pollutants. However, adsorbed compounds are less available to sunlight and microorganisms, and are thereby not degraded at the same rate as in a water solution. This means that their lifetime is prolonged, and if the soil is disturbed in some way (for example during construction work) the pollutant may reappear.

Soil is a complex mixture of various inorganic minerals as well as organic matter, and its composition is different from location to location (see also Chapter 4). While the amount of organic matter is substantial at the surface, very little is found deeper down where for instance the groundwater is present. In addition, the size of the particles in soil is an important factor, as small particles have a much larger surface per gram and thereby much better qualities as a sorbent. This makes it difficult to generalize about how efficiently different types of compounds are sorbed. As most minerals contain charged groups at their surface, charged compounds and easily protonated/deprotonated compounds are normally efficiently adsorbed. An example is the herbicide paraquat (see Figure 3.9), which is toxic to the lungs (Section 10.2.5) but has been considered relatively safe to use from an environmental point of view. Paraquat is an efficient herbicide that kills only the plants whose leaves it contacts. The paraquat that ends up in the soil is strongly sorbed and will not wash away.

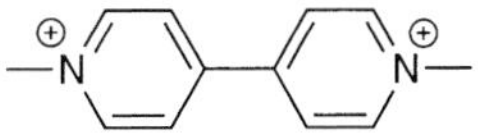

Figure 3.9. The paraquat ion has two positive charges.

Most organic pollutants are devoid of electric charge, but can still be adsorbed to inorganic materials if they contain polar groups that are attracted by ion–dipole and dipole–dipole forces. However, more important is their sorption to organic matter in soil, and this is strongly correlated with the solubility of the sorbate. Neutral compounds with high water solubility (having small logP values) are poorly sorbed by organic matter, while lipophilic compounds in general are sorbed well. The soil particles are not always retained in the soil, and very small organic particles (approximately 1 μm in diameter), so called organic colloids (e.g. humic substances), behave as if they are part of the water solution. Colloids do not separate from water by gravitation, and may be seen as microparticles or macromolecules that can still function as sorbents. Organic colloids may in this way substantially increase the "water solubility" of pollutants.

3.3.4 Persistence

A chemical is called persistent if it stays unchanged for a very long time in the environment. Most inorganic chemicals can be regarded as persistent, they are at the most transformed to other inorganic compounds. The life-time of an organic chemical may vary considerably, many are rapidly degraded by for example microorganisms and sunlight, while others are difficult to get rid of. Many compounds developed for specific purposes have been designed to be persistent, as this was considered to be an advantage, and examples are most chlorinated pesticides used in the 1950s and 1960s. Persistence is not really a chemical property, it is a lack of other properties that results in the preservation of a compound. A half-life of a chemical in the environment increases if it is poorly soluble in water and if it is strongly sorbed in the soil or at the bottom of a sea, because the microorganisms and sunlight will not get at it. Another important property that persistent chemicals should not have is reactivity. It should in fact be as inert as possible to both chemical and biological degradation. Saturated chemicals are therefore generally more persistent than the corresponding unsaturated analogues, although the aromatic hydrocarbons are the most stable. The persistence of aromatic hydrocarbons will increase with increasing number of substituents, especially halogens.

3.3.5 Reactivity

A compound's reactivity is perhaps the most important factor when it comes to the ability of a chemical to cause a toxic effect in an organism. Even if volatility and solubility as well as other properties of a compound are important for bringing it to the target, it normally has to trigger or precipitate a chemical reaction in some way to induce the toxic effect. However, it should be remembered that chemical reactivity in general can be anything from an explosion to subtle interactions between a ligand and a receptor in a nerve cell. Also, it is not automatically the most reactive chemicals that are the most dangerous, as such compounds often are well-known and their hazardous properties may be evident from their behaviour. Apparently non-reactive compounds are often considered to be relatively harmless, but may for example slowly be converted inside the body to a form that is carcinogenic. It is therefore important to have a basic understanding of the different forms of reactivity, and how they can affect an organism as well as be related to the chemical structure of the reactive compound.

Explosives are compounds or mixtures of compounds that rapidly may react with the generation of large amounts of heat and pressure, if a certain amount of initial energy (activation energy) is provided. For very sensitive explosives it is enough if they are scraped or struck, while others require an ignition spark or a shock wave from a detonator. If the reaction spreads through the explosive material less than 1,000 m/s it is called a deflagration, and if the speed is higher it is a detonation. In most cases, explosions can be regarded as an internal combustion, without external oxygen.

3.3.5.1 Explosives

Compounds that have an intrinsic instability may be prone to rapid decomposition which converts them into other more stable compounds and releases large amounts of energy. Such compounds cause explosions or detonations, and although they are not the main subject of this book it should be recognized that hazards with explosive chemicals are of major concern in many workplaces where chemicals are used. In the text, comments will be made about functional groups and combinations of functional groups that frequently are associated with explosives, but not in a comprehensive way.

3.3.5.2 Acids and bases

Acids and bases are reactive and will affect biological systems in various ways depending on their strength and concentration. Strong acids and bases in high concentrations will simply disintegrate biological tissue and destroy it, while weaker acids and bases in lower concentrations give milder effects. Formic acid, hydrofluoric acid, sodium or potassium hydroxide, and nitric acid are examples of acids and bases that are particularly harmful because they break up the skin and make the chemical penetrate deeper. A substantial part of all accidents that happen in workplaces with chemicals and that lead to bodily harm involve acids and bases. Besides the strength and concentration, the damage that exposure to acid or base will cause to a person also depends on the route. Some organs are extremely sensitive (e.g. the eyes) while other parts of the body are more resistant. The external effects of acids and bases are fairly well-known to anybody working with chemicals, and will not be further discussed in this text. However, it should be stressed that accidents with acids and bases are very common and that such compounds never should be handled unless protective measures (*always wear protective glasses*) have been taken. The internal effect of acids will be discussed in Chapter 10.

The acid-base concept is defined in a number of different ways:

According to the *Brønsted-Lowry* definition, an acid is a proton donor and a base is a proton acceptor, and this is the way we normally think of acids and bases.

A *Lewis acid* is a compound or an ion that can form a new covalent bond by accepting a pair of electrons while a *Lewis base* is a compound or an ion that can form a new covalent bond by donating a pair of electrons. All Brønsted-Lowry acids/bases are also Lewis acids/bases, in addition electrophiles (section 3.4) are also Lewis acids while nucleophiles (section 3.4) are Lewis bases.

The *hard and soft acid-base* theory (HSAB) states that hard acids favour binding to hard bases, while soft acids favour binding to soft bases. A hard acid is typically a small positively charged ion (e.g. lithium and sodium ions) while a hard base is a small negatively charged ion (e.g. chloride and hydroxide ions). Soft acids and soft bases ware typically larger neutral molecules with high polarizability. The hard-hard interaction depends on the attraction of opposite charge, while the soft-soft interactions, which are important for the reactions between nucleophiles and electrophiles discussed in section 3.4, take place between filled orbitals and empty orbitals.

3.3.5.3 Reducing and oxidizing agents

Oxidation is the loss of electrons while reduction is the gain of electrons. The two are always coupled, so the oxidation of a substrate is accompanied by the reduction of another chemical, the oxidation agent. Oxidations are for example the introduction of a heteroatom into a hydrocarbon and of unsaturations in saturated compounds while reductions are the opposite, as illustrated below:

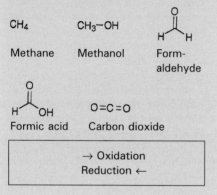

CH$_4$ CH$_3$—OH

Methane   Methanol   Form-
                                aldehyde

Formic acid   Carbon dioxide

→ Oxidation
Reduction ←

Compounds or reagents used for oxidations and reductions will normally react with and damage biological tissue. Strongly oxidizing and reducing agents, for example sodium dichromate and potassium hydride, are highly reactive and will damage whatever tissue they contact. During oxidations and reductions, acid or base will be formed as secondary products and this will aggravate the effect. However, oxidations and reductions may also be more selective and for example change the oxidation stage of a metal ion needed for a certain enzymatic reaction (for an example, see subsection 10.1.4.2), or generate highly reactive radicals (*vide infra*), and we get back to some examples of selective toxicity caused by oxidations and reductions.

3.3.5.4 Excited molecules

Molecules can be excited to reactive states by the input of energy in some form, most commonly electromagnetic radiation in the form of sunlight. Sunlight is a spectrum of wavelengths of which the visible region is only a part. The most powerful is the ultraviolet light (UV-light), with wavelength between 10 and 400 nm, and this is able to dissociate molecules (e.g. $O_2 + hv^-$ > 2 O, by UV-light with the wavelength 242 nm) or ionize them to produce free electrons (e.g. $H_2O + hv^-$ > $H_2O^+ + e^-$, by UV-light with the wavelength 98 nm). Such reactions take place in the upper parts of the atmosphere, and the reactive species formed actually serve to clean the atmosphere of air pollutants. However, the photochemical reactions can also be severely disturbed by pollution, and this will be discussed in Chapter 12. Little UV-light penetrates the atmosphere all the way down to the earth, and both humans as well as other organisms have developed protections against this potentially very dangerous radiation. Nevertheless, exposure to sunlight is associated with a increased risk of skin cancer, and this is caused by photoexcited reactions that take place in the cells of the skin. For example, UV-light can excite thymine, a

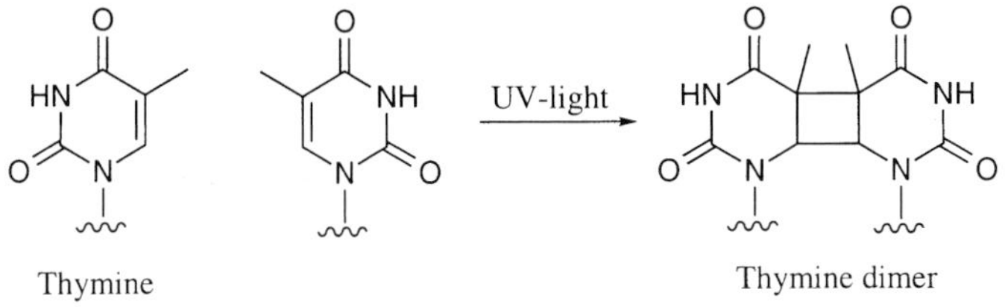

Figure 3.10. The formation of thymine dimers after UV radiation of our cells.

component of our genetic material, to a form that readily reacts with another thymine and forms a thymine dimer (as shown in Figure 3.10).

Another example, which will be discussed in Chapter 10, is the phototoxicity of some plant metabolites (e.g. the psoralenes) and antibiotics (tetracyclines), the toxic effect resulting from the sunlight exciting the chemicals in the skin (they do not have to be applied directly to the skin) to reactive forms.

3.3.5.5 Electrophilic compounds

The fact that biochemistry is based on molecules containing nucleophilic functionalities will for obvious reasons make us sensitive to electrophilic compounds that react with and bind covalently to for example proteins and genetic material. In general, such nucleophilic substitutions or additions will irreversibly transform functional biomolecules to forms that will not work or even be harmful. Important examples are allergens that provoke the immune system to react to the body's own proteins, and carcinogens that transform genes to forms that can no longer control the rate of cell division. Electrophilic compounds are involved in many of the toxic effects that we are most concerned about, e.g. allergies, tumours and teratogenic effects, discussed in some more detail below.

3.3.5.6 Radicals

Radicals are organic compounds with unpaired electrons, and as a rule they are very reactive. According to fundamental chemical principles (Hund's rule), electrons prefer to be paired, and the search for another electron to pair with is the driving force for the reactivity of radicals. While some radicals, for example the hydroxyl radical, are so reactive that they will react with the first molecule they encounter, others are less reactive and may even survive the time it takes to diffuse around a cell. (The reactions of radicals will be discussed shortly in Section 3.5.) Radicals may be formed by the homolytic cleavage of a covalent bond, although this normally requires so much energy (e.g. an open flame) that it is not relevant in biochemical reactions. However, radicals may also be formed by the addition or abstraction of an electron to or from a compound by a reducing or oxidizing agent, for instance by the enzymatic systems that metabolize most of the unnatural chemicals we are exposed to, and this will be further discussed in Chapter 7.

3.3.5.7 Non-covalent binding

A compound may react with other compounds in ways that produce toxic effects by forming non-covalent bonds (electrostatic forces). Such reactions are normally reversible, especially when weak dipole–dipole forces or hydrogen bonds are involved, although their effects to an organism may not be so. Ionic bonds may in rare cases be involved in toxic effects, an example is the reaction between calcium and oxalate (the anion of oxalic acid) yielding an insoluble product that may form stones that block the kidney. Many of the classical toxic compounds acting for example on the nervous system bind to receptors in the membranes of the nerve cells and trigger a reaction that may result in for example cramps or paralysis. The non-covalent bonds are the forces that have been discussed earlier in this Chapter, and this is the way that most pharmaceutical agents act, by binding to receptors and enzymes and either stimulating or blocking them. A compound that fits perfectly in for example a receptor can be seen as a key, if the receptor is the lock, and it is evident that the compound must have its functional groups in the right places in space in order to fit. The stereochemistry of the compound is therefore extremely important, and for pharmaceutical agents it is well-known that one isomer and one enantiomer is the most active. As we shall see in Chapter 10, non-covalent binding is also the mechanism by which some toxic chemicals used in workplaces act, and it should be recognized that it is more difficult to find general relationships between structure and activity for such compounds.

3.4 Reactions between nucleophiles and electrophiles

Probably most important chemical reactions causing damage that, for example, may lead to cancer, are nucleophilic substitution and addition. A nucleophile is an electron donor, with at least one unused electron pair that can participate in a chemical reaction. It may be neutral or have a negative charge. An electrophile is an electron acceptor, positively charged or neutral and with the capability in some way to accommodate the electrons donated by a nucleophile. The biochemical components, e.g. proteins and nucleic acids (the topic of the following Chapter), are good or reasonably good nucleophiles, while organisms contain virtually no electrophiles, and it is obvious that the exposure of an organism to electrophiles, or chemicals that will be converted to electrophiles by its metabolism, will create problems.

3.4.1 Nucleophilic substitutions

In a nucleophilic substitution or addition, a nucleophile will donate (or "push") an electron pair and is consequently a Lewis base, while the electrophile act as an electron sink and will accept (or "pull") the electron pair, and therefore is a Lewis acid. The actual reaction is the flow of electrons from the source to the acceptor, from nucleophile to electrophile, under the influence of electrical charge, and can be compared with the flow of water in a stream due to the influence of gravity. Although the nucleophilicity and electrophilicity of the reactants are not the only rate determining factors, as discussed below, a good nucleophile will normally react fast with a good electrophile while a poor nucleophile will react slowly with a poor electrophile. However, fast and slow are relative concepts and, while a very good electrophile may react upon contact with biological tissue and for example give rise to chemical burns, poor electrophiles may react selectively with components of genetic material and cause cancer.

The flow of electrons in a nucleophilic substitution or addition is indicated by curved arrows, a full-headed curved arrow indicates the movement of an electron pair while a half-headed curved arrow indicates the movement of a single electron in a radical reaction (*vide infra*). An illustrious example from inorganic chemistry is the reaction between water and hydrogen chloride (HCl) which takes place if hydrogen chloride gas is bubbled through liquid water, and produces hydrochloric acid:

Figure 3.11. The reaction between water and HCl.

The base of the arrow always emanates from the site of electron density, either an unshared electron pair or a bond, and the head always points to the site that will accept the electrons. They will either form a new unshared electron pair of an atom or a new bond between two atoms. Nucleophilic substitutions take place with electrophiles that contain a suitable leaving group that will act as the electron sink, parting with the additional electron pair initially introduced by the nucleophile. In nucleophilic additions the electrophile is either positively charged, e.g. a carbocation, or has a polarized multiple bond activated by for example a carbonyl group. The latter addition includes additions to carbon–carbon double or triple bond conjugated with electron-withdrawing groups, also called Michael additions. Reactions between nucleophiles and electrophiles normally also include proton transfers, protonations of anions or unshared electron pairs or deprotonations producing anions or unshared electron pairs, which are important parts of the reaction but are often taken for granted and not shown explicitly.

Nucleophilic substitutions may be characterized as S_N1 or S_N2, depending on the number of molecules that are involved in the rate-determining step of the reaction. In S_N1 (Substitution Nucleophilic monomolecular) reactions, the electrophile reacts alone in an initial slow step to form a highly reactive cation that is extremely electrophilic and immediately reacts with the nucleophile. The reaction rate is therefore determined by the concentration of the electrophile alone, the concentration of the nucleophile does not matter (as long as it is there). S_N1 reactions are favoured by factors that stabilize the reactive intermediate formed, e.g. the presence of electron-donating groups. An example is the substitution of a chlorine for a methoxy group during the transformation of *tert*-butylchloride to methyl-*tert*-butylether in methanol:

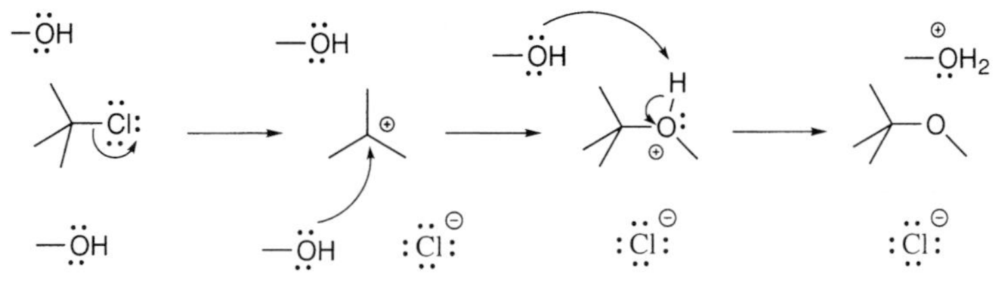

Figure 3.12. The S_N1 reaction between *tert*-butylchloride and methanol.

The rate-determining step involves only one molecule, *tert*-butylchloride, and as soon as the trimethylmethyl cation is formed it directly adds the nucleophile methanol. After the addition the oxygen of the added methanol has a positive charge, because it has donated one electron pair to the cation. This is taken care of by the abstraction of the proton from the same oxygen by another methanol molecule, acting as a base, and the two electrons of the oxygen–hydrogen bond make a second unshared electron pair on the ether oxygen. Methanol is not known to be a good nucleophile, but that does not matter in a S_N1 reaction. Any nucleophile will do, simply because the true electrophile (the trimethylmethyl cation) formed from the substrate is so very reactive. In that sense one can say that electrophiles that react via a S_N1 mechanism are less dangerous, because they are not choosy when it comes to selecting a nucleophile. Organisms always consist of large amounts of water that will be sufficiently nucleophilic to take care of such electrophiles.

In a S_N2 reaction the reaction rate is instead determined by the concentration of 2 molecules, both the electrophile and the nucleophile, and it is called a bimolecular reaction. The two have to collide and undergo a reaction in which both participate. An example is the reaction between a hydroxide ion and methylchloride:

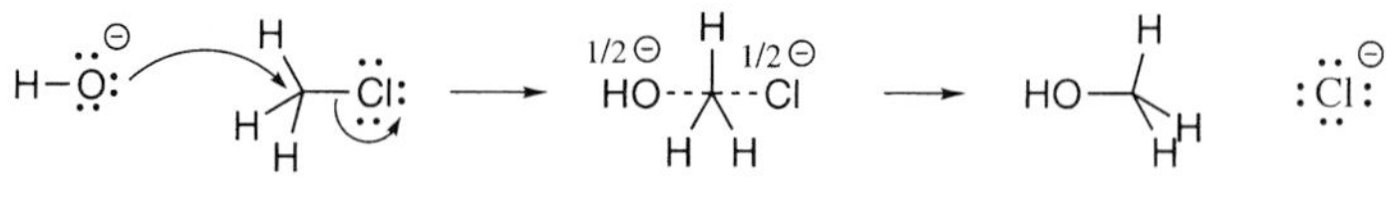

Figure 3.13. The S_N2 reaction between methyl chloride and a hydroxide ion.

The hydroxide ion, the nucleophile, is attracted by the positively charged carbon in the electrophile, which can accommodate the electrons donated by the nucleophile by pushing the electrons between the carbon and the chlorine (in the carbon–chlorine bond) towards the chlorine. In the middle of the reaction one reaches a high-energy transition state that appears to have five bonds to the central carbon. This collapses to the products, in which a new bond has been formed between the oxygen and the carbon while the carbon–chlorine bond has been broken. Chlorine leaves the electrophile as a negative ion, together with the two electrons that bound it to the carbon. It is called the leaving group in the reaction and as such it ultimately takes care of the electrons donated by the nucleophile. As the nucleophile takes part in the rate-determining step of S_N2 substitutions, the nucleophilicity of the nucleophile will be an important factor for the reaction. A S_N2 electrophile will react fast with a good (but perhaps not at all with a poor) nucleophile, and may therefore be highly selective for nucleophilic groups in proteins and genes. It should be mentioned that nucleophilic substitutions seldom proceed via pure S_N1 or S_N2 mechanisms, and it is common that mixed mechanisms are observed in which one or the other predominates.

3.4.2 Nucleophilic additions

Nucleophilic additions, for example to carbon–carbon double bonds activated by electron withdrawing groups (EWGs), are very common among the toxic chemicals. Carbon–carbon double bonds are normally not electrophilic, but if it is polarized by its substituents and one of them has the ability to accept an extra pair of electrons, they become good electrophiles. The most common EWG that activate carbon–carbon double bonds for nucleophilic additions is the carbonyl group, conjugated with the double bond as for example in α,β-unsaturated aldehydes and ketones (α and β because the unsaturation is located between the carbons α and β to the carbonyl group). Nucleophilic additions to α,β-unsaturated carbonyl compounds as well as α,β-unsaturated nitriles and nitro compounds are called Michael additions.

The addition of for example esters of acetoacetate to unsaturated aldehydes, amides, esters, ketones and nitriles was described already in 1887, by the American chemist Arthur Michaels who has given the reaction its name.

The simplest α,β-unsaturated aldehyde is acrolein, a powerful electrophile and a highly toxic chemical. As can be seen in Figure 3.14, the carbonyl group is polarized because it contains an electronegative oxygen atom, which actually causes the electrons in the carbon–carbon double bond to be slightly drawn towards the carbonyl group. Three reasonable resonance structures can be drawn for acrolein, the one without electric charge is of course the most stable but the other two are not too bad because the negative charge is positioned on the electronegative oxygen. This indicates that both the carbonyl carbon and the β-carbon have substantial positive charge, and it it to these carbons that nucleophiles will add (see Figure 3.14).

Besides S_N1 and S_N2 substitutions and nucleophiles additions there is an array of other reactions that electrophiles may undergo, although this lies outside the scope of this text.

3.4.3 The electrophile

Electrophiles involved in nucleophilic substitutions and additions relevant for toxicity contain functionalities of the general types discussed in this Section. In addition, other features (e.g. steric and electronic factors) will influence the reactivity of an electrophile, and this will be discussed in Section 3.6.

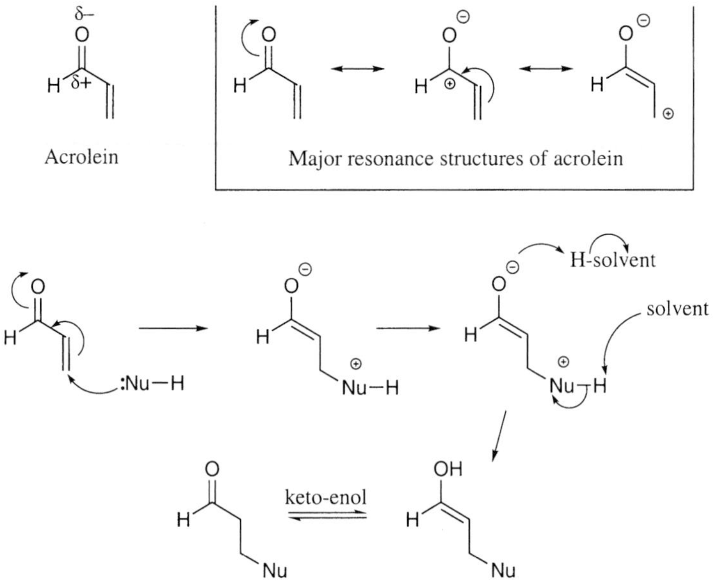

Figure 3.14. The Michael addition to acrolein.

3.4.3.1 Strained cyclic hydrocarbons and heterocycles

In epoxides, aziridines and thiiranes (three-membered ring with oxygen, nitrogen or sulphur), there is a combination of polarized bonds that will attract a nucleophile, and a ring strain that would be relieved if the ring opened up. The ring strain is caused by the fact that saturated hybridized carbon atoms have 109.5° between their sp^3 orbitals while the formal angles in an equilateral triangle is 60°, so the bonds are somewhat "banana shaped" and this increases the energy of three-membered rings. The energy that the system will gain by opening the ring is the driving force for the reaction (see Figure 3.15). Cyclopropanes in most cases are not electrophilic, because they lack the heteroatom to polarize a bond in the ring. In some cases, electron withdrawing substituents on the cyclopropane ring may activate it and facilitate a nucleophilic attack. An example is the microbial antibiotic duocarmycin A, which possesses potent biological activities (e.g. antitumour activity) because the cyclopropane ring can react with certain positions in the genetic material. In four-membered rings, e.g. oxetanes, the ring strain is smaller and normally not enough to make these compounds electrophilic. However, if the polarization of the carbon–oxygen bond increases because of the presence of an EWG, as is the situation in β-propiolactone, also oxetanes will be good electrophiles (β-propiolactone is a well-known chemical carcinogen that reacts with genetic material).

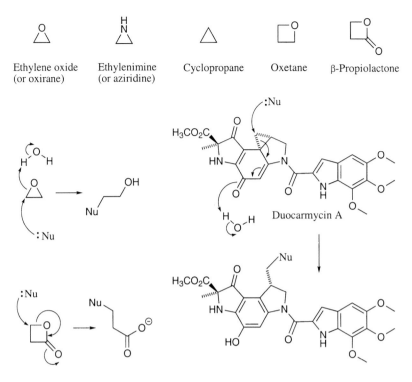

Figure 3.15. Electrophiles containing strained cyclic hydrocarbons and heterocycles.

3.4.3.2 Weak single bonds between a sp³ carbon and a leaving group

Atoms or groups (neutral or positively charged) bound to a sp^3 carbon with a weak, polarized bond have a tendency to leave together with the electron pair of that bond, either as an anion or as a neutral species. Such atoms or groups are called leaving groups and we have already observed that chlorine may act as a leaving group in S_N1 and S_N2 reactions. In general, the quality of a leaving group is correlated with the acidity of its corresponding acid, and good leaving groups are the bases of relatively strong acids having negative pKa values. (The pKa value of an acid is a measure of its acidity, at the pH corresponding to the pKa at which the acid is ionized to 50 %. For example, the pKa of acetic acid is 4.8, and in a water solution having pH 4.8 acetic acid is 50 % present as CH_3COOH and 50 % as CH_3COO^-. If the pH is 3.8 it is 90 % CH_3COOH and 10 % CH_3COO^-, while if the pH is 5.8 it is 10 % CH_3COOH and 90 % CH_3COO^-.) More basic leaving groups (having weaker corresponding acids with positive pKas) can be regarded as fair (up to pKa 10) to poor (pKa over 10). In Figure 3.16, some examples of leaving groups of different qualities are shown.

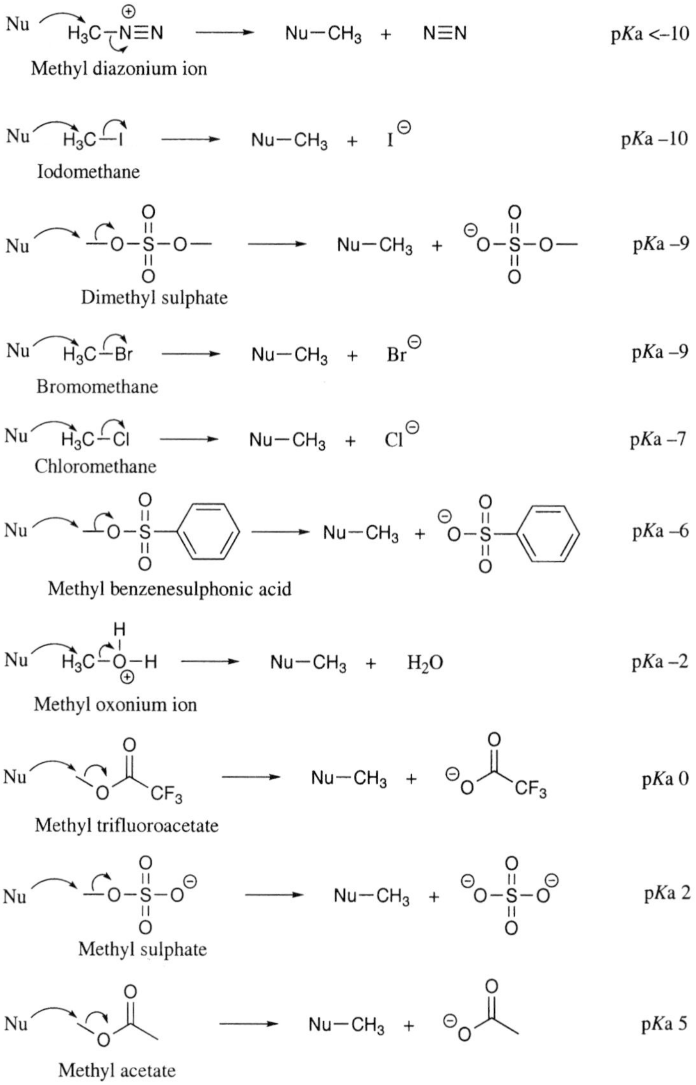

Figure 3.16. Some important leaving groups and the pKa values of their acids.

Dimethyl sulphate is obviously a better electrophile than methyl sulphate, because methyl sulphate is a better leaving group than sulphate. (Dimethyl sulphate is an efficient methylating agent used in chemistry to put methyl groups on various nucleophiles.) However, methyl sulphate is not a poor electrophile and electrophiles having sulphate as a leaving group in activated positions may be reactive enough to cause toxic effects. Examples of this will be given in Chapter 8. The hydroxyl group in alcohols is a poor leaving group, the pKa of the corresponding acid

(H_2O) is 16. However, it the oxygen is protonated (methyl oxonium ion in Figure 3.16 is protonated methanol) it becomes an excellent leaving group. This trick can be performed in the laboratory, and also by enzymes carrying out reactions that one would not expect to proceed with a reasonable speed. Acetate is normally not a leaving group in electrophiles relevant to chemical toxicology, but if it is attached to a saturated nitrogen instead of carbon it will become a good leaving group because the nitrogen–oxygen bond is so much weaker from the beginning.

For electrophiles with a weak single bond between a sp^3 carbon and a leaving group, substitution competes with elimination, and the ratio depends on the position of the leaving group (primary, secondary or tertiary sp^3 carbon). There are few electrophiles in biochemistry, and those that are there are used in a safe way. One example of a fair leaving group used in nature is pyrophosphate, present in for example dimethylallyl pyrophosphate and isopentenyl pyrophosphate that take part in the enzyme-catalyzed biosynthesis of terpenoids as shown in Figure 3.17. The fully protonated acid pyrophosphoric acid is a strong acid, but the anions formed after deprotonation are also acidic. The geranyl pyrophosphate formed in the reaction shown in Figure 3.17 is a monoterpenoid, which can undergo the same reaction again to form a sesquiterpenoid. Another example of an electrophile used by organisms is the cofactor SAM, discussed in Chapter 7 (Section 7.4.5).

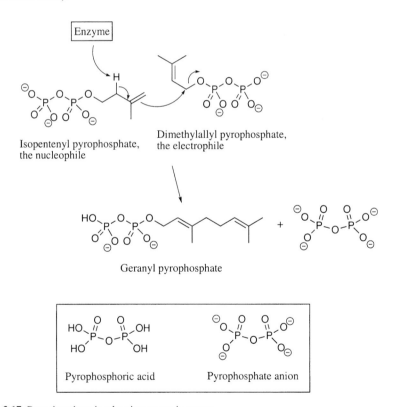

Figure 3.17. Pyrophosphate is a leaving group in nature.

3.4.3.3 Polarized double or triple bonds

We have already noted that carbonyl groups may react with water to form hydrates (Figure 3.2), and this is in fact a nucleophilic addition. The same reaction could in principle also take place with a polarized triple bond, for example in nitriles, although they are much less reactive than ketones. If the positive end of a polarized double or triple bond has a suitable leaving group, the addition may be followed by an elimination making the overall reaction look like a substitution. Carboxylic acid chlorides (see Figure 3.18), bromides and anhydrides are examples of such electrophiles, which often are very reactive. The product is an acyl derivative and the reaction is often called an acylation (in contrast to alkylations).

> The reaction between a nucleophile and an electrophile in general is sometimes called an alkylation, which by definition is a reaction in which a new carbon-carbon bond to an alkyl group is formed. Acylations are reactions during which acyl groups (R–CO–) are introduced into molecules.

Also polarized double bonds in heteroallenes (two adjacent double bonds with at least one heteroatom) can react with nucleophiles, as shown by the toxic methyl isocyanate. Other isocyanates are used for the preparation of polyurethanes used for example in upholstery. The bubbles formed in this material are caused by water reacting with the isocyanate to form an unstable carbamic acid that decomposes into an amine and gaseous carbon dioxide (see Figure 3.18).

However, most important are the conjugated additions discussed above (Figure 3.14), with an EWG activating a carbon–carbon double or triple bond. In general, the more efficiently the electrons are withdrawn from the carbon–carbon multiple bond by an EWG, the more reactive is it towards nucleophiles. Because EWGs attract electrons, they will stabilize negative charge on an adjacent carbon, and compounds with the general formula CH_3–EWG are therefore slightly acidic and can be ionized by a base (see Figure 3.19). By comparing the acidity (pKa value) of the methyl protons of various CH_3–EWGs, one can get an impression of how efficient they are in activating carbon–carbon multiple bonds (see Figure 3.19). The combination of two EWGs on the same unsaturated carbon will naturally increase the electrophilicity of the carbon–carbon multiple bond.

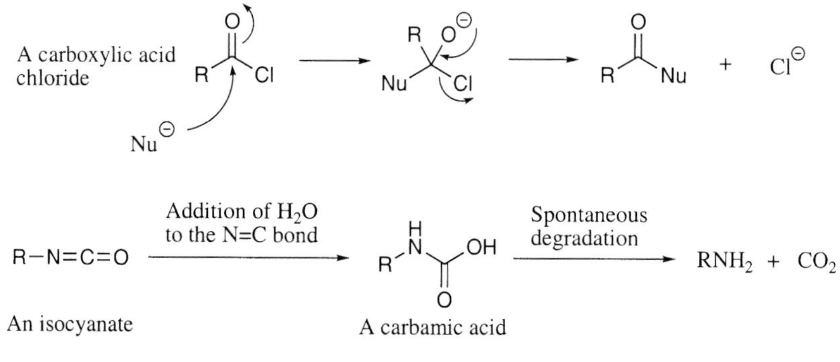

Figure 3.18. Additions to polarized multiple bonds.

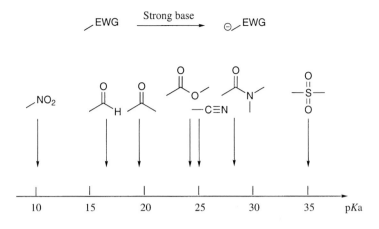

Figure 3.19. Approximative pKa values for selected CH$_3$–EWGs.

3.4.4 The nucleophile

The nucleophilicity of a nucleophile depends on its ability to donate electrons and to form a new covalent bond with the electron sink. In general the best nucleophiles have a negative charge, which not only provides it with the free electron pair necessary for the reaction but also enhances the attractive force to the partially positively charged centre of the electrophile. However, nucleophilicity does not depend on the presence of a full negative charge, the partial negative charge caused by polarized bonds and conjugations is sufficient for turning many organic compounds into good nucleophiles. Most nucleophiles contain a heteroatom, in most cases oxygen, nitrogen, or sulphur, which provides the unshared electron pair that forms the new covalent bond to the electrophile. Table 3.4 list a number of nucleophiles and their relative nucleophilicity towards the electrophile iodomethane (in methanol as solvent).

Obviously, the same heteroatom is a better nucleophile when it is negatively charged. Compounds having the same heteroatom with the same charge are more nucleophilic the more basic the unshared electron pair is. The methoxide ion is a stronger base compared to the phenoxide ion and acetate ion (methanol is a weaker acid that phenol and acetic acid), and consequently a better nucleophile. Triethylamine is a stronger base than ammonia and therefore a better nucleophile. When comparing nucleophilic atoms of the same column of the periodic table, nucleophilicity increases with increasing atom number. This is due to the fact that the electrons in a larger atom (e.g. sulphur) are less tightly associated with the nucleus compared to those of a smaller atom (e.g. oxygen), the electrons are more polarizable and may interact more efficiently with the types of electrophiles that we are mainly concerned with.

Table 3.4. The relative nucleophilicity of various nucleophiles towards iodomethane in methanol.

CH$_3$OH	(methanol)	1
F$^-$	(fluoride ion)	500
CH$_3$COO$^-$	(acetate ion)	20,000
Cl$^-$	(chloride ion)	23,000
(CH$_3$CH$_2$)$_2$S	(diethyl sulphide)	220,000
NH$_3$	(ammonia)	320,000
PhO$^-$	(phenoxide ion)	560,000
Br$^-$	(bromide ion)	620,000
CH$_3$O$^-$	(methoxide ion)	1,900,000
(CH$_3$CH$_2$)$_3$N	(triethylamine)	4,600,000
CN$^-$	(cyanide ion)	5,000,000
I$^-$	(iodide ion)	26,000,000
(CH$_3$CH$_2$)$_3$P	(triethyl phosphine)	520,000,000
PhS$^-$	(thiophenoxide ion)	8,300,000,000

3.5 Radical reactions

As has been mentioned, most radicals are highly reactive due to the presence of an unpaired electron. They will react with other compounds by abstraction of an atom, by addition to a multiple bond, or by combination with another radical (see Figure 3.20). The first and the second reactions generate one radical for each consumed, and radical reactions are often called chain reactions. As the concentration of radicals is low the third reaction is unlikely to happen, but it is important because it will effectively take radicals out of circulation. The fishhook arrows used in Figure 3.20 show the change in position of single electrons, not electron pairs.

Radicals and radical reactions have attracted considerable attention by toxicological chemists because molecular oxygen readily participates in radical reactions, forming reactive and toxic oxygen species. This will be further discussed in Chapters 7 and 10.

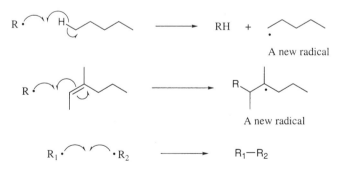

Figure 3.20. Principle reactions of radicals.

3.6 Factors modulating reactivity

The reactivity of a compound naturally depends on a number of factors, and large differences in the reactivity may be observed in a group of compounds that contain a certain reactive functional group. Most important are the electronic and steric effects, which are related to the reactive chemical, but the conditions under which the reaction is carried out may also influence the outcome. In the human body the temperature and pH normally do not change, but a compound would react differently if dissolved in the water solution of a cell or present inside the lipophilic membranes (i.e. different solvents). In addition, a large number of enzymatic conversions in organisms, as well as spontaneous chemical transformations, will take place with most compounds. As the products formed may be more reactive and more toxic, it is important to have an idea about which conversions and transformations are likely to take place.

3.6.1 Electronic effects

The electronic effects are divided into inductive (or field) effects and resonance effects. Consider for example compounds that are reactive because they are acidic. Weak organic acids are carboxylic acids and phenols. As a class of compounds carboxylic acids are depicted as R–COOH, and phenols as Ar–OH, but the acidity of individual carboxylic acids and phenols depends greatly on what is R and Ar. Compare the acidity of acetic acid and trichloroacetic acid (see Figure 3.21). The latter is almost 10,000 times stronger because the three electronegative chlorines makes the α-carbon partially positively charged and this stabilizes the negative charge of the carboxylate group. Trichloroacetic acid is therefore more willing to give up its acidic proton and become an anion, compared to acetic acid, because of the inductive effect of its α-substituents.

A similar difference is observed between phenol and 4-nitrophenol (see Figure 3.22). To some extent this is caused by the inductive effect of the nitro group, but the main contribution

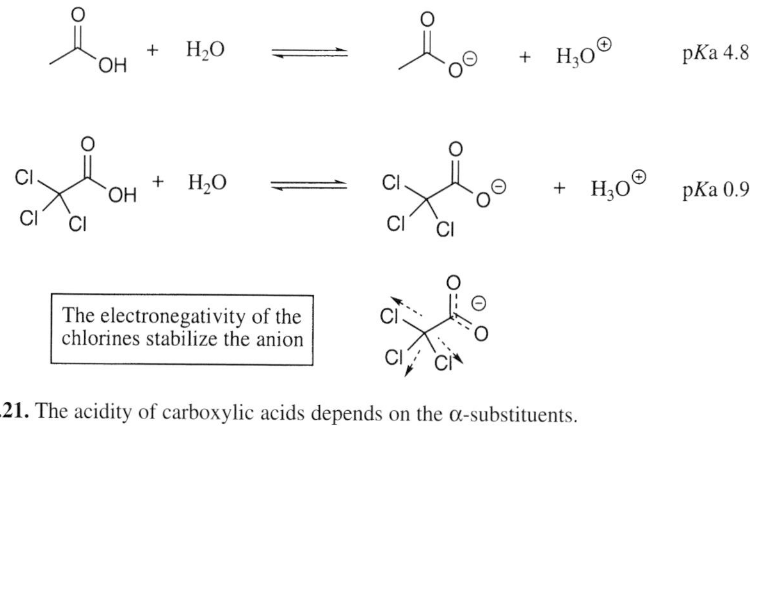

Figure 3.21. The acidity of carboxylic acids depends on the α-substituents.

When a benzene ring contains two substituents it is common to describe their relative position with the prefixes *ortho-*, *meta-* and *para-* (or *o-*, *m-* and *p-*). 2-nitrophenol, 3-nitrophenol and 4-nitrophenol can consequently be named *ortho*-nitrophenol, *meta*-nitrophenol and *para*-nitrophenol. However, with more than two substituents in the ring we have to return to numbers.

comes from its ability to accommodate an extra pair of electrons by resonance (i.e. the resonance effect). In Figure 3.22 the additional resonance form of the anion of 4-nitrophenol (compared to phenol itself) is shown, and this ability to distribute the negative charge over several (electronegative) atoms will stabilize the anion and facilitate its formation. (In Figure 3.22, the nitrogen atom in the nitro group is shown with a double bond to one of the oxygens and a single bond, made from the

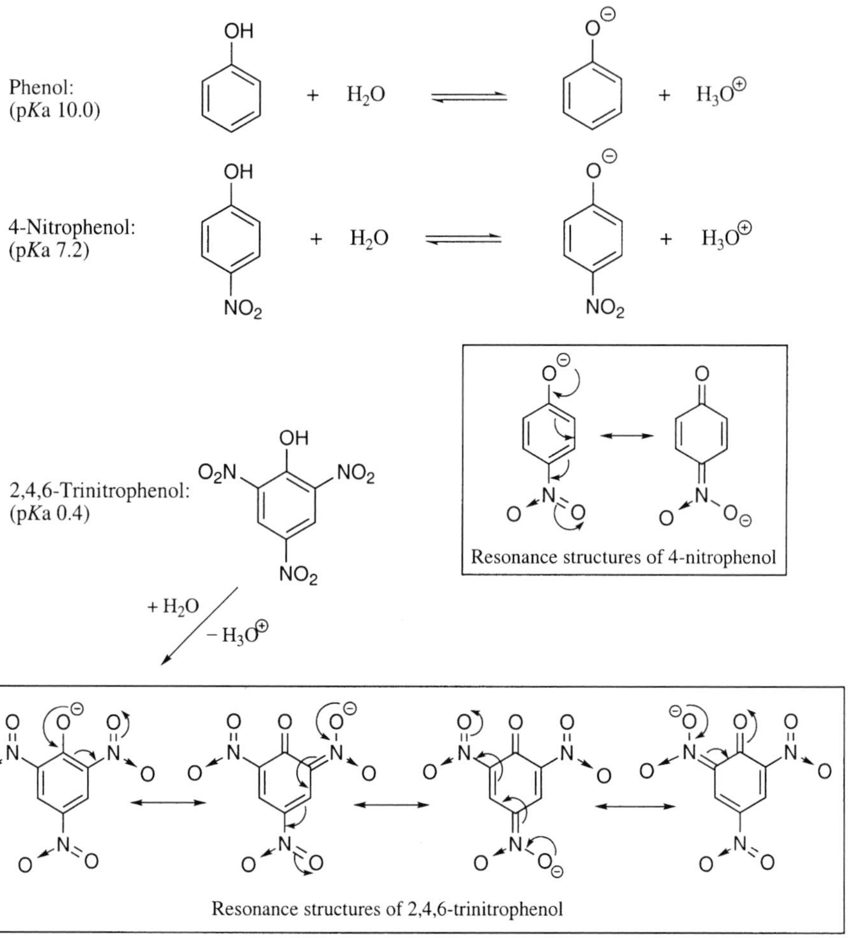

Figure 3.22. Nitro groups will increase the acidity of phenols.

unshared electron pair of the nitrogen, to the other. In reality both nitrogen–oxygen bonds are identical and both oxygens have a half negative charge, similar to the situation in carboxylates.) If the phenol is substituted with several nitro groups in suitable positions, the resonance effect is additive. 2,4-Dinitrophenol is a considerably stronger acid with a pKa value of 4.0 (more acidic than acetic acid), while the pKa of 2,4,6-trinitrophenol is 0.4. The latter is a strong organic acid, and in Figure 3.22 it is indicated how the charge of the anion is distributed on four oxygen atoms thanks to resonance. Note that the nitro groups have to be in the correct positions (2, 4 and 6). In the 3 or 5 position they can not participate in any resonance structures and will only give a weaker inductive effect (the pKa of 3-nitrophenol is 9.3).

The nitro group has a strong electron-withdrawing effect, but also for example aldehyde, keto and cyanide substituents will have an effect in the same direction. Other substituents connected via a heteroatom (e.g. ethers, amines, halogens) may instead donate electrons and have an opposite effect.

Electronic effects also influence the reactivity of electrophiles and radicals, although things get a bit more complicated. Electrophiles must have an electron sink, and are polarized by definition. Electronic effects by substituents can influence the polarization, and also stabilize or destabilize the high-energy intermediates or transition states formed during the reaction. An example is the difference in electrophilicity between bromoacetone, 2-methyl-3-bromopropene and 2-methyl propylbromide (see Figure 3.23). Bromoacetone is a highly reactive chemical, a violent lacrimator that has been used as a war gas; 2-methyl-3-bromopropene is a good electrophile that is carcinogenic; while 2-methyl propylbromide is a poor electrophile. The two former have a double bond adjacent to the carbon that holds the bromine, the leaving group, and the p orbitals of the double bond will interact with and stabilize the *p*-like orbital holding both nucleophile and leaving group in the transition state, and thereby lower the energy for the transition state of the reaction (see Figure 3.23).

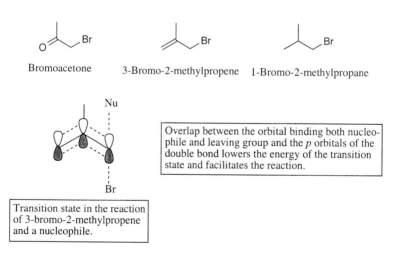

Bromoacetone   3-Bromo-2-methylpropene   1-Bromo-2-methylpropane

Overlap between the orbital binding both nucleophile and leaving group and the *p* orbitals of the double bond lowers the energy of the transition state and facilitates the reaction.

Transition state in the reaction of 3-bromo-2-methylpropene and a nucleophile.

Figure 3.23. Electronic effects may increase the electrophilicity.

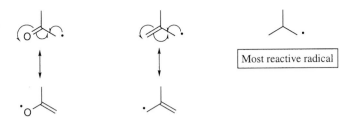

Figure 3.24. Resonance stabilized radicals are less reactive.

On the contrary, the 2-methyl propyl radical is more reactive compared to the corresponding radicals of acetone and 2-methylpropene (see Figure 3.24), because the unpaired electron in a radical is stabilized by the electrons of an adjacent double bond as indicated.

It should again be stressed that it is not necessarily the most reactive species that is the most dangerous, especially in limited amounts, as such compounds will react with for example the water of our bodies. Instead it is reactive species that are stable enough to get around and react selectively with targets such as genetic material that we should be most concerned about. However, it depends on the amount of material that one is exposed to as well as how one defines "dangerous".

3.6.2 Steric factors

In addition to electronic factors, nucleophilic substitutions are very sensitive to steric effects that block the nucleophile from coming in contact with the electrophile. The major effect is

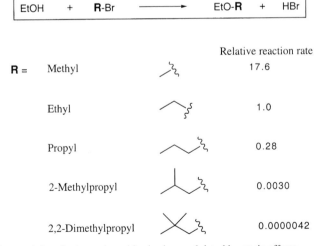

Figure 3.25. The reactivity of primary bromides is also modulated by steric effects.

observed with sterically hindered electrophiles, but nucleophiles may also be too big. For S_N2 reactions, the general and important rule is that an electrophile reacts faster the less substituted it is on the carbon with the leaving group, e.g. bromomethane reacts faster than bromoethane which in turn is a better electrophile than 2-methyl bromopropane and 2,2-dimethyl bromopropane. However, steric effects not only affect the proximate positions, but may also have an effect further away. An example is shown in Figure 3.25, showing the relative rates for the reaction between ethanol (the nucleophile) and various bromides (the electrophiles) in the solvent ethanol. For comparison the rate of the reaction with ethyl bromide has been set to 1.

Although the steric effect in electrophiles may appear to be easily predicted, this is not the case. The problem how to compile all the different chemical properties of an electrophile in a way that the risks associated with it can be estimated is addressed in Chapter 13.

3.6.3 Transformations and conversions that increase reactivity

Although transformations and conversions are not inherent in a molecule, there are in principle no compounds that are inert in an organism or in the environment. The multitude of chemical and enzymatic reactions that may take place will sooner or later change the structure of the compound and convert/transform it to another compound. If the new compound formed is more toxic we can say, from the point

> In principle, a conversion of one chemical to another in the biosphere is enzymatical (catalyzed by enzymes, *vide infra*) and takes place inside an organism, while a transformation is a chemical reaction and can take place both inside and outside organisms.

of view that is interesting in this text, that a chemical or enzymatic activation has taken place. It is crucial to understand and be able to foresee the reactions that a compound is likely to undergo inside organisms, subjected to various enzymatic systems, or in the abiotic environment, subjected to for example irradiation from the sun, if one aims at predicting the toxic and ecotoxic effects of a chemical, and this will be the theme of Chapters 7, 8 and 9.

4 The chemicals of nature

An ecosystem can be defined as a functional unit of life on earth, with animals, plants, microorganisms and minerals that together provide everything necessary for their co-existence. In principle it is not possible to identify isolated ecosystems as the whole planet is one gigantic ecosystem. However, for practical reasons it can still be of interest to discuss for example a forest or a lake, with natural boundaries to the surroundings, as an ecosystem. In an ecosystem, the species depend on each other in sensitive balances that have evolved with time. If the conditions in an ecosystem change, e.g. the climate, individual species will be favoured or disfavoured, and eventually the change will lead to the establishment of a different ecosystem with new balances. Small changes may be dramatic and even lead to the extinction of species, because they are in most cases highly specialized and depend heavily on the ecological equilibria. This has led to big differences in colour, size and function of the organisms of the world, but however different they may appear they all share the basic chemical features of life. They are all composed by cells, one cell in unicellular organisms and many in multicellular organisms (approximately 10^{15} in humans), and all living cells use the same basic chemical components and reactions for the primary life processes. As the toxicity of chemicals is caused by their interference with the chemical components and reactions of a cell, the basic principles for chemical toxicology are the same for all organisms and can to some extent be generalized. Beside the primary life processes, there are of course a number of secondary processes responsible for various specialized functions that differ completely between the species. As the effect of many toxic chemicals depends on their interference in such secondary processes, one species can be much more sensitive to a certain chemical than another. In the coming Chapters several examples of such differences will be encountered and the reasons for the differences in sensitivity will be discussed, but in this Chapter we shall focus on the chemical constituents of nature. As mentioned in Chapter 1, the species *Homo sapiens* is central for our interest, and will get more attention than other species.

4.1 The cell

A human being consists of 10 organ systems that have special tasks and functions to perform but collaborate closely with each other. A brief description is given in Table 4.1.

The organ systems are composed by organs and organ subunits, which in turn are made up by tissues (epithelial tissue, connective tissue, muscle tissue and nerve tissue), and cells. A human body contains approximately 200 significantly different cell types, although they can be classified according to their basic functions as either epithelial cells, connective cells, muscle cells or

Table 4.1. The 10 organ systems of humans.

Organ system	Major organs or tissues	Major functions
Musculo-skeletal	Bone, ligament, joints, muscle	Support, protection, motion
Integumentary	Skin	Protection, defence, regulation
Circulatory	Heart, vessels, blood	Transportation
Respiratory	Throat, trachea, bronchi, lungs	Exchange of gases, regulation
Digestive	Stomach, intestines, pancreas, liver	Digestion, absorption, excretion
Urinary	Kidneys, bladder	Regulation, excretion
Nervous	Brain, spinal cord, peripheral nerves	Detection, coordination, control
Immune	Lymphocytes, thymus, bone marrow	Regulation, surveillance, defence
Endocrine	Hormone secreting glands	Regulation, coordination
Reproductive	Testes, ovaries	Reproduction

nerve cells. The cell is the lowest common denominator for life, and although cells of simpler organisms (e.g. bacteria) are quite different from human cells it is nevertheless natural to have the cell as a starting point for our discussions about the effects of harmful chemicals.

4.1.1 Human cell types

Epithelial cells. The epithelial cells are found on all surfaces that cover organs and organ subunits, and form the border between different macroscopic components and functions. They are responsible for the exchange of compounds between the body and the environment, for example in the gastrointestinal tract (between gut contents and the blood) and the lung (between the inhaled air and the blood), as well as between different compartments of the body. The skin, which forms the boundary between the body and the rest of the environment and protects us, is also composed by epithelial cells. They are characterized by a relatively rapid division rate, something that makes them sensitive to agents that affect their genetic material and more prone than other cells to be transformed to cancer cells (see Chapter 11).

Connective cells. A major function of connective cells and connective tissue is to give support to various structures in the body, giving the organs and the organ subunits their form and shape. In most organs the connective tissue make up the bulk of the tissue, forming a cellular net onto which other cells may attach and be anchored. This is done by secreting a matrix consisting of a polymer in a polysaccharide gel into the extracellular space. Bone cells secrete calcium phosphate which is crystallized to bone, while ligaments connecting muscles are produced by connective cells secreting the proteins collagen and elastin. Besides that, connective cells perform a variety of functions, examples of cell types that are included in this category are the adipose cells, storing fat, as well as the red and white blood cells.

Muscle cells. The muscle cells have the ability to contract, and if a muscle cell is attached at both ends the contraction may generate motion. Besides the most obvious function of muscle tissue, to enable the body to move around, it should be remembered that the heart consists of muscle tissue, and that the tubes through which body fluids like the blood flow are surrounded by muscle cells that can regulate the diameter of for example a blood capillary and thereby the flow rate.

Nerve cells. Nerve cells can receive and transmit signals from and to other cells as well as from the environment. For example, muscle cells that contract do so when they receive a signal from a nerve cell. One of the central functions for nerve tissue is therefore to control and coordinate the actions of other cells, and indirectly of organs (via for example the stimulation of a gland cell producing a hormone). The signals are electrical within nerve cells and chemical between the cells, and a more detailed discussion about the normal function of nerve cells and how this is affected by neurotoxic chemicals is found in Chapter 10.

The fact that all cells in a human body originate from one single cell, the ovum after conception, deserves a moment of reflection. All the different shapes and functions must be preprogrammed in this cell and this information must be made available at exactly the right time. This is one of the many wonders of life, and we will return to it in Chapter 11.

4.1.2 The basic components and functions of a cell

The cell is the basic unit of life, and all organisms consist of one or more cells. By definition, a cell has the ability to absorb nutrients from its surroundings, to use them for the production of energy and the molecules it needs for its various activities, and to multiply. As has been discussed above, cells differ tremendously in size, appearance and function. Bacteria for example are differently organized and between 1,000 and 1,000,000 times smaller than the average human cell, although the basic chemical machinery that sustains life is roughly the same. Several bacteria have been so thoroughly investigated that one knows more or less in detail how they work, and in bacteria life can be said to consist of a few thousand, say 3000, chemical reactions. These will together perform all functions discussed below and provide all things necessary for the normal life of the bacterium. Human cells are not only bigger but of course also considerably more complicated than bacteria, and all cellular functions are not known on the molecular level. The number of chemical reactions in our cells is therefore much greater than 3000, but it is still finite and will in the near future be determinable.

The most important parts of a mammalian cell are indicated in Figure 4.1. The plasma membrane surrounds the cell and provides a boundary across which the flow of at least some chemicals can be regulated by the cell. This is also where the receptors that convey chemical signals from other cells are situated. Inside the plasma membrane is the cytoplasm, a thick water solution containing several cellular structures called organelles, and the nucleus (in eukaryotic cells, the cells of animals, fungi and plants). The nucleus is the largest structure of a mammalian cell, it is surrounded by the nuclear envelope and it contains the genetic material. The genetic material (chromatin) is organized in 46 strands (in a human cell nucleus), which is condensed to chromosomes that can be observed in a light microscope at the time of cell division (see also Chapter 11). In simpler organisms (bacteria) the cell nucleus is missing (prokaryotic cells) and the genetic material is present in the cytoplasm To facilitate the movement of molecules and

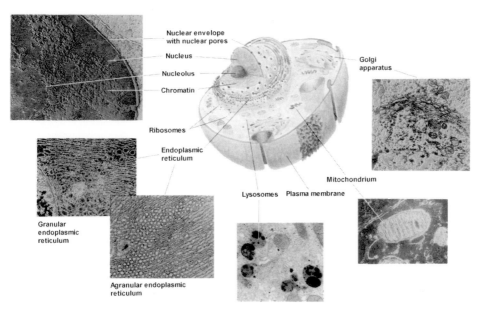

Figure 4.1. An average mammalian cell.

molecular complexes between the cytoplasm and the nucleus, the nuclear envelope has a number of openings called nuclear pores. The nuclear envelope is made up by two membranes separated by a small space. The endoplasmic reticulum is a large organelle also made up by two membranes with a space between them that traverse the cell, and the intermembrane space of the nuclear envelope and the endoplasmic reticulum is interconnected. Parts of the endoplasmic reticulum is granular (rough), containing ribosomes that synthesize proteins, or agranular (smooth), in which synthesis of steroids and fatty acids as well as the metabolic conversions of exogenous chemicals take place. All cells contain many mitochondria, small oval-shaped organelles in which energy is produced, and lysosomes, which are responsible for the re-cycling of non-functional cellular components by the action of its degrading enzymes. The Golgi apparatus is another membrane structure, which function is to concentrate and modify proteins to be secreted from the cell.

The size of a average mammalian cell is 10–20 μm, which make them invisible for the naked eye but only just (the limit for our eyes is approximately 100 μm). The three dimensional form of cells, which can be anything from a sphere to an extended structure is largely maintained by filaments and microtubules, and especially the latter provide a cytoskeleton to which organelles can be anchored. Transportation inside a cell is normally no problem for molecules that are dissolved, as molecules rapidly diffuse around the cell. However, for larger structures there are active cellular transport systems that for example make use of the cytoskeleton, and some cells (nerve cells which can have a length of 1 m) depend on such systems (which can transport vesicles up to 40 cm per day) for their function.

Before we take a closer look at the most important molecular materials of the cell, the membrane, the proteins, and the genetic material, we need to briefly discuss chemicals in general and biochemicals in particular.

4.2 Chemicals

Chemicals can be classified in many different ways, but in this context it is convenient to differentiate the chemicals that are endogenous to organisms from those that are exogenous.

4.2.1 Endogenous vs exogenous chemicals

Endogenous chemicals have a natural place in the normal biochemical processes of the organism, being either essential or non-essential compounds. Compounds or elements that an organism can not survive without and not make (sufficient amounts of) itself, that consequently have to be provided from the environment, are called essential. For humans we know of approximately 50 essential compounds and elements, e.g. water, molecular oxygen, 9 amino acids, some unsaturated fatty acids (e.g. linoleic acid, linolenic acid and arachidonic acid), approximately 20 vitamins, and minerals. Approximately 25 elements are assumed to be essential to life. Some of these are present in large amounts (> 1 % of the body weight, e.g. carbon, hydrogen, nitrogen, oxygen and phosphorus) and part of molecules which compose the main building blocks used in the macromolecules of cells, others (e.g. the trace metals) are only present in minute amounts and are important for the function of enzymes. Non-essential but still endogenous chemicals are the remaining 11 amino acids, many carbohydrates, and all other compounds that take part in the biochemical processes. An example of a non-essential but endogenous chemical is glucose, which we use to generate energy (our brains burn more than 100 g glucose per day). Although we depend on glucose for the generation of energy, we may survive on a glucose-free diet because our metabolism can prepare glucose from other chemicals present in our bodies in adequate amounts, but if it is present in the food we will of course use it as it is.

Shortage of essential compounds and elements will result in a less than optimal capacity of that organism, although the effect does not have to be dramatic, while too rich a supply of essential compounds and elements will also be harmful and may give toxic effects. An example is molecular oxygen, which we absolutely need for our survival but which is harmful in high concentrations for longer times. Our bodies are designed to live in an environment containing approximately 20 % oxygen in the air we inhale, this will enable our blood to transport oxygen safely and only leave a low concentration of free oxygen in our tissues. However, if we breathe air containing more oxygen for a longer time, the blood will not be able to bind all of it and the amount of free oxygen that can participate in potentially harmful reactions (Section 7.2.3) increases.

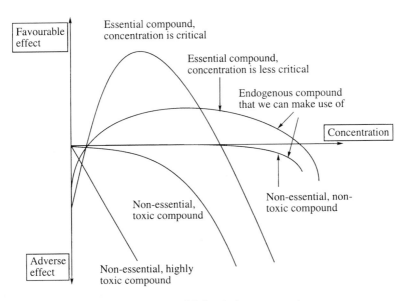

Figure 4.2. The effects of essential and non-essential chemicals on an organism.

Exogenous chemicals. Exogenous chemicals, or xenobiotics (from the Greek: "xenos" and "bios" make "stranger to life"), are chemicals that do not occur naturally in an organism. Exogenous compounds may in rare cases be inert, like molecular nitrogen, but should in principle be considered to be potentially toxic.

The differences between essential, other endogenous and exogenous chemicals are indicated in Figure 4.2. For essential chemicals there is always an optimal concentration for the organism, above which the favourable effect will diminish and eventually become a toxic effect. In some cases the maximal favourable effect is obtained within a broad concentration range while the concentration of the essential compound is more critical for others. Non-essential endogenous compounds such as glucose can be imagined either to give a favourable effect, because they relieve the metabolism, or to give no effect at all (in reasonable concentrations), because the organism will cope with the situation anyway. Exogenous compounds will never give a favourable effect. If they are non-toxic, as molecular nitrogen, they will give an adverse effect in high concentrations (> 90 % molecular nitrogen in the respiration air is dangerous because the concentration of oxygen is too low).

4.2.2 Natural vs synthetic compounds

Quite often one encounters the terms "natural" and "synthetic" compounds, which are used to distinguish between compounds that are formed by organisms (natural) and by man in artificial ways (synthetic). Some believe that natural compounds are less toxic that unnatural, since

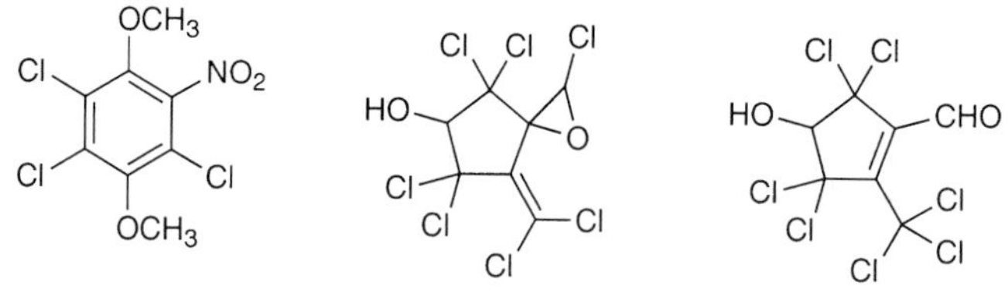

Fungal metabolites

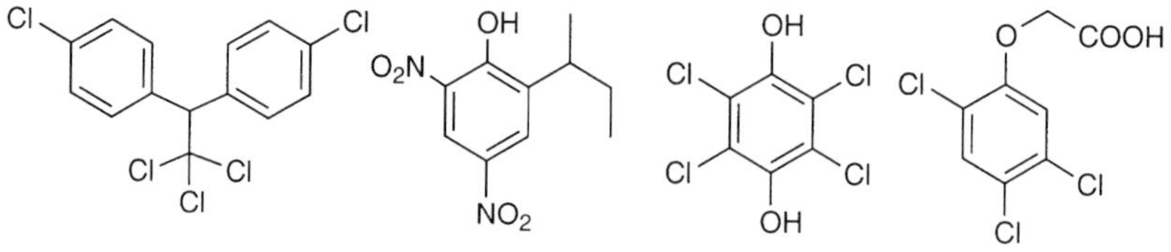

Banned pesticides

Figure 4.3. Natural compounds can be as toxic as synthetic.

the human species has evolved on earth in the presence of natural compounds and has had the possibility to adapt to their presence. Synthetic compounds have only existed for a short time, and it is obvious that organisms have not had the same chance to learn how to deal with them (although microorganisms have shown a remarkable ability to become resistant to newly developed antibiotics). To some extent this is of course true, and natural compounds are for instance more easily biodegradable compared to many man-made compounds (e.g. DDT and PCB) because of the co-evolution of organisms and natural products. However, the most toxic compounds that we known of are natural, produced by organisms that possibly use them to resist or control competitors for food and space. The production of biologically active and potentially toxic compounds in nature is enormous, and as can be seen in Figure 4.3 where the structure of some fungal metabolites are compared with those of a few notorious pesticides, it is not always obvious from the structure which compounds are natural and which are synthetic!

Many of the hazardous synthetic compounds that cause problems today are halogenated, and traditionally the presence of halogen atoms in an organic compound has been associated with a synthetic origin. However, the natural production of halogenated compounds exceeds synthetic production by far, albeit the latter in some cases are poorly biodegradable.

We also have to be aware that the borders between endogenous and exogenous as well as natural and synthetic compounds are not permanent, and that new scientific discoveries may change the status for a compound. A recent example is nitric oxide (NO), which for decades has been known as a noxious gas formed for example when copper metal is dissolved in nitric acid or in a petrol engine when N_2 is oxidized by O_2 at high temperature. Besides damaging the lungs (causing pulmonary oedema, Section 10.2.5), it has been shown to affect the ozone in the stratosphere (Section 12.4) and to be a component in acid rain. However, recently nitric oxide has been found to have several important physiological functions, e.g. as a transmittor of nerve signals and as an inhibitor of platelet aggregation. Nitric oxide also dilates the capillaries delivering blood to for example the heart muscle, and has been used in the form of nitroglycerin (which is converted in the body to nitric oxide) to treat vascular spasms (e.g. angina pectoris). In addition, nitric oxide is one of the components produced by the immune system (the macrophages) and used to kill infecting microorganisms.

4.3 Biochemicals

4.3.1 Water

The earth contains approximately 1.4 billion km^3 water, making the water molecule (H_2O) the most abundant molecule on earth. Water is the medium in which life originally formed and evolution started, and is the solvent in which biomolecular processes take place. As far as we know, the evolution of any form of life anywhere in the universe would depend on the presence of water. Water has a number of qualities that makes it exceptional in this respect. For example, the density of water is highest at +4 °C, something that not only protects water-living organisms in cold climates but also is important for the transformation of stone to dust (shattering by freezing water in cavities). The intermolecular forces between water molecules and between water molecules and polar compounds as well as ions dissolved in water are exceptionally strong, giving water an unexpectedly high boiling point and enabling it to transport a variety of chemicals in biological systems. Due to the high dielectricity constant of water, ions dissolved in water will efficiently be shielded from each other and can stay in solution. It absorbs IR radiation, and is important for the natural green house effect that keeps the temperature on earth at a favourable level. The heat capacity of water is exceptional, and seas and oceans function as buffers against variations in temperature. So is its heat of evaporation, which makes it suitable for cooling organisms by transpiration. The surface tension of water is the highest known for liquids (except liquid mercury), important for the interaction of water with other chemicals as well as for the formation of rain drops in the atmosphere. (Several of the chemical properties of water that arise from these qualities have already been discussed in Chapter 3.)

Water is considered to be the solvent of life, in which the molecular reactions on which life is based take place. However, a cell contains approximately 80 % water and 20 % other material, and the chemical properties of the thick water solution present inside a cell cannot be compared with those of pure water. Water is an essential chemical, and the only hazard with water recognized by most people is the potential to drown in it. However, the consumption of large quantities of pure water (distilled or desalted) may be dangerous, as it may bleed the body of minerals.

4.3.2 Carbohydrates

This class of organic compounds received its name because the elemental composition of many (but not all) of its members correspond to $C_x(H_2O)_y$. Take for example glucose (blood

β-D-glucose   Sucrose

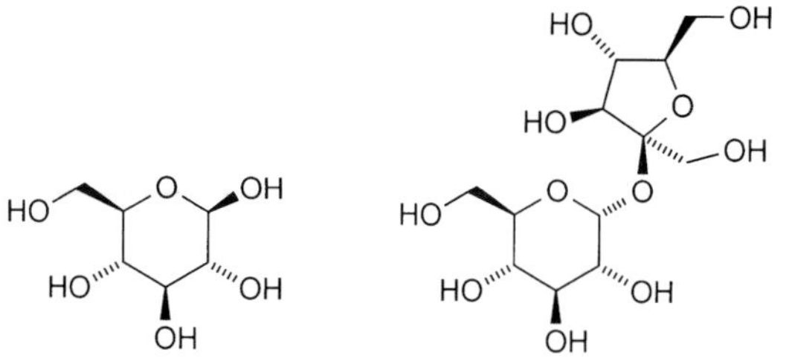

Figure 4.4. The structures of glucose and sucrose.

sugar) and sucrose (table sugar) with the molecular formulas $C_6H_{12}O_6$ or $C_6(H_2O)_6$, and $C_{12}H_{22}O_{11}$ or $C_{12}(H_2O)_{11}$, respectively. The simpler members of the carbohydrates are also called saccharides, and the monosaccharides, with the general formula $C_xH_{2x}O_x$ (x=3, 4, 5, 6, 7 or 8) (for example glucose) are the monomers of more complicated carbohydrates. Disaccharides are made from two monosaccharides, for example sucrose (consisting of one glucose and one fructose), maltose and lactose, while trisaccharides are made from three monosaccharides. Oligosaccharides contain between four and ten monosaccharides, while polysaccharides (e.g. starch, cellulose and glycogen) contain a large number. (As can be seen in Figure 4.4, we can indicate the direction in space of chemical bonds in a molecule. In this text, a bold wedged bond indicates that a bond arises from the plane of the paper while a hashed bond indicates that it is directed from the plane of the paper downwards.) In addition, as most natural compounds, monosaccharides are chiral (the mirror image of the molecule is not superimposable on the molecule itself) and are essentially produced in the biological world as only one enantiomer (mirror image). For the vast majority of carbohydrates it is the D-series, shown for glucose and sucrose in Figure 4.4.

The carbohydrates are formed by plants and green algae which convert water and carbon dioxide into carbohydrates and molecular oxygen, using the energy of sunlight to make this energetically unfavourable reaction proceed. This is called photosynthesis, and generates the fuel that other organisms (including ourselves) depend on (see Figure 4.5).

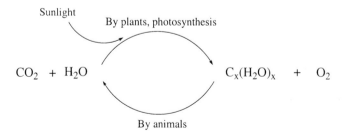

Figure 4.5. The energy consumed by organisms on earth originally comes from the sun.

Besides storing chemical energy in for example starch and glycogen, carbohydrates are also used by organisms as construction materials (e.g. cellulose in plants, polysaccharides in the cell walls of bacteria and chitin in the exoskeletons of insects and crustacea). Cellulose is actually the most abundant organic compound in the world, containing more than half of all carbon bound in organic molecules. Carbohydrates are also used by our cells as "flags", linked to proteins and lipids and positioned in cell membranes. Such signals can be recognized by the immune system and used for identification, and our different blood types (A, B, O, etc), for example, are determined by different carbohydrates bound to the membrane of the red blood cells. A person having type A blood cannot be given type B blood because his immune system will identify the type B blood cells as foreign and destroy them. Additional important functions of carbohydrates are as components of the DNA and various metabolic cofactors (*vide infra*).

4.3.3 Fatty acids

A fatty acid is a long unbranched carboxylic acid with 12–20 carbon atoms (see Figure 4.6 for examples), essentially derived from the hydrolysis of natural fats, oils and waxes. Fats (solid at room temperature) and oils (liquid) are triesters of fatty acids with glycerol, while waxes are esters of a fatty acid with a long-chained alcohol. Fatty acids can be saturated, like palmitic acid and stearic acid, or unsaturated, like oleic acid. Polyunsaturated fatty acids, e.g. linoleic acid and linolenic acid, contain two or more carbon–carbon double bonds. Unsaturated fatty acids are less stable compared to the saturated, and take part in oxidative reactions (lipid peroxidation) that will be discussed in Chapters 7 and 10. Linseed oil and other drying oils contain (the triglyceride of) linolenic acid, and linseed oil paint will dry not because a solvent evaporates as in normal paints but because the linolenic acid reacts with molecular oxygen in the air and polymerises.

Just as carbohydrates, fatty acids (in the form of triglycerides) are used to store chemical energy, and because fatty acids are more reduced their energy content is larger (complete oxidation of fatty acids yields approximately 9 kcal/g while carbohydrates yield 4 kcal/g). Fatty acids are also important components, as phospholipids and glycolipids, of biological membranes

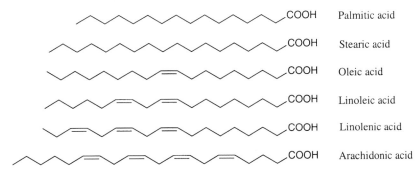

Figure 4.6. Common and important fatty acids.

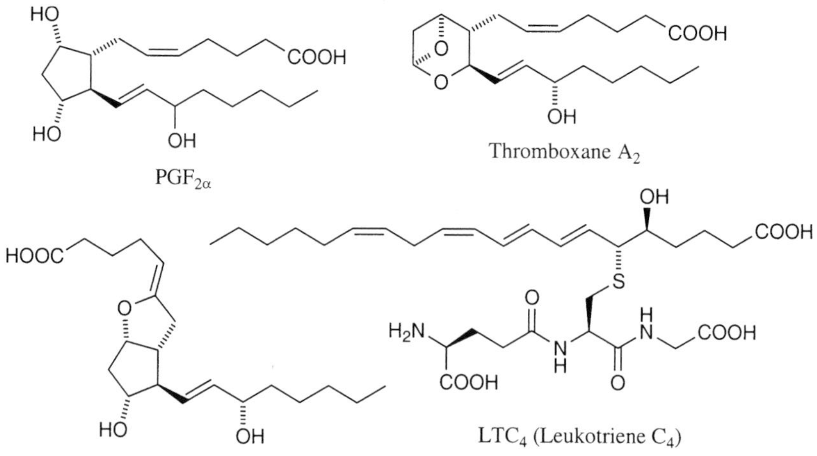

$PGF_{2\alpha}$ Thromboxane A_2

LTC_4 (Leukotriene C_4)

Prostacyclin

Figure 4.7. Some physiologically important compounds derived from arachidonic acid.

(*vide infra*), and some are essential compounds with important functions in our biochemistry. Arachidonic acid for example, is used by our bodies as a starting material for the biosynthesis of the prostaglandins (e.g. $PGF_{2\alpha}$, which stimulates contraction especially of uterine smooth muscle, and prostacyclin, which dilates blood vessels), the leukotrienes (e.g. LTC_4, which stimulates the contraction of especially lung smooth muscle), and the thromboxanes (e.g. thromboxane A_2, which triggers the constriction of injured blood vessels and the aggregation of blood platelets) (see Figure 4.7).

4.3.4 Terpenoids

The terpenoids are in general not directly involved in the basic metabolism that provides us with energy, but are instead products of what is called secondary metabolism (see also Chapter 7). They are found in all organisms, and are for example responsible for the smell of flowers. In humans, the major class of terpenoids are the steroids, represented in Figure 4.8 by cholesterol. Most of the cholesterol produced in a human (an adult contains approximately 150 g) is present as a component of the cell membrane, and the membranes of nerve tissue are particularly rich in cholesterol (containing 10 %). Cholesterol is also used as a precursor in the synthesis of the sex hormones (e.g. testosterone in men and progesterone in women), the glucocorticoid hormones (e.g. cortisone) which for example take part in the inflammatory process and regulate the metabolism of carbohydrates, cholic acid which acts as an emulsifying agent in the intestines, and vitamin D (formed in a photochemical reaction, see Figure 4.8). The amounts of vitamin D (actually a group of structurally related compounds) that are made from cholesterol are not sufficient, and it is still an essential compound. Cholesterol is formed by the terpene biosynthetic

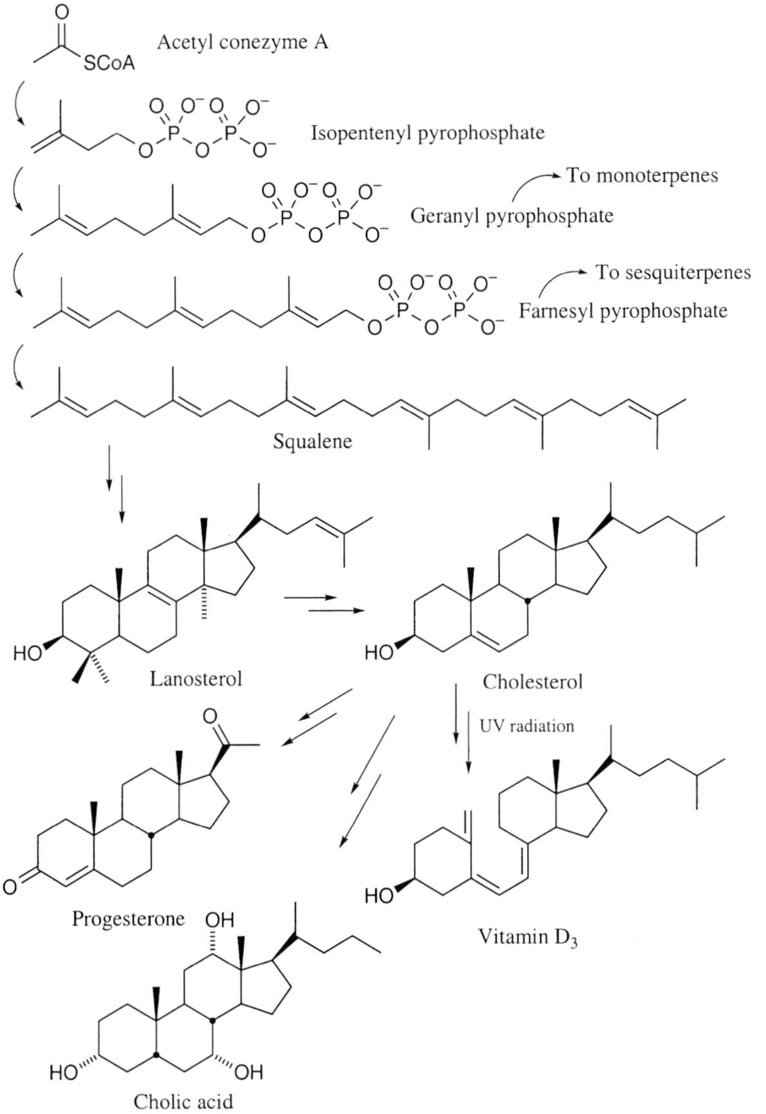

Figure 4.8. Some important steroids and their formation.

pathway via lanosterol, an extremely elegant procedure that synthetic organic chemists so far only can dream of mimicking.

In addition, vitamin A is a terpenoid (formally a diterpene with 20 carbon atoms), while in vitamin E and vitamin K a major part of the molecule is a terpene. The four vitamins (A, D, E and K) discussed in this subsection are relatively nonpolar compared to the other vitamins and fat-soluble.

4.3.5 Amino acids

Amino acids, or more correctly α-amino acids (as the amino group is positioned on the carbon α to the carboxylic acid), are the building blocks of peptides (up to approximately 20 amino acids), polypeptides (> 20 amino acids) and proteins (one or several polypeptides and a molecular weight exceeding 5000 g/mol). The general structural formula of an α-amino acid is shown in Figure 4.9, and we should note a couple of important features. As discussed in Chapter 2, amines are basic and the nitrogen will be protonated by an acid while carboxylic acids are acidic and will be deprotonated to form a carboxylate anion in the presence of a base, and the α-amino acid moiety is at physiological pH actually present as a zwitterion, with one positive and one negative charge (i.e. no net charge). The solvent water is acidic and basic enough to both protonate amines and deprotonate carboxylic acids. As zwitterions amino acids is more polar and water soluble than one would expect from its normal structure. The α-amino acid moiety is also the link that connects the amino acids in peptides and proteins by forming an amide bond (called peptide bond) between two amino acids. The α-carbon is chiral in most α-amino acids, and mainly L-amino acids are found in biological systems.

The structures of the 20 amino acids common in proteins are shown in Figure 4.10. The α-substituents have different size, polarity and charge, and the combination of amino acids, result-ing in a chain with a combination of different α-substituents, is what gives the peptide or pro-tein its characteristics. In Figure 4.10 they are presented according to their chemical features, first those with a nonpolar side chain of different sizes, then those with a polar but neutral side chains that are able to form strong hydrogen bonds, followed by weakly acidic (cysteine and tyrosine) and weakly basic (histidine) amino acids. The latter are only deprotonated/protonated to a minor extent at pH 7, but it will nevertheless influence their behaviour. The last group are more or less completely deprotonated (glutamic acid and aspartic acid) or protonated (lysine and arginine) at pH 7, and consequently have a net charge. In addition to size and polarity, we can note that some amino acids have chemical moieties that readily should participate in certain chemical reactions, something that we shall return to in the next Chapter.

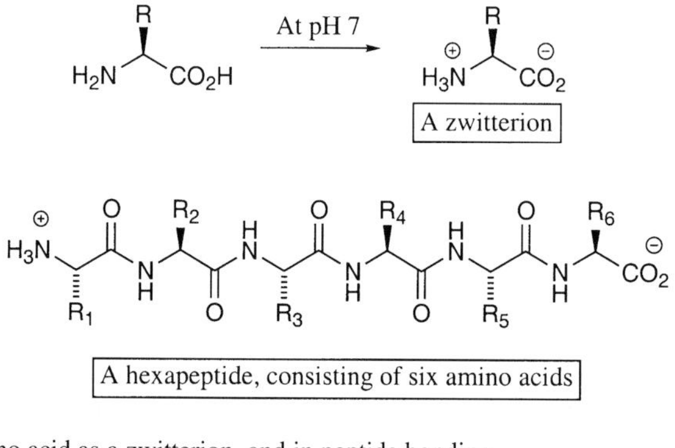

Figure 4.9. The amino acid as a zwitterion, and in peptide bonding.

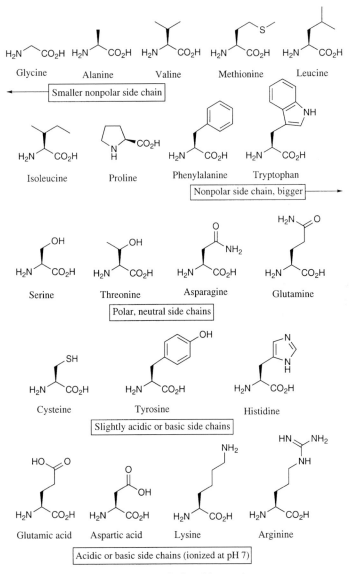

Figure 4.10. The amino acids (L-isomers) commonly found in proteins.

4.3.6 Nucleosides

A nucleoside a molecule composed of one of five heterocyclic aromatic amine bases (adenine, cytosine, guanine, thymine or uracil, derived from purine or pyrimidine) and one of two monosaccharides (ribose or 2-deoxyribose), linked together with a β-*N*-glycoside bond (see

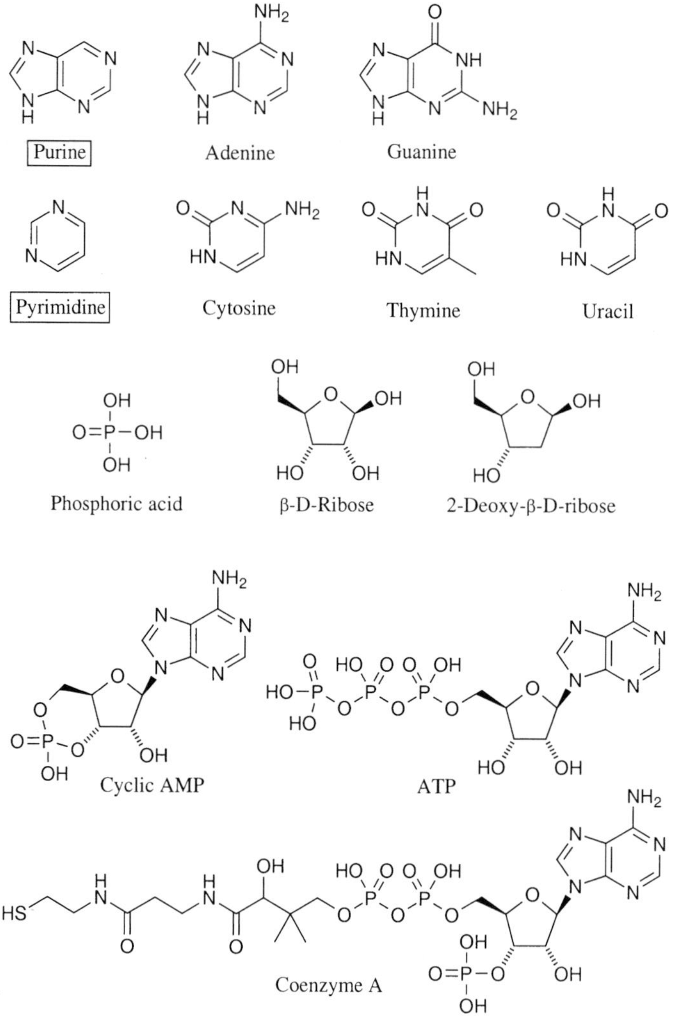

Figure 4.11. The components of nucleosides, and some important nucleoside derivatives.

Figure 4.11). Nucleoside monophosphates (called nucleotides) have important functions as monomers in DNA and RNA (*vide infra*) and as secondary messengers that relay the message delivered to the cell membrane by hormones into the cell (e.g. cyclic AMP [AMP = adenosine monophosphate]), and are part of coenzymes such as coenzyme A and ATP (ATP = adenosine triphosphate).

4.4 Biomacromolecules and cellular constituents

4.4.1 Cell membranes

The major functions of the membranes in a cells have already been mentioned, and in Chapter 6 we will discuss the transport of chemicals through membranes in more detail. Cell membranes are basically made up by a double layer of amphipathic phospholipids, ordered in the sense that the lipophilic parts of the molecules are inside the double layer while the hydrophilc parts are on the outside in contact with the surrounding water solution. It is also called a lipid bilayer (see Figure 4.12). This is how amphipathic compounds spontaneously arrange themselves in water, as it minimizes the interactions between lipophilic (parts of) molecules and water and maximizes the number of stabilising hydrogen bonds that can be formed. The phospholipids are diesters of phosphoric acid, with diacylglycerol (the fatty acids of the diacylglycerol is normally palmitic, stearic or oleic acid) and for example the alcohol cholin (resulting in lecithin) or ethanolamine (resulting in cephalin). The fatty acid esters will give that part of the molecule lipophilic character, while the phosphate group is charged and hydrophilic. Amphipathic compounds in a membrane can move relatively freely in two dimensions, but will not be able to change sides.

Besides phospholipids, membranes also contain approximately equal amounts of proteins, although the proportions are different in different membranes (varying from 1:4 to 4:1). The

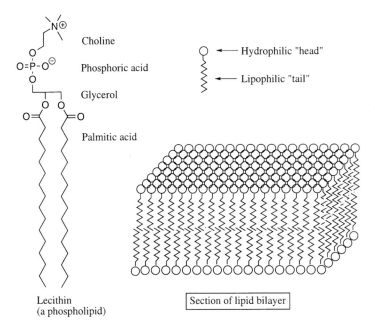

Figure 4.12. The lipid bilayer of cell membranes.

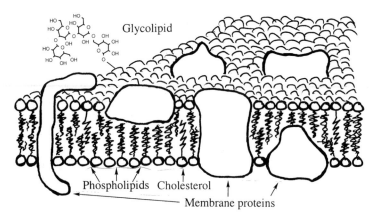

Figure 4.13. A "true" membrane.

proteins may be associated with the inside or the outside of the membrane or they may penetrate it, and their various functions will be discussed in the next Section. In addition, the membranes contain small amounts of cholesterol, which together with the nature of the fatty acid in the phospholipid determine the fluidity of the membrane, and glycolipids, which reach out signalling carbohydrate structures into the extracellular space. A cell membrane is consequently very complicated, and not completely understood. Figure 4.13 shows how one can imagine a cell membrane with all its different constituents, although it should be emphasised that this is only a relatively simple picture and that large variations between different membranes exist.

4.4.2 Proteins

The proteins may be described as the workhorses of cells, with a number of different functions. For example, they catalyze most biochemical reactions (enzymes), transform exogenous chemicals so that we can get rid of them (described in Chapter 7), make cells respond to signals (receptors), regulate the flux of chemical through membranes (described in Chapter 6), transport oxygen (haemoglobin), make cells mobile (actin and myosin), recognize foreign structures (antibodies), give cells and tissues their structure and form (collagen, elastin and keratin), etc. Peptides have functions as transmittors of signals between nerve cells and between cells of the immune system, and as hormones. Many peptides and proteins are also highly toxic, e.g. bee and snake venoms, mushroom poisons, as well as diphtheria and botulin toxins.

Proteins are composed of a large number of amino acids, normally more than 100, often as a combination of several polypeptides. The chain of amino acids arranges itself in various regions of the polypeptide as certain conformations are more stable, for example as an α-helix, and will fold in space to form a three-dimensional structure whose form ultimately depends on the number and order of the various amino acids in the chain (see Figure 4.14). In addition, the form of a polypeptide is often stabilized by cross links between cysteine side chains that form disulphide

Figure 4.14. The folding of a protein.

bonds. In principle, if a polypeptide is stretched out it will spontaneously fold back into its original form because this represents the most stable conformation. However, many proteins are only stable under special circumstances, for example when they are situated in a membrane, and will not fold to a functional three-dimensional form in a water solution. Besides the polypeptide(s), many proteins (most enzymes) also need a cofactor. This can be a simple metal ion or more complicated organic molecules (called coenzymes).

In enzymes and receptors only a small part of the protein is directly involved in the chemical conversion, or recognition of a messenger, and this is called the active site. In the active site, a few critical amino acids positioned in different locations in the polypeptide, are brought together by the folding of the polypeptide. Evidently, chemicals may interfere with the function of such proteins by binding to or reacting with components of the active site, but it should be remembered that the interaction of a chemical with another part of the protein may affect its overall structure and thereby indirectly its active site.

4.4.3 Nucleic acids

The nucleic acids DNA (deoxyribonucleic acid) and RNA (ribonucleic acid) are biopolymers of nucleotides, joined together as a chain of phosphodiesters as indicated for DNA in Figure 4.15. While the function of DNA is to be the data base of the cell and provide the information about how and when proteins should be prepared, as well as enable the cell to pass on the information to future cell generations, RNA is used to transcribe and translate the information provided by DNA. DNA makes use of the four bases adenine (A), cytosine (C), guanine (G) and thymine (T) (in RNA thymine has been exchanged for uracil [U]), the monosaccharide is deoxy-

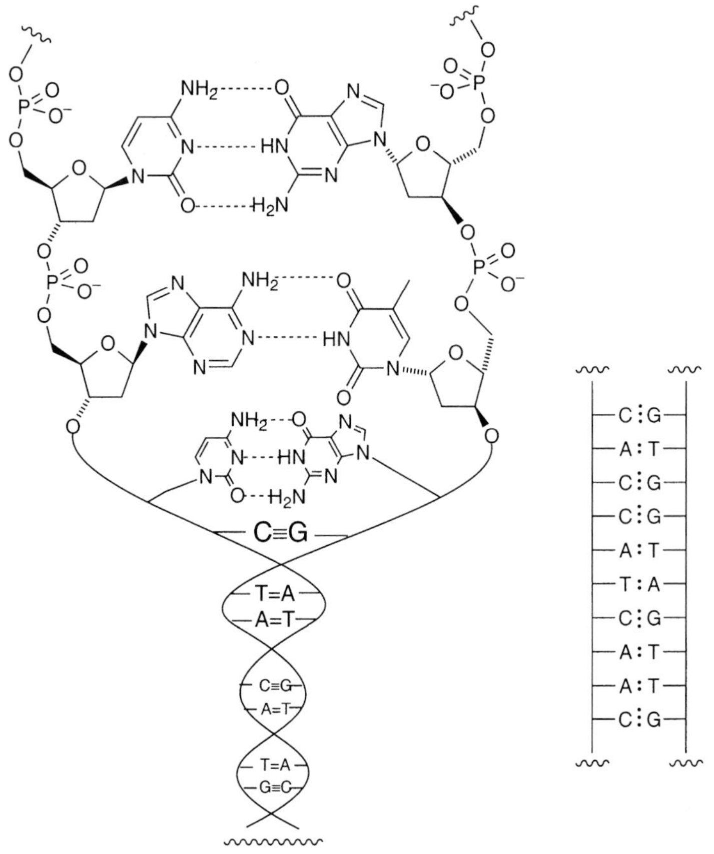

Figure 4.15. A schematic picture of DNA.

ribose (ribose in RNA), it occurs as two complementary and paired strands twisted into a double helix (as indicated by Figure 4.15) (RNA is single-stranded). In DNA, adenine will only form a base pair with thymine while guanine only will form a base pair with cytosine, making A–T, T–A, G–C and C–G the only base pairs observed. Each one of our cells contain approximately 6 billion base pairs, on 46 pairs of strands. The double-stranded nature of DNA makes it a very stable molecule, and also facilitates the replication of DNA that precedes a cell division as each strand can serve as a unique template for the cell to make two identical copies of its DNA.

The information stored in DNA is called the genetic information, and the DNA of a cell or an organism is called the genetic material. The genetic information is encoded by the sequence of the bases in DNA, and as it is a chemical memory it is sensitive to chemical interference. Genotoxic chemicals (toxic to the genetic material) will affect the genetic information, and this can for example lead to the loss of cellular control systems that facilitate the development of cancer cells. An obvious target for genotoxic chemicals is the DNA bases, which have to fit exactly into the double helix and bind the complementary bases with the correct set of hydrogen bonds. As the bases are relatively good nucleophiles they may react with electrophils to forms that can not function in DNA. In addition, the double helix is sensitive to agents that causes breaks in the phosphodiester-deoxyribose polymer, and we shall return to this in Chapter 11.

4.5 Basic environmental chemistry

The chemistry of the earth can be described with one word: Complex. So far we have discussed the elements of biochemistry, or the chemistry of organisms, which constitutes what is called the biosphere of the earth. In addition, we can identify several other distinct "spheres", the atmosphere (the air stratum surrounding earth), the hydrosphere (oceans, seas, lakes, rivers, etc), the lithosphere (stone and rock material of the earth's crust) and the pedosphere (the soil, formed from the weathering of the lithosphere and the site for much of the life on earth). The biosphere is the part of the world that is inhabited by living organisms, approximately 3 km of the earth crust, some 10 km deep in the oceans and 25 km up in the atmosphere. The elemental composition of the earth is completely different compared to that of organisms and of cosmos, as can be seen in Table 4.2 where the 12 most common elements in various materials are shown.

The 12 most common elements of our world together comprise more than 99.5 % of the total mass, and for example carbon, a basic element in biochemistry, can be regarded as a trace element. Oxygen is the most common, comprising approximately 50 % of the mass. Most of it is present in the lithosphere and hydrosphere, while the atmosphere contribute with surprisingly small amounts of this element.

Table 4.2. The amounts (weight %) of the 12 most common elements of the world (the explorable parts, including the earth crust, the atmosphere and the hydrosphere), the earth (excluding atmosphere and hydrosphere), man and cosmos. Estimated values.

World		Earth		Man		Cosmos	
O	49.5	Fe	36.9	O	63	H	74.6
Si	25.8	O	29.3	C	19	He	23.7
Al	7.6	Si	14.9	H	9	O	0.81
Fe	4.7	Mg	6.7	N	5	C	0.28
Ca	3.4	Al	3.0	Ca	1	Fe	0.14
Na	2.6	Ca	3.0	P	1	N	0.09
K	2.4	Ni	2.9	S	0.6	Ne	0.08
Mg	2.0	Na	0.9	K	0.2	Si	0.08
H	0.9	S	0.7	Cl	0.2	Mg	0.07
Ti	0.4	Ti	0.5	Na	0.1	S	0.04
Cl	0.2	K	0.3	Mg	0.04	Ar	0.02
P	0.1	Co	0.2	Fe	0.01	Al	0.01

4.5.1 The atmosphere

There is no sharp limit between the atmosphere surrounding the earth and the rest of the cosmos, but for practical reasons the limit is set to 1000 km. The first 10 km above the sea level is called the troposphere, the stratosphere is the space between 10 and 50 km, the mesosphere between 50 and 100 km while the termosphere constitute the space between 100 and 1000 km. The pressure at the end of the troposphere (10 km altitude), the absolute limit for humans without technical equipment, is 25 % of that at the sea level, while at the end of the atmosphere (1000 km altitude) it is only 0.000,000,000,01 %. Nevertheless, the N_2 and O_2 present in the termosphere absorb the most energetic radiation (with wavelengths between 120–220 nm) from the sun. The energy release is enormous, and the "temperature" in the termosphere is above 1000 °C, but because of the low pressure this value can not be compared with 1000 °C at sea level. Most of the ultraviolet radiation with a wavelength above 220 nm, still very dangerous to man, passes both the termosphere and the mesophere, but is absorbed in the stratosphere by O_2 and ozone (O_3). Some chemicals, e.g. the freons, hamper this process and increase the influx of UV light, and this will be discussed in Section 12.2.1.

In view of the central functions of the atmosphere in regulating the influx of radiation and the outflux of heat, combined with the facts that it contains only a small part of the total mass of the world and that the transport of chemicals in the atmosphere is very efficient (*vide infra*), it is not surprising that several of the environmental threats discussed in Chapter 12 are linked to the atmosphere. An example of how sensitive the atmosphere is to chemical pollution is a massive volcanic eruption, a natural event that occurs regularly, and which in a short time (within a year) may have a significant impact on the whole world mainly due to the release of particles. The composition of the atmosphere appears simple (all Figures including those of Table 4.3 are given for dry air at earth level 1992 [the average concentration of water in air close to the earth is approximately 0.05 %]), more than 99.96 % is N_2 (78.08 %), O_2 (20.95 %) and argon (Ar) (0.93 %). However, the remaining 0.04 % is a complex mixture (called the trace gases) and contains several compounds (natural as well as anthropogenic, released by human activities) that are causing some of the negative effects discussed primarily in Chapter 12. In Table 4.3, the concentration of the various trace gases in the atmosphere is shown.

4.5.2 The hydrosphere

Water as a chemical has been discussed in Section 4.3.1. It is not only the solvent in which the life processes take place, but also participates in many biochemical reactions of which perhaps the photosynthetic reactions producing oxygen and carbohydrates are the most important. The total amount of water in the world is 1.4×10^9 km^3. Measured in litres, the amounts are 1.4×10^{21}, an almost inconceivably large number that still only represents the approximate number of water molecules in one drop of water. Its distribution in the various parts of the hydrosphere (and other spheres) is shown in Table 4.4.

Table 4.3. Approximate concentrations (in ppb, litres per 1,000,000 m^3 air) of the trace gases in the atmosphere (dry air at earth level in 1992). "Stable" signifies a half-life of at least 10 years, "unstable" a half-life of days while "very unstable" compounds have a half-life of minutes or shorter. Incr. = increasing, Fluc. = fluctuating.

Compound	Structure	Conc. in ppb	Notes
Carbon dioxide	CO_2	360,000	Increasing
Neon	Ne	18,000	Noble gas
Helium	He	5,200	Noble gas
Methane	CH_4	1,800	Increasing
Krypton	Kr	1,100	Noble gas
Hydrogen	H_2	500	
Nitrousoxide	N_2O	300	Incr., stable
Carbon monoxide	CO	100	Fluctuates
Xenon	Xe	87	Noble gas
Ozone	O_3	50	Fluctuates
Nitrogen dioxide	NO_2	50	Fluct., very unstable
Nitric oxide	NO	50	Fluctuates
Sulphur dioxide	SO_2	10	Fluct., unstable
Chloromethane	CH_3Cl	0.6	
Dichlorodifluoromethane	CCl_2F_2	0.5	Incr., stable
Carbonyl sulphide	COS	0.5	
Trichlorofluoromethane	CCl_3F	0.3	Incr., stable
Ammonia	NH_3	0.3	Fluctuates
Formaldehyde	HCHO	0.3	Fluct., unstable
1,1,1-Trichloroethane	CH_3CCl_3	0.1	
Tetrachloromethane	CCl_4	0.1	
Nitric acid	HNO_3	0.1	Fluct., unstable
Tetrafluoromethane	CF_4	0.07	Stable
Carbon disulphide	CS_2	0.05	Fluct., unstable
Dimethyl sulphide	CH_3SCH_3	0.05	Fluctuates
Peroxyacetylnitrat	$CH_3CO_3NO_2$	0.02	Fluctuates
Methanethiol	CH_3SH	0.02	Fluctuates
Hydroperoxyl radical	$HO_2\cdot$	0.004	Very unstable
Bromotrifluoromethane	$CBrF_3$	0.002	
Hydrogen peroxide	H_2O_2	0.001	Unstable
Sulphur hexafluoride	SF_6	0.0005	
Hydrogen sulphide	H_2S	0.0001	
Hydroxyl radical	$OH\cdot$	0.00004	Very unstable

Table 4.4. The distribution of water in the world.

	% of total
Oceans and seas	97.2
Snow and ice	2.1
Subsoil water	0.7
Lakes and rivers	0.009
(Atmosphere	0.0009)
(Organisms	0.00004)

The mixing of various natural waters takes much longer compared to the mixing of the air-masses of the atmosphere (*vide infra*), and the detailed chemical composition of different waters may vary considerably. Besides a longer mixing time, the hydrosphere weighs almost 300 times more than the atmosphere. While pollutants released into the atmosphere may result in global effects, those released into the hydrosphere are more likely to result in local (within a range of 100 km) or regional (within a range of 1000 km) damages.

4.5.3 The lithosphere and the pedosphere

The lithosphere goes down approximately 100 km from the face of the earth and consists mainly of rock, but more interesting is the pedosphere, the top layer (in average a few decimetres or metres deep) of weathered stone that we call soil, because it is here that we find most life-forms. Soil is composed of complex mixtures of organic and inorganic chemicals, in varying amounts, but the components minerals, decayed or decaying dead organisms (organic material), living organisms, water and air are present in most soil. Water and air fill the pores, and may together constitute 50 % of the volume of soil. The minerals (as clay particles) are formed by erosion (flowing water and winds) and by the action of freezing water on stones and rocks, and provide many essential elements as inorganic salts that can be dissolved in the water and absorbed by organisms. The organic rests of dead organisms provide the basis for the organisms living in soil. As indicated by Table 4.5, the number of organisms living (on average) on a square metre of soil is enormous and the amounts of organic material continuously added to the soil are large.

The various organisms have adopted to each other's presence, and to a large extent they live separately from each other, although parasites and symbiotic relations have also emerged. An example of an extremely important symbiosis is the one between plants and fungi, that exchange nutrients to the benefit of both organisms. Any negative influence on the growth conditions for the fungus (e.g. by the presence of an antifungal chemical) would seriously affect the plant as well. The debris from dead organisms can either be consumed by living organisms, e.g. microorganisms which in principle will convert it to carbon dioxide, water and some other end products, or be transformed by humification to humic substances (e.g. humic acids) and humin. The humic acids, which may constitute anything from one to 20 % of a soil, play an important role as they form complexes with metal ions (due to the presence of free –COOH, –OH and

Table 4.5. Typical numbers of various organisms found in the uppermost 30 cm layer of 1 m^2 soil.

Organism (group)	No.
Bacteria	60,000,000,000,000
Fungi	1,000,000,000
Other microorganisms	500,000,000
Nematodes	10,000,000
Algae	1,000,000
Mites	150,000
Springtails	100,000
White earthworms	25,000
Earthworms and fly larvae	200
Millipedes	150
Beetles	100
Snails, spiders, wood-lice, centipedes	50
Vertebrates	0.001

–NH groups). This makes important ions (e.g. Ca^{2+}, Mg^{2+} and K$^+$) available for the organisms. In addition, due to the polarity of the humic acids, they also retain the water present in soil. They are not defined molecules but instead polymers made up from biochemicals available (e.g. proteins and carbohydrates), and they are constantly formed and degraded by the microorganisms. A final point that should be stressed is that the metabolic capacity of the organisms in the soil is enormous, and will in many cases have a significant impact on pollutants that end up in the soil. This will be further discussed in Chapters 9 and 12.

4.5.4 The turnover of chemicals in and between the spheres

Most important for the possibilities of chemicals to give rise to ecological effects and to be subjected to biological conversions and chemical transformations, is their transportation within and between the different spheres. The turnover of chemicals in the lithosphere is of course very slow, while in the pedosphere it is comparable to that of the hydrosphere. As already has been indicated, it is much faster and more efficient in the atmosphere, because there are relatively small amounts of air which are transported efficiently by the weather systems. The atmosphere can be divided into 2 hemispheres, the southern and the northern. The mixing time (the time it takes until the air is homogenous after a discharge) is less than 2 months within a hemisphere but almost 2 years between the hemispheres. As most of the emission into the atmosphere, at the moment, takes place in the northern hemisphere, there is a substantial difference between the 2 hemispheres in the concentration of pollutants with relatively short half-lives (the time it takes until the concentration is 50 %) (see Figure 4.16). The concentration of carbon monoxide, for

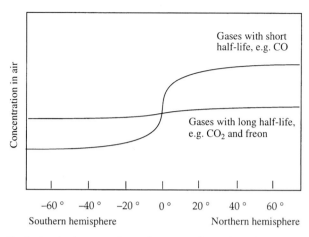

Figure 4.16. The distribution in the atmosphere of gases continuously emitted for years in the northern hemisphere.

example, a compound that only will stay a couple of months in the atmosphere, is approximately 50 ppb in the southern and 150 ppb in the northern hemisphere. However, the concentration of compounds that are more stable (having a half-life of years, e.g. carbon dioxide and the freons) will slowly be equalized in the whole atmosphere.

Hydrophilic compounds in the atmosphere will be washed away by the rain (water in the atmosphere is rapidly turned over with a half time of a few weeks) and be transported to the oceans and the soil. From the soil a major flow for chemicals is with the excess water to rivers and eventually to the oceans, although substantial amounts of chemicals (if they are naturally volatile or become volatile after conversion in the soil) are also transported back to the atmosphere. Also from the oceans there is a flux of chemicals to the atmosphere, approximately 0.03 % of the water of the oceans evaporate yearly and with it any volatile chemical. Volatile, lipophilic chemicals (e.g. the freons and methane) will largely stay in the atmosphere, while non-volatile hydrophilic chemicals will end up in the oceans. Non-volatile chemicals that are insoluble in water may be adsorbed on various clay particles and deposited in the mud at the bottom of oceans and seas or in the deeper layers of the soil.

5 Toxicological concepts

A general definition of the term toxicology is "the study of adverse effects of chemicals on living organisms", and the organism in focus is of course man. Toxicologists around the world are constantly generating new knowledge about chemicals that makes the user more aware of the potential hazards associated with them, and thereby also more able to reduce the risk for themselves as well as for the environment. The "adverse effects" studied can be anything from irritation to sudden death and chemicals anything from minerals to proteins, making toxicology a truly interdisciplinary science. In this Chapter we will not go into toxicology in any depth, we will just define some concepts that we need to use in later Chapters, as well as give an impression about how difficult it is to determine how hazardous is a certain chemical.

5.1 The toxicity of different chemicals

In Table 5.1 the toxicity, expressed as the LD_{50} value (a measure for the short-term fatal toxicity indicating the dose, in mg per kg body weight of the test organism [rats in this case], that would kill approximately every second individual exposed) of 13 compounds is compared. The compounds chosen have completely different chemical structures and properties, some are familiar and present in most homes while others are classical poisons.

Table 5.1. Approximate LD_{50} values (rats, oral intake) for 13 chemicals.

Chemical	LD_{50} (mg/kg)
Ascorbic acid (vitamin C)	12,000
2-Propanol (isopropanol)	6,000
N-Acetyl-4-aminophenol (acetaminophen)	2,400
Acetylsalicylic acid (aspirin)	1,500
2,4,5-Trichlorophenoxyacetic acid [(2,4,5)-T]	300
Acrylamide	120
1,1,1-Trichlor-2,2-bis(4-chlorophenyl)ethane (DDT)	110
Nicotine	50
Arsenic(III)oxide	40
Aflatoxin B$_1$	5
Mercury(II)chloride	1.4
Strychnine	0.24
2,3,7,8-Tetrachlorodibenzodioxin (TCDD)	0.022

Ascorbic acid is an essential compound that we can not live without, a daily intake of approximately 50 mg is recommended although it is commonly used in prophylaxis against common colds and other diseases (we will return to ascorbic acid in Chapter 10). 2-Propanol is a very common solvent present in many household products (e.g. windscreen washer liquid). *N*-Acetyl-4-aminophenol and acetylsalicylic acid are common medicaments used as antipyretics, to treat inflammations and to alleviate headaches, and show low acute toxicity if not abused. 2,4,5-Trichlorophenoxyacetic acid was a popular herbicide (weed-killer) in the 1960s and 70s, until it (and similar phenoxycarboxylic acids) came under suspicion for causing various illnesses. Initially this was blamed on the presence of TCDD (*vide infra*) as an impurity, as it can be formed in significant amounts during the synthesis of 2,4,5-trichlorophenoxyacetic acid, but this may not be the whole truth. Acrylamide is a high volume chemical, used for example for the preparation of polyacrylamide and to seal tunnels and sewers. 1,1,1-Trichloro-2,2-bis(4-

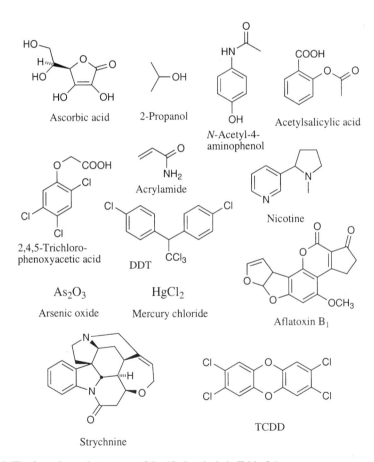

Figure 5.1. The formulae and structures of the 13 chemicals in Table 5.1.

chlorophenyl)ethane is the famous insecticide discussed in Chapter 1 that has been banned in many parts of the world, not for its acute toxicity but because it is poorly degradable in the environment. Nicotine is the stimulating principle of tobacco, but also quite toxic. Arsenic oxide and mercury chloride are examples of toxic inorganic chemicals, that are spread in large quantities in the environment. Aflatoxin B_1 is a mycotoxin formed by a mould that may attack food. It is highly toxic although we are mostly concerned about it because it is one of the most potent carcinogenic compounds known. Strychnine is an alkaloid isolated from a plant, and a classical poison used in detective stories. Finally, 2,3,7,8-tetrachlorodibenzodioxin (also known simply as dioxin or TCDD) is an extremely toxic chemical formed as a by-product during the synthesis of 2,4,5-trichlorophenol and 2,4,5-trichlorophenoxyacetic acid (discussed in detail in Chapter 12).

Even though Table 5.1 does not include the least toxic (e.g. water with an approximate LD_{50} value of 500,000 mg/kg) or the most toxic (e.g. the botulin toxins which are neurotoxic proteins capable of killing a man at doses around 1 µg) compounds, the differences are still striking and the acute toxicity of TCDD is approximately 500,000 times higher compared to ascorbic acid. As can be seen in Figure 5.1, the chemical structures of the 13 compounds differ a lot, giving them different chemical properties (e.g. solubility and reactivity) and thereby different abilities to interfere with the normal chemical processes that keep an organism alive. More in detail how these interferences create toxic effects on the molecular level will be discussed in Chapter 10-12, at least for some of the compounds. However, besides such intrinsic properties of a chemical, there are also a number of other factors related to exposure that will modulate the toxic effect of a chemical, which somewhat surprisingly may vary substantially. Before we explore those factors, we need to define a number of toxicological concepts.

5.2 Toxicological concepts

5.2.1 Acute and chronic effects

Acute effects of a chemical are observed immediately or shortly after an acute exposure, when the victim has received the dose at one single occasion or as multiple doses within 24 hours. When volatile chemicals are tested, animals normally inhale air containing a certain concentration of the compound for 4 hours. Chronic effects are then the result of chronic exposure, that goes on continuously for a long time (at least 3 months). Between these two, when the exposure goes on for more than 24 h but less than three months, the terms subacute (multiple dose short-duration, less than 1 month) and subchronic (multiple dose long-duration, 1-3 months) are used. Note that a toxic compound can give both acute and chronic effects, by different mechanisms. An example is benzene, which after a single exposure will give an acute effect on the CNS, as most organic solvents do, while it may give leukaemia after chronic exposure. Most acute toxic effects are immediate, they develop rapidly after an exposure, but some are not perceived immediately and are called delayed. An example is a tumour, which in principle can be caused by an acute exposure but may have a latency period of many years.

Chronic effects can be caused, for example, if a compound is taken up faster than it is excreted, and will result in an accumulation of the compound in the body that eventually will produce the effect. Heavy metals, e.g. mercury and cadmium, are not easily excreted, and also very small daily intakes can lead to accumulated amounts in the liver and kidneys that can affect their normal function. However, several other mechanisms for chronic effects also exist, and it can be a question of a decreasing capability to withstand the effects of a compound (e.g. as a result of poisoning, ageing or sickness).

5.2.2 Reversible and irreversible effects

Reversible effects will disappear if the exposure stops, for example, inhalation of solvent vapours may result in dizziness which will pass away if the solvent is removed and the inhaled amounts are allowed to eliminate. Irreversible effects on the other hand never disappear, and the nervous tissues are especially sensitive as they not are regenerated if damaged. Carcinogenic and teratogenic effects are other examples of irreversible effects. Note that one compound may give both reversible and irreversible effects at the same time.

5.2.3 Local and systemic effects, target organs

Local effects of a chemical are observed on the part of the body that first came in contact with it, examples of compounds that give local effects are strong acids and bases that rapidly burn the skin for example, and other highly reactive chemicals. Systemic effects on the other hand require that the chemical is taken up by the body, distributed, and eventually reaches the organ (the target organ for that chemical) that is affected. The target organ may be injured because, for various reasons, it receives most of the compound, or because it is more efficiently converted to its final toxic form in this organ. However, it may also be especially sensitive, and then the amounts are not critical. An example is DDT (see Figure 5.1) which due to its lipophilicity is distributed mainly to the adipose tissue (the body fat) although it is a nerve poison. A primary injury on one organ may also lead to a situation where other organs are affected by a secondary mechanism. For example, if the kidneys are injured and malfunctioning, this will change the liquid balance in the body in a way that puts extra strain on the heart and lungs, and the effect first noticed may come from this. The central nervous system (CNS) is the most commonly affected target organ, followed by the circulatory system, the liver (where most of the metabolism of exogenous compounds take place), the kidneys, the lungs and the skin. The muscles and the bones are seldom, although not never, affected by chemicals. Metabolic activation is often required for a chemical to be toxic, indicating that it is the metabolized form that is toxic, while metabolic inactivation indicates that the compound is rendered less toxic by the metabolic conversions (see Chapter 7).

5.2.4 Independent, additive, synergistic and antagonistic effects

Important but difficult to estimate is the total effect from a mixture of chemicals, because normally at workplaces where chemicals are handled several chemicals are used at the same time and in practice people are exposed to mixtures. Although almost nothing is known about this, we can say that two compounds given simultaneously act together to give a toxic effect in one of the following ways:

1. Independently (independent effect), if the compounds exert their own toxic effects (both target organ and potency) independent of each other.

2. Additively (additive effect), if the compounds each give the same type of effect and together give an effect that is the sum of the individual effects. Additive effects are commonly observed for compounds with similar chemical structure, and an example is toluene and *p*-xylene. Both compounds affect the CNS as most organic solvents do, and the effect of 100 mg of any of them is similar to the effect of a 100 mg mixture containing 50 mg of each.

3. Synergistically (synergistic effect), when the two compounds give an united effect that is stronger than the additive effect. Ethanol and carbon tetrachloride both produce liver damage, although with different potencies, but together the effect is much stronger. This is caused by the induction by ethanol of the enzymes converting carbon tetrachloride to liver-damaging metabolites. The influence of one compound on the metabolism of another is a very common mechanism for synergistic effects. As tobacco smoke contains several compounds that have this ability, smokers may be especially sensitive to synergistic effects.

4. Potentiating (potentiating effect), if one of the compounds by itself does not give a toxic effect but will enhance the effect of another. 2-propanol is not known to be toxic to the liver, but just as ethanol it will enhance the liver toxicity of carbon tetrachloride and by the same mechanism.

5. Antagonistically (antagonistic effect), when one compound opposes the effects of another. Antidote taken to counteract the effects of a poison is an example of antagonistic effect, which is very common in the pharmaceutical sciences. An antagonistic effect may be functional, when the two compounds give neutralizing effects (e.g. one compound raises the blood pressure while another lowers it) resulting in no observable net effect. It can also be chemical, when the two compounds react with each other and are transformed to something less toxic, and chemical antagonism is used to treat poisonings by metal ions (for example arsenic and mercury) that can be chelated and masked by 2,3-dimercapto-1-propanol (see also Chapter 12). Dispositional antagonism is when one compound decreases the concentration and/or duration of another in the target organ, for instance by influencing its uptake, metabolism or excretion. The parallel case leading to a synergistic effect has already has been mentioned, and further examples will be given in Chapter 6. Finally, receptor antagonism is observed when two compounds bind to the same receptor and the response is weaker when the two are together compared to the sum of their individual effects. A receptor antagonist (blocker) is a compound that binds to a receptor without stimulating it but blocking it for other active compounds, and we shall come back to this mechanism in Chapter 10.

5.2.5 Tolerance

The development of tolerance to the toxic effects of a chemical means that previous exposure to the chemical has made the individual more resistant to it. Tolerance is more rare than generally thought, and definitely nothing to count on. It can in principle develop as a result from the induction (increased production) of detoxifying proteins, or from the depletion of enzymes that converts the compound to a more toxic form. Cadmium will induce the production of metallothioneins (discussed in Chapter 6), proteins which complex cadmium and take it out of circulation, while the reactive metabolites of carbon tetrachloride will destroy the enzymes that activate it to a reactive form and thereby not be metabolized so efficiently.

5.3 Factors modulating toxicity

5.3.1 The purity and physical state of the toxicant

The presence of impurities in a chemical is in principle not relevant if we want to discuss the toxicity of only one compound. However, completely pure chemicals do not exist. As small amounts of a very toxic impurity will change the outcome of a toxicity test significantly, the purity of a compound being evaluated is an important and relevant property that has to be assessed. A number of inconsistent examples in the toxicological literature, e.g. the carcinogenic effects of various aromatic amines (see Chapter 7 and 8), have later been explained by the fact that compounds of different purity had been used. The physical state of a compound will influence the principal route of exposure, which in turn will determine how efficient the compound is taken up by a man (discussed in Chapter 6). In workplaces, the most important route for uptake of chemicals into the body is via the lungs, and any compound that can exist as a vapour or an aerosol will normally be more dangerous in this state. Obvious examples are volatile compounds, but also non-volatile solids that can exist as either heavy crystals or a granulate or finely ground like flour. The latter will raise dust during handling and easily be inhaled compared to the same compound in a non-dusty solid form. If swallowed, solids that are not easily dissolved in the intestines will be better absorbed if they are finely ground (having a large surface area) compared to a granulated form. Arsenic oxide is an example, and the LD_{50} value given in Table 5.1 actually depends a lot on the form given orally to the rats. Some compounds will be safer and easier to handle as solutions in a suitable solvent, although, as also will be discussed in Chapter 6, some solvents will facilitate the uptake of a solute through the skin and thereby enhance its potential ability to give rise to toxic effects.

5.3.2 The exposure

The route of exposure is an important modulator, and the same amount of a toxicant administered to the same victim by different routes may give completely different effects. In workplaces the principle routes are via the lungs (by inhalation), via the skin, and via the intestines (after accidental swallowing), but when the toxicity of chemicals are assessed during animal experiments also other routes of exposure are used. As much of the toxicological data available in the literature are based on animal experiments, it is important to know about the techniques used, whatever one's feelings towards animal experiments may be.

ivn – intravenous administration means that the compound to be tested is injected into a vein of the animal. It enters directly into the circulation, and will rapidly be spread throughout the whole body.

ihl – inhalation of the chemical to be tested, which is volatile.

ipr – intraperitoneal administration is when a compound is injected into the peritonal cavity (into the belly but beside the intestines and not into a blood vessel. From there, the compound is taken up by the blood and distributed to the rest of the body. Although ivn and ipr administration does not have any apparent relevance for the exposure encountered in workplaces and in the environment, they have been used frequently and there are a lot of ivn and ipr data in the literature.

scu – subcutaneous administration means that the compound was injected into the skin, between epidermis and dermis (Section 6.2.1).

orl – oral administrations is used when the chemical to be tested is given via the mouth of the animal, together with food or simply forced down with a syringe. The compound will pass the stomach and eventually end up in the intestines where it can be taken up by the blood and delivered into the body.

d (skn) – dermal (skin) administration is used when the uptake through the skin is investigated. The compound to be tested is simply applied on the outside of the skin and has to penetrate through the epidermis down into the dermis where there are blood vessels. How fast this is done depends greatly on the structure of the chemical (discussed in Chapter 6).

These are the most common routes for the administration of chemicals tested in animal experiments, given in order of efficacy. The natural ways to absorb chemicals (natural = without needles) are by inhaling air containing pollutants, by swallowing them, or by spilling them on the skin, and ihl, or, and d correspond to the routes that we see in workplaces. Which one is actually used in an experiment depends on what kind of information is sought, and also practical reasons. If, for example, the same amount of the same compound was given to three animals by ipr, ivn, and or administration, the blood concentrations in the three animals as a function of time would look something as in Figure 5.2.

If a certain blood concentration is required to trigger a toxic response, this may be reached by using ivn administration but not by the two other routes. In general, ivn administration is the most efficient, followed by inh, ipr, scu, or and d. In addition to affecting the rate of absorption,

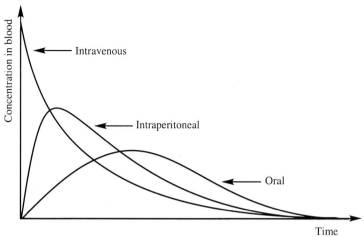

Figure 5.2. Different routes for the administration will produce different effects.

the route of exposure also influences the possibilities of the victim to convert the compound by the metabolic processes (discussed in Chapter 7).

5.3.3 The victim

The major reason for choosing animals like rats and mice for assessing the toxicity of chemicals is that they are mammals like humans, and that results from animal experiments should in principle be transferable to us. This is true in many cases, and chemicals that are toxic to rats are probably also toxic to man and vice versa. However, there are numerous exceptions, and we need to be aware about the reasons for such differences.

5.3.3.1 Species-related differences

Even though most mammalian species are very similar in function, and generally believed to be approximately equally sensitive to toxic chemicals, there are nevertheless a number of cases known where this is not true. A well-known example of how man is more sensitive to a chemical compared to other animals is methanol, or wood alcohol. Approximately a tea-cup of methanol may be fatal to a man, and it is much more toxic compared to the apparently very similar compound ethanol (this will be further discussed in Chapter 10). However, in mice and rabbits for example there is no big difference in the toxicity of the two compounds, and this is caused by the fact that they are able to convert both compounds to carbon dioxide and water while we can only do this with ethanol. Differences in the metabolism of exogenous compounds are the major reason for species-related differences in toxicity, although there are also others. Another example of species-related differences is shown in Table 5.2.

Table 5.2. The toxicity of TCDD towards various mammalian species.

Species	LD_{50} (mg/kg, oral intake)
Guinea pig	0.0005
Dog	0.001
Monkey	0.002
Rat (male)	0.022
Rat (female)	0.045
Mouse	0.11
Frog	1.0
Hamster	1.2

TCDD is an environmental toxicant of great concern during the last decade or so. Its presence as a byproduct in various circumstances will be discussed in Chapter 12. TCDD is extremely toxic to some species (see Table 5.2) while others are less sensitive. In addition, it is carcinogenic. For the layman there may not be any striking differences between a guinea pig and a hamster, but the former is nevertheless more than 1000 times more sensitive to TCDD. It is believed that also TCDD is metabolized differently in various species, but it is not yet known in detail how TCDD is toxic at the molecular level. Neither is it known in what position the human species should be placed in Table 5.2, because no fatal poisoning with TCDD has been reported for man, but we probably do not belong among the most sensitive species. Species-related differences are more a rule than an exception, although it is unusual with differences as large as with TCDD.

5.3.3.2 Sex-related differences

Small differences are normally observed between males and females of one species. A major reason is that the need for sex hormones differs and males and females therefore differ slightly in their secondary metabolism.

5.3.3.3 Age-related differences

In general, infants and old people are more sensitive to toxic chemicals. In infants, some metabolic systems as well as biological barriers (see Chapter 6) are poorly developed, and old people may be more sensitive as for instance their immune system is getting less efficient.

5.3.3.4 Individual differences

Individuals of the same species, same sex and same age have different health and nutritional status, and that will modulate the sensitivity for toxic chemicals. In addition, as individuals we all look slightly different and we all have our own variants of the human metabolic system with

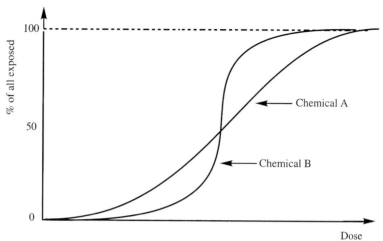

Figure 5.3. The effects of different doses of two chemicals on a group of individuals.

small but significant differences in for example the balances between competing metabolic pathways. As a result, a larger group of individuals of the same species, sex and age exposed to a toxic chemical by the same route will not be affected in the same way by exactly the same dose. The individual variations become apparent if the fraction of the exposed group afflicted is plotted against the dose, as in Figure 5.3. As the dose increases from zero a few sensitive individuals are affected long before the majority, while other indviduals withstand surprisingly high doses. Such graphs are normally S-shaped, although the shape of the S can vary a lot.

However, for some effects, e.g. chemical allergy (when certain individuals have developed an extreme sensitivity for a chemical, see Chapter 10) or in cases of chemical idiosyncrasy (when a hereditary abnormality increases the sensitivity for a chemical), it may look completely different.

5.4 Acute toxicity

5.4.1 Measures for acute toxicity

From such dose–effect relationships (see Figure 5.3), one can extract values for the dose that will give rise to a certain effect in X % of the exposed individuals (same species, sex, age and exposure route). Most common is to enter the diagram at 50 % affected and read the dose for that, but any percentage of the exposed group can of course be chosen. If the effect studied is death, then one can read the Lethal Dose on the y-axis, and the value is called the LD_X-value (X being the number in %). The median lethal dose LD_{50} is therefore the dose that will kill 50 % of the individuals in the exposed group, while LD_{10} indicates the dose that will kill the 10 % most sensitive. As species and exposure route are the two most important modulators of lethal doses, it is normal to state at least these two facts with the value, e.g. LD_{50} = 25 mg/kg (or, rat). If it is interesting to assay volatile compounds and exposure via the lungs, the toxicity is expressed instead as LC values (Lethal Concentration) measured in mg/m³ in the inhaled air. For natural reasons, few (although nevertheless shockingly many!) LD_{50} or LC_{50} values are known for man, most have been obtained with rats and mice. Besides LD and LC values, one also encounters TD and TC values for acute toxicity in the literature. T stands for Toxic, and TD_X (TC_X) values give the dose (concentration) of a compound that will give a toxic effect (that has to be defined) in X % of the exposed group. In addition, for many compounds a LD_{lo} (or LD_{low}) value is reported in the literature, and such values state the lowest known dose that has caused death in an individual. LD_{lo} values are often reported for man, as a result of a fatal accident when the dose involved could be estimated, and are therefore important.

Measuring acute toxicity with LD_{50} values obtained from animal experiments (involving something in the order of 100 animals) has been criticized as unethical and cruel, and this is certainly the case. In addition, animal experiments are always expensive and the information obtained is actually rather limited unless a detailed (and very expensive) *post mortem* investigation is carried out with the dead animals. It is therefore possible to replace animal experiments for LD_{50} determinations with an array of simpler, more ethical and less expensive *in vitro* assays based on mammalian cells, at least if a new standard for preliminary testing of acute toxicity was to be established. However, animal experiments for more specific toxic effects testing (e.g. tumours in certain organs) are not easily replaced, and will continue to be used for some time. Whole animals, with their metabolic and excretion systems intact, do provide more information about uptake, distribution, metabolism, excretion, recovery etc., which so far is difficult to simulate in *in vitro* studies. In spite of its various disadvantages, LD_{50} values are still used when the acute toxicities of chemicals are compared, simply because there are so many LD_{50} data in the literature.

5.4.2 Specific and non-specific toxicity

Toxic effects can be highly specific due to the interaction of the toxicant with only certain proteins or certain stretches of the DNA helix. Examples are compounds that bind to receptors in for instance CNS or that inhibit enzymes, and such compounds are in many cases very toxic and give dramatic effects. The recently discovered antitumour antibiotic calicheamicin γ_1^{Br} (see Figure 5.4) is also an example of a compound with a specific toxic effect. It will become associated with DNA through hydrogen bonding between the carbohydrate based tail and the minor groove of DNA in a way that positions the sulphur bridge so it can be attacked by a nucleophilic group in DNA (see Figure 5.4). This results in a conformational change in the enediyne ring that promotes its closure to an aromatic system (relieving the substantial ring strain in the enediyne ring). This change leaves 2 unpaired electrons, i.e. the product formed is a diradical which is highly reactive and will attack both strands of the DNA polymer and cleave it. This is a severe injury that in most cases will kill the cell, although it may lead to chromosome mutations (see Chapter 11).

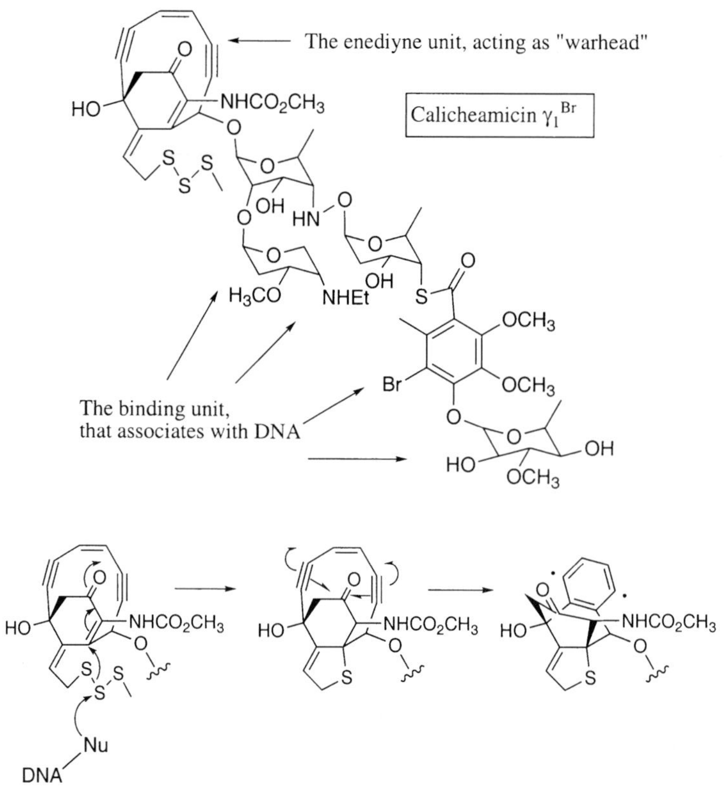

Figure 5.4. The antitumour antibiotic calicheamicin γ_1^{Br} and its interaction with DNA.

Non-specific toxic effects on the other hand do not rely on the combination between toxic chemicals and specific targets, instead the effects are caused by a reversible lipophilic association with the membranes of the cell or cell organelles. Compounds giving non-specific effects are therefore less dependent on their reactivity and more on the lipophilicity, although the same compound can of course give both specific and non-specific effects. A typical non-specific effect is the dizziness caused by organic solvents, because they are gathered in the cell membranes in neurones in the CNS and disturb the normal function of the receptors in the nerve cell membranes. It has been suggested that there are receptors that mediate this effect, but so far no such receptors have been identified. Instead it is believed that the lipophilic molecules are collected inside the membrane which will expand and become more fluid. This will disturb the proteins associated with the membrane, and even change the conformation and function of integral proteins. Besides effects on the nervous system the cardiac function may also be affected, and despite the reversible nature of the interactions even non-specific toxicity may be fatal.

In general, there is a correlation between lipophilicity (measured for example as the logP value) and non-specific toxicity (see Figure 5.5).

Figure 5.5. The general correlation between lipophilicity and non-specific toxicity.

However, extremely poorly water soluble compounds will have problems reaching their targets (e.g. CNS), at least within a limited time period, and will not show the acute toxicity expected by their lipophilicity.

5.4.3 Selective and non-selective toxicity

A chemical that shows selective toxicity is more toxic to one organism than to another, or to one organism than to another organ in the same organism. Non-selective chemicals are consequently equitoxic to both organisms (organs, cells). The mechanism for selective toxicity is either that the compound is accumulated in one organism (organ, cell) due to differences in absorption, distribution, metabolism and/or excretion, or that it interferes specifically with cellular or biochemical components that are only present or of importance in one organism (organ). Selective toxicity is of course of great importance when antibiotics, pesticides and herbicides are developed. Examples are the penicillins which kill bacteria without affecting the cells of the person infected, because they block the construction of the cell wall (which mammalian cells do not have) of bacteria. DDT is an efficient insecticide with relatively low toxicity for humans that handle it, because it is poorly absorbed through the skin, but approximately equally toxic if injected. Norbormide (see Figure 5.6) has found use as a rodenticide, because of its selective toxicity towards rats. Its LD_{50} value (oral intake) is 3.8 mg/kg for rats, while the corresponding value is 2,200 mg/kg for mice and 1,000 mg/kg for monkeys, dogs, cats, and rabbits. Norbormide acts irreversibly on the smooth muscle of peripheral blood vessels, leading to contraction and subsequent necrosis, but the molecular target is not known and the reason for the exceptional sensitivity of rats is not understood. Glyphosate (see Figure 5.6) is both selective, killing plants but not affecting microorganisms or mammals (LD_{50} for rats is 4,900 mg/kg

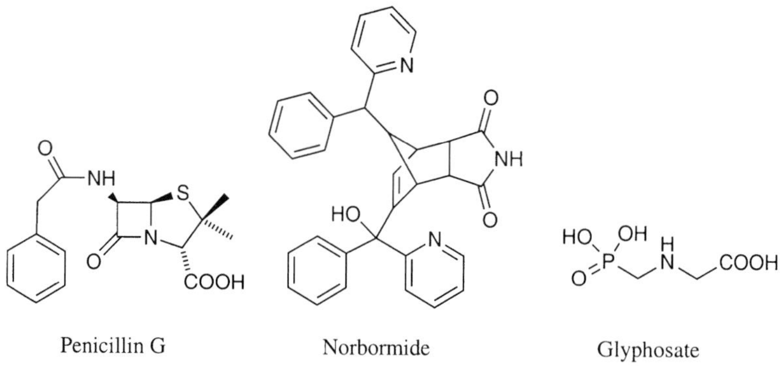

| Penicillin G | Norbormide | Glyphosate |

Figure 5.6. Examples of chemicals exhibiting selective toxicity.

orally), and non-selective, killing all plants except conifers. It apparently interferes with the biosynthesis of aromatic amino acids, although the exact mechanism for its action has not yet been elucidated.

5.5 Interpretation of results from toxicological testing

In principle, all toxicity testing is carried out to obtain knowledge that can be used to protect man from exposure to hazardous compounds, or to assess the risks connected with exposure. However, the assessment of any hazard or risk is difficult, there is no sharp limit between hazardous and non-hazardous chemicals, and a judgement will not be based on objective information alone (see also Chapter 1). For example, many industrial chemicals are of great economical importance, and such factors are normally considered when their use is being regulated. The adverse effects of cigarettes and alcoholic beverages are well documented, but society has chosen not to ban such products because they are considered, at least by many, to be means of enjoyment and the risks are (again by many but not all) acceptable. Also the risks involved with handling a hazardous chemical are difficult to estimate, for example which level of exposure should be regarded as acceptable in a workplace. Hazard (qualitative) and risk (quantitative) assessments of chemicals often to some degree have an element of economical/political decisions. Man is very seldom the test-organism during toxicological testing, although the results should be applicable to Man. Therefore, the results have to be translated from whatever test organism was used, and this is of course not trivial. We have already discussed how different species and even individuals may be differently sensitive to toxic effects, and in cases where results from assays other than animal experiments are to be evaluated the estimations will be even more precarious. This is normally accounted for by a safety margin if for example an exposure limit should be determined from data from an animal experiment. If a volatile chemical is considered to be harmless to rats breathing air containing 1000 mg/m^3, man would be

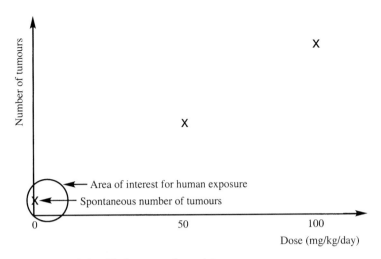

Figure 5.7. Dose–response relationship from a carcinogenicity assay.

considered to resist 10, 100 or 1000 times less (depending on the character of the toxic effect, 10 times for non-selective toxicity and 1000 times for, for example, carcinogenicity). However, the safety margins are no guarantee that humans will not be affected, something that we have learned the hard way several times over the years.

The problems with translating data from animal experiments to humans are also linked to the way the experiments are normally designed. It is important to obtain information about two things: If the chemical assayed possesses a certain toxicity; and how potent it is. To obtain the necessary correlation between dose and response it is required that several doses are tested, one being 0 mg/kg (the so called negative control). For economic and ethical reasons, as few animals and as few doses as possible are used, and in order to obtain unambiguous effects as high a dose as possible is employed. The situation may look as in Figure 5.7, in which the incidence of tumours in animals exposed to 0, 50 and 100 mg/kg/day of a chemical (that for example is considered as a new preservative in food) is plotted against the dose. The results clearly suggest that the chemical is carcinogenic.

However, man is unlikely to be exposed to the same massive doses. Instead the expected human daily intake is perhaps less than 1 mg/kg. If we look at that part of the diagram, there are no data to suggest whether low doses are carcinogenic. It can of course be assumed that the apparent linear dose–response relationship indicated in Figure 5.7 can be extrapolated all the way to zero (case b in Figure 5.8), and that low doses actually are slightly carcinogenic. However, we do not know if this is the case from the data shown in Figure 5.7. Other possibilities are that low doses are more carcinogenic than expected (case a in Figure 5.8), or less (case c or d in Figure 5.8).

Many carcinogenic compounds appear to behave as indicated by graphs c and d in Figure 5.8, because of the endogenous defensive systems that to some extent can protect us from reactive compounds (further discussed in Chapter 7). Such systems have a limited capacity and will

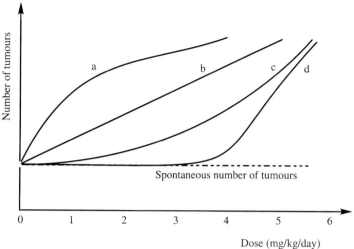

Figure 5.8. Possible dose–response relationship at low doses.

only be able to neutralize small amounts of reactive chemicals (the threshold dose). If we want to know for certain about the low-dose toxicity, it would be very expensive and consume large numbers of animals. Say that the preservative discussed above in low doses (1 mg/kg/day) gives cancers in 0.1 % of the exposed people, that would mean that 1 individual among 1,000 exposed would be affected. That is an unacceptable incidence and the preservative should be banned. However, at this dose it is not a potent carcinogen, and in order to statistically establish the effect in an animal experiment several thousands of animals would have to be used. As this is not possible in practice, unrealistically high doses are tested instead with the result discussed above, and we shall return to this topic in Chapter 11.

6 Uptake, distribution and elimination of chemicals

In order for a toxicant to give rise to a systemic toxic effect, it has to be taken up by the organisms and distributed before it reaches its target organ. As has been mentioned in Chapter 5, there are a number of different routes for chemicals to enter our bodies, but two organ systems, the digestive (intestines) and the respiratory (lungs), have been designed for the uptake of chemicals. In addition, chemicals can in principle enter the body wherever they contact it, through the skin or the mucous membranes in the nasal cavities and the mouth, in the stomach or the rectum, and even via the ear as the father of Hamlet so tragically experienced. In places of work, people are mainly exposed to chemicals as air contaminants, and uptake via the lungs normally accounts for 80–90 % of the total uptake.

6.1 Transport through biological membranes

Any uptake of a chemical from the outside into the body requires that it passes through biological membranes, which one way or the other consist of layers of cells that have to be passed. Cells are surrounded by a cell membrane, briefly described in Chapter 4, and the cell membranes constitute a major barrier for the chemicals that are transported.

6.1.1 Diffusion

Diffusion is the transport of molecules in the gas phase or the liquid phase (solution) due to the continuous and rapid movement of the molecules. The efficiency of diffusion as a transport mechanism depends on the temperature and the weight of the molecule, at 37 °C the average velocity of water molecules is 2,500 km/h while a glucose molecule, being approximately 10 times heavier, only moves 850 km/h. In spite of its velocity, a molecule will not get far (especially not in solutions) before it has a collision with another molecule which changes its direction, so it will mainly move back and forth within the same volume. With time this movement will result in a transport in all directions, driven by concentration differences in the system and with a homogenous liquid or gas as the end point. The rate of transport caused by diffusion is therefore extremely sensitive to distance, and this will in practice limit the size of cells. An average cell is 10–20 μm in diameter, and for most compounds the transport from the cell membrane to the centre of the cell by diffusion takes place in seconds. However, if the cell was the size of a football (20 cm) it would instead take years!

The transport by diffusion through the membrane of a cell in our bodies also depends on two additional things; the concentration difference on the two sides of the membrane and the resist-

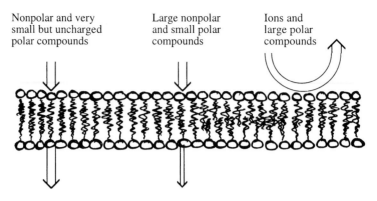

Figure 6.1. Diffusion of different species through a synthetic membrane.

ance of that particular membrane to let this particular molecule through (or the permeability of the molecule in membrane). A substantial concentration gradient will always promote diffusion, while no net diffusion will take place if the concentration is equal on both sides. The diffusion resistance of the membrane is linked both to the membrane itself, which is similar in all cells as long as the double layer of amphipathic molecules is considered, and to factors of the diffusing species. If one compares how easily chemicals in a water solution diffuse through a synthetic membrane, consisting only of amphipathic compounds and no proteins, it is obvious that very small neutral molecules (e.g. molecular oxygen and water) are not really bothered by the membrane but diffuse straight through even if they are polar. They are simply believed to fit into cavities in the membrane created by its continuous movement. The bigger and the more polar a compound is, the more difficult it is to diffuse through a membrane. Ions will not be able to pass a synthetic membrane (see Figure 6.1), simply because they are covered by several layers of stabilizing water molecules (Section 3.2.3). In order to leave the water solution and move into the lipophilic core of the membrane the ion would have to get rid of this water and that costs a lot of energy. Compounds containing for example carboxylic acid or amine functions may be ionized at physiological pH, although most organic acids and bases are not completely ionized which makes it easier to pass membranes. Most lipophilic organic compounds, also reasonably large (up to approximately 500–1000 g/mol), will rapidly pass membranes, while hydrophilic, if they are not as small as water or methanol, will not.

Two additional things are important to remember concerning the diffusion of exogenous compounds through biological membranes: It is by far the most important mechanism for transportation, and with time almost anything will diffuse through a membrane.

6.1.2 Passive transport by proteins

For hydrophilic compounds that can only diffuse through biological membranes with difficulty, there are in all cells special proteins, membrane transport proteins, that may facilitate the

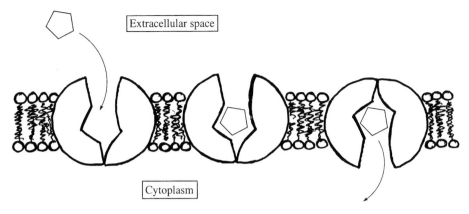

Figure 6.2. Passive protein-assisted transport.

transport. There are two principal mechanisms for protein-assisted transport, passive transport that only takes place with a concentration gradient (or a electrochemical gradient for ions), and active transport (see Section 6.1.3) that can also operate against a concentration gradient. Certain membrane transport proteins, carrier proteins, may transfer hydrophilic compounds that the cell either needs (e.g. sugars, amino acids and ions) or wants to dispose of, in and out of the cell. Carrier proteins only transport one type of compound, which must fit into the protein as a key in its lock, and the specificity that this type of transport shows is characteristic and important, but not absolute. As the transporting proteins have a limited capacity, it is also characterized by its ability to become saturated. The principal function of these transmembrane proteins is that they can adopt two conformations, that are available for the transported molecule on different sides of the membrane. In both conformations, the solute can bind to the protein as well as be released, and if there is a conformational change in the protein between the binding and the release the solute will have been transported across the membrane (see Figure 6.2). As there is no preference for binding or releasing on either side, it is the concentration of the solute on the different sides of the membrane that will determine the flow in each direction. Most mammalian cells use this mechanism to transport glucose from the extracellular fluid in which the concentration of glucose is relatively high. Note that the same system also transports the sugars mannose, ribose and xylose, but not, for example, fructose.

Other membrane transport proteins, channel proteins, form transmembrane pores in the cell membranes through which hydrophilic molecules may pass. The pores are gated by other proteins attached to the cell membrane so that they can open and close as a response to some stimulus, for example an external signal. In their open form, they provide channels of water through the membrane through which hydrophilic compounds may diffuse from one side or the other without coming in contact with the inner part of the membrane, or be passively transported by the flow created by hydrostatic or osmotic pressure (see Figure 6.3). Another name for this transport is filtration. The size of the pores may differ in different cell types, in most cells they only let small molecules pass. In mammalian cells it is mainly ions that are trans-

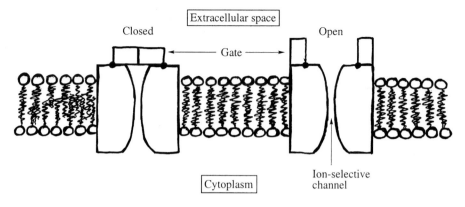

Figure 6.3. Passive transport with channel proteins.

ported in this way. The pores are then called ion channels, and they are to a large degree selective for a certain ion by the size of the pore and its internal properties. As the regulation of the flow of certain ions through the membrane is very important for cells, especially nerve cells, the interference of exogenous chemicals with the pores may give rise to toxic effects (see Chapter 10).

Exogenous compounds may also facilitate diffusion, as demonstrated by the ionophores. They increase the membrane's permeability for ions, either by disguising the ion in a lipophilic envelope and helping it to pass the membrane as a mobile ion carrier (as the insecticide and nematicide valinomycin, Figure 6.4), or by forming unnatural channels through which an ion can diffuse (as the gramicidins, used as antibiotics).

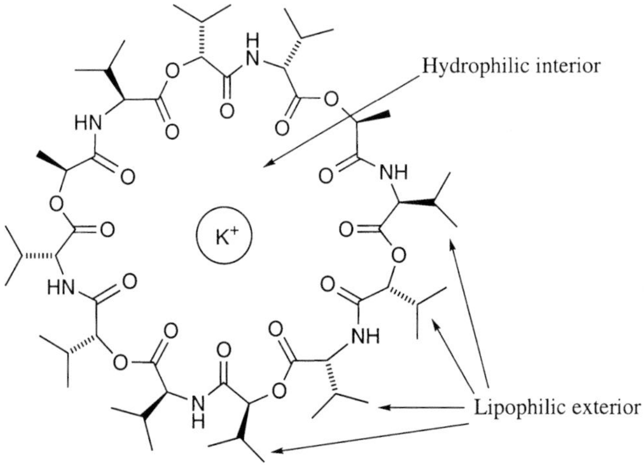

Figure 6.4. Valinomycin.

As mentioned above, passive transport only takes place in the direction of a concentration gradient (or the electrochemical gradient for ions), and can only result in dilution of a chemical. On the other hand, it shows a high degree of specificity and it is relatively simple, not requiring the input of energy for its function (although the gating of the ion channels is an active process). Passive transport was not developed for exogenous chemicals, so when they are transported through membranes by this mechanism it is simply because they happen to fit into the carrier proteins, or the channel proteins.

6.1.3 Active transport by proteins

Contrary to passive transport, active transport requires energy in the form of ATP (primary active transport) or an ion gradient (secondary active transport) and it is able to concentrate a chemical or an ion against a concentration gradient (see Figure 6.5). The most important example of a primary active transport is the plasma membrane Na^+–K^+ pump, that generates and maintains the sodium ion (Na^+) and potassium ion (K^+) gradients across plasma membranes.

The Na^+ and K^+ gradients serve several functions, for example to maintain the osmotic balance and regulate the cell volume, and we shall come back to it in Chapter 10. An example of a secondary active transport is the resorption of glucose in the kidney tubules (see Section 6.5.3), where the glucose concentration is low. This is coupled with and driven by the simultaneous uptake of Na^+, of which the concentration is low inside the cell and high outside, and the carrier

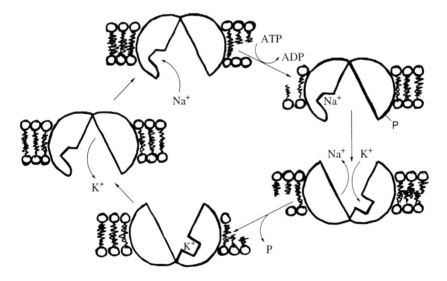

Figure 6.5. The plasma membrane Na^+–K^+ pump.

protein will only transport glucose and Na⁺ together. The Na⁺ that gets into the cell in this way is pumped out again by the Na⁺–K⁺ pump, so a low concentration of Na⁺ is maintained inside the cell.

Besides the absorption of chemicals through cell membranes by active transport, compounds are also excreted from the body by the same mechanism. The brain has active transport systems for getting rid of organic acids and bases, and so have the liver and kidneys (*vide infra*).

6.1.4 Exocytosis and endocytosis

The need for transportation also of larger molecules, such as proteins, polysaccharides and polynucleotides, or even particles, is satisfied by the transport mechanisms exocytosis and endocytosis. As depicted in Figure 6.6, they involve the formation of vesicles surrounded by a membrane, like a miniature cell. These are either formed inside a cell and can by fusing with the plasma membrane be emptied into the extracellular space (exocytosis), or by budding a piece of the plasma membrane with a volume of extracellular fluid (containing whatever the cell desires) into the cell where it is dissolved (endocytosis). Endocytosis can be divided into pinocytosis, when a volume of fluid is ingested (pino = drink), and phagocytosis, when particles are ingested (phago = eat). Almost all mammalian cells make frequent use of pinocytosis for their transport

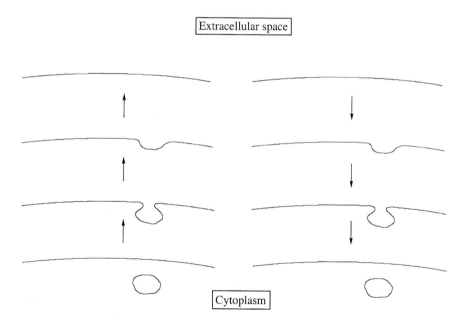

Figure 6.6. Exocytosis (left) and endocytosis (right).

needs, and the terms endocytosis and pinocytosis sometimes replace each other, while only specialized cells use phagocytosis.

Insulin-producing cells export the insulin by exocytosis. Cells that need cholesterol for membrane synthesis produce and place receptors for LDL (low-density lipoproteins) in the plasma membrane. LDL, which essentially is a complex between one protein and 2000 cholesterol molecules, is attracted by the receptors, concentrated on the outside of the cell and engulfed by pinocytosis. The newborn obtain immunity to various diseases by absorbing immunoglobulins (proteins that recognize for example pathogenic microorganisms). Phagocytosis is used by the macrophages and the neutrophiles of the immune system. They "eat" bacteria and other cells recognized as alien as well as damaged and dead cells of our own. Although exocytosis and endocytosis do not play a central role for the transport of exogenous compounds, it is involved in for example the absorption of the botulin toxins (toxic proteins, discussed in Section 5.1) and we shall also came across a few other examples.

6.2 Absorption of chemicals in man

6.2.1 Absorption via the skin

The skin is involved in many important tasks, it constitutes a barrier and helps to protect the organism, it regulates the body temperature, etc, but it is not involved in the absorption of nutrients. Therefore, it does not contain any specialized transport systems, and the only way for exogenous compounds to be absorbed by penetrating the skin is to diffuse through it. In addition to diffusion directly through the cells that make up the biological membrane, it is in some cases also possible to diffuse between the cells. The space between the cells contains fat, which can be dissolved by organic solvents leaving a skin that is considerably more permeable to chemicals. The skin consists of several layers, but for our discussions a division into dermis and epidermis is adequate. The principal difference between the two is that the epidermis, which is the outermost, is an epithelial tissue containing no blood vessels, while the dermis is a connective tissue rich in vessels (see Figure 6.7). For a exogenous chemical it is therefore

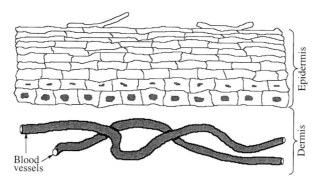

Figure 6.7. The skin.

sufficient to penetrate the epidermis; arriving at the dermis it will efficiently be swept away by the blood.

The epidermis is composed of 10–20 layers of cells, of which the innermost layer is constantly dividing, generating new cells that push those above closer to the surface. However, as soon as the cells lose contact with the basal layer they stop dividing and begin to differentiate. They fill themselves with keratin, a protein, dissolve the cell nucleus and dry as they are slowly pushed upwards. When they approach the surface they are dead, flattened, highly keratinized and relatively dry cells, containing only 5–10 % water. Eventually they are sloughed off by wear (we all lose approximately 600,000 skin cells per day). The process of renewing the epidermis takes a few weeks, and for products that penetrate the epidermis very slowly this continuous shedding serves as protection. The thickness of the skin may vary substantially in different parts of the body and the thickness will adjust to the actual need, caused by different amounts of keratin in the cells of skin experiencing different workloads. In addition, the skin contains hair follicles and different glands which make up approximately 1 % of the surface of the skin.

The absorption of exogenous chemicals through the skin depends on both the skin, its general condition and thickness, and on the chemical, how efficiently it can diffuse through the epidermis. The epidermis of the skin of arms and legs is more than 10 times thinner than that of the palms, and exogenous chemicals will consequently be absorbed faster in such areas. However, even more important is the water content of epidermis as the diffusion rate increases rapidly in moist skin, and the water content may increase dramatically if for example rubber gloves or boots are worn. In addition, if the skin is diseased or damaged by reactive chemicals, it may also be easier to penetrate for exogenous compounds. As discussed above, molecules that are small and nonpolar will diffuse rapidly through biological membranes, and the epidermis is no exception. Many organic solvents, e.g. methanol, will diffuse through the skin so efficiently that toxic amounts may be absorbed as a result of accidents when solvents are spilled. Although water molecules normally are able to diffuse through the membrane of a living cell rapidly, water will not diffuse through the epidermis as efficiently. However, it will make it swell, something that everyone can observe after having taken a bath, and as mentioned above this makes the skin more permeable to other chemicals.

Many other chemicals also increase the permeability of the skin, and facilitate more rapid absorption of others. An example is the polar organic solvent dimethyl sulphoxide (DMSO), which is used in veterinary medicine to increase the absorption of certain remedies administered dermally. The mechanism is again that DMSO makes the outermost cells in the epidermis swell and thereby easier to diffuse through, and in addition it will loosen the cells from each other.

It is important to remember that there is no sharp line dividing compounds into one group that efficiently diffuse through the epidermis and are absorbed, and compounds that will not be able to penetrate the skin. Most compounds will eventually get through, although it takes time. Synthetic gloves and boots are often used as a protection against chemical spills when larger amounts of chemicals are handled, and although some synthetic materials are efficient barriers for certain types of compounds, they should not be regarded as fail-safe. Just as chemicals can penetrate the skin they will eventually also penetrate synthetic materials, and whatever diffuses through a rubber glove will encounter skin that is humid and very easily penetrated. Good advice is therefore not to wear such equipment for too long, and to wash it thoroughly if it has been in contact with hazardous chemicals. A special risk with synthetic gloves and boots is the possibility that a chemical is spilled inside a glove or a boot, and in such cases also quite polar compounds may be absorbed efficiently. This happens frequently, for example during the preparation and use of pesticides, and is responsible for many unnecessary accidents that could easily be avoided.

Most absorption of exogenous chemicals via the skin depends on accidental spills combined with carelessness. If one is observant and immediately takes care of spills on the skin or on the clothes by removal of the contaminated garments and careful washing of the skin, and uses protective equipment in a sensible way, it could essentially be eliminated.

6.2.2 Uptake by the gastrointestinal tract

Even though this only accounts for a small part of the uptake of chemicals in workplaces, it is of course of interest in other contexts like accidents, suicides, etc. The gastrointestinal tract, all the way from the mouth to the anus, can be regarded as an approximately 8 m long tube that passes through a human body. Its contents are still on the outside, but its task is to convert whatever is put in at one end to something that can be taken up and used. The whole construction is controlled by the nervous system, although hormones regulate some functions. After mechanical treatment in the mouth, the food and drink passes the oesophagus and reaches the stomach where it is degraded both chemically (by hydrochloric acid) and enzymatically (by pepsin which will hydrolyze peptide bonds in proteins). The resulting mush (called chyme) is then passed into the small intestine (divided into the duodenum, jejunum and ileum), and the low pH in the stomach (approximately 2) is neutralized with bicarbonate secreted by both the pancreas and the liver. The pancreas also secretes enzymes that degrade proteins, polysaccharides, fats and nucleic acids, while the liver in addition secretes bile that emulsifies fats, and the cells of the small intestine secrete a lot of different enzymes (see Figure 6.8). A continuous conver-

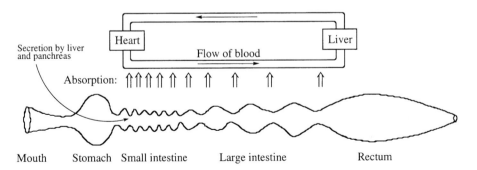

Figure 6.8. A schematic view of the gastrointestinal tract.

sion of the nutrients to absorbable units takes place and when they reach the end station, the rectum, via the large intestine, only non-digestible material is left. Also important for this process is the fact that the intestines harbour hundreds of different species of bacteria, which in number of cells exceed the number of cells of a human being by a factor of 10. In some ways, the bacterial flora of the intestines can be regarded as a separate organ in the body, with a huge and different metabolizing capacity as well as the ability to produce toxins of its own, and we shall see how this can create problems.

Chemicals are absorbed all the way, from the mouth to the rectum, although most is absorbed in the first part of the small intestine. The total internal surface of the small intestine available for absorption is approximately 300 m² (as big as a tennis court!), created by numerous villi (see Figure 6.9) that stick out like arms into the contents of the intestine, the villi in turn being covered by microvilli (small projections that increase the surface even more, see Figure 6.9). The small intestine absorbs approximately 8.5 l fluid material per day in an adult. The inside of a villus is vascular, and the blood vessels are only separated from the outside by a thin monocellular layer of endothelial cells with microvilli. It is obvious that any molecule that passes through the intestine and is able to diffuse a little, will be adsorbed. It is so efficient because the surface is large, the diffusion barrier is low (compared with the skin), and the time spent in this environment is relatively long. In addition, the endothelial cells lining the intestines are packed with membrane transport proteins that pick up hydrophilic compounds (e.g.

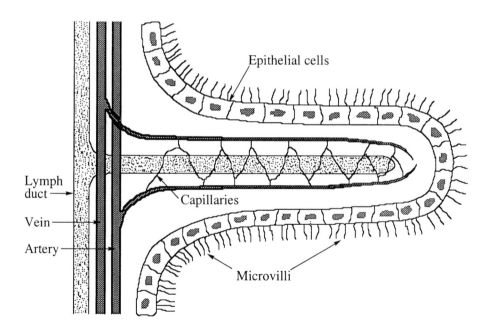

Figure 6.9. The structure of a villus.

monosaccharides and free DNA bases) and ions (e.g. amino acids and minerals) that can not diffuse by themselves. Even endocytosis is used to engulf some proteins, for example the special protein (called intrinsic factor) that binds vitamin B_{12} and helps it to pass the membrane, and the extremely toxic botulin toxins.

Exogenous compounds that for any reason end up in the intestines can be expected to be absorbed, most likely by diffusion in the small intestine but also by being structurally similar to the endogenous compounds and fitting into one of the many specific membrane transport systems. For example, thallium, cobalt and manganese all use the carrier for iron, while lead is sufficiently similar to calcium to make use of its transport system. Poorly absorbed are chemicals that are more or less insoluble in the contents of the intestines, because absorption in the intestines requires that the chemical is dissolved. Examples are paraffin and metallic mercury, which both pass the intestines almost unaffected. Mercury is well-known as a very toxic heavy metal to which we shall return several times. If mercury vapour is inhaled or salts of mercury are ingested it gives severe toxic effects, but ingestion of comparable amounts of metallic mercury is relatively harmless. In this context we can also note a case of dispositional antagonism, which is taken advantage of when intoxications are treated with active carbon. Toxicants are adsorbed to the carbon, which is insoluble and stays in the intestines.

As indicated above, the environment in the gastrointestinal tract shows several distinct features. One is the low pH in the stomach, which is not found anywhere else in the body. Although absorption directly from the stomach into the blood is small, it is evident that acidity will affect this uptake. Many carboxylic acids are ionized at pH 7 but are protonated and neutral in the stomach, while amines are protonated and ionized. It has actually been shown that some amines may be transported from the blood, where they are neutral, to the stomach, as the reverse cannot take place. Besides being corrosive, acid is also an efficient catalyst for many chemical reactions, and the pH in the stomach is optimal for the transformations of nitrite (encountered for example as a preservative in meat) and amines or amides to N-nitrosamines or N-nitrosamides. Some of these, e.g. the dimethylnitrosamine formed from dimethylamine (see Figure 6.10), are

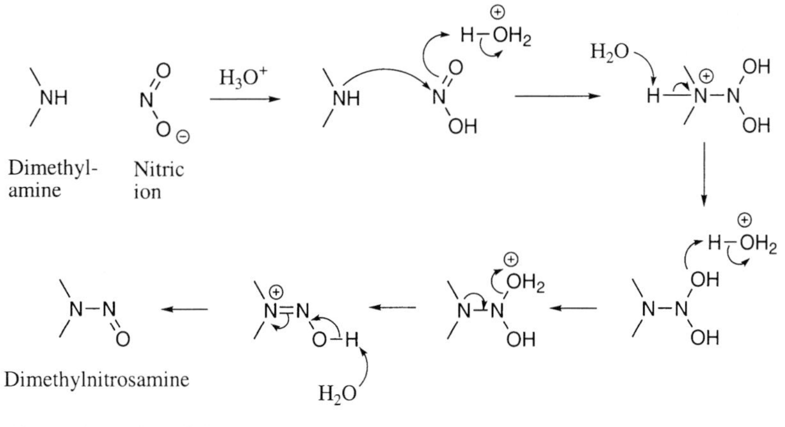

Figure 6.10. The formation of dimethylnitrosamine.

stable and will be absorbed in the intestines, but converted enzymatically to potent carcinogens in for instance the liver, and we shall return to *N*-nitrosamines in Chapters 8 and 11.

The most dangerous nitrosamines are those formed from secondary amines, with two alkyl groups and one hydrogen attached to nitrogen. Secondary amines are found in many foodstuffs (amino acids are secondary amines), and several have found important uses in industry, so it is not unlikely that people are exposed to nitrite and secondary amines at the same time. However, it is not known if the formation of nitrosamines in the stomach of humans has any significant impact on tumour frequency.

The metabolizing capacity of the bacterial flora of the intestines is enormous, and has a different profile compared to our own metabolic system. While our biochemistry is oxidative (aerobic), the bacterial flora in the intestines, especially in the colon, can be regarded as a reductive (anaerobic) enclave separated from us by a protective intestinal mucous layer. Over the colonic mucous layer, which is approximately 1 mm thick, there is a redox potential of several hundred meV, positive in our cells and negative on the outside, and this adds another element to absorption of exogenous compounds in the intestines because reactive radicals are easily generated in such an environment. Some years back scientists could show that extracts of the faeces of some people had quite high mutagenic activity, and the compounds responsible were isolated and characterized. The mutagenic principles turned out to be fecapentaenes, e.g. fecapentaene 12 which is the major component (see Figure 6.11), and it is suspected that they are involved in the genesis of colon tumours.

It has been shown that the fecapentaenes are formed from the plasmalopentaenes (a mixture of different fatty acid esters, the stearic acid ester shown in Figure 6.10 being a major component), by various bacterial strains of *Bacteroides* in the colon. The plasmalopentaenes are believed to be protective compounds, produced by the epithelial cells of the intestines to act as a

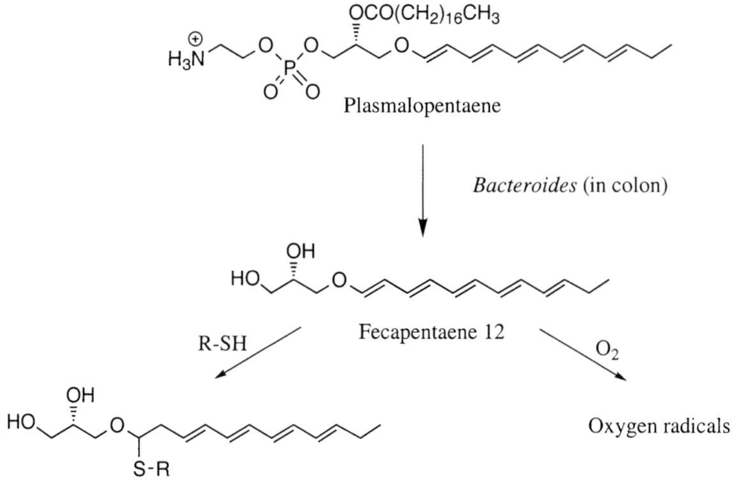

Figure 6.11. Fecapentaene 12 and its precursor plasmalopentaene.

scavenger of radicals formed at the redox border. They are therefore important endogenous metabolites, and their conversion to fecapentaenes depends on the presence of the *Bacteroides* strains in the colon. As nutritional factors are known to influence the composition of the bacterial flora it is possible that the formation of the potentially carcinogenic fecapentaenes is a piece in the cost–cancer puzzle. It is not understood in detail how the fecapentaenes exert their mutagenic activity. Fecapentaene 12, a very potent mutagen, can react with nucleophiles (e.g. thiols, see Figure 6.10) that add to the first carbon of the enol residue, but it has also been shown that the fecapentaenes generate reactive oxygen radicals (discussed in Chapter 7). The hydrolysis of the enol moiety would, after keto–enol tautomerism, yield an unsaturated aldehyde that could be suspected to be the electrophilic form of the fecapentaenes, but this is apparently not the case.

From the intestines, the blood flows directly to the liver via the portal vein, and the liver is the first organ that compounds absorbed in the intestines contact. This is "intentional", and the liver converts nutrients into suitable forms before they are distributed to the rest of the body (the so called first-pass effect). Also exogenous compounds absorbed in the intestines (but not necessarily exogenous compounds absorbed in the skin or the lungs) will primarily be metabolized in the liver, which is perfect if they thereby are rendered less toxic. However, as we shall return to in the following Chapter, this is not always the case.

6.2.3 Uptake via the lungs

As already noted, the lungs are the principal absorption organs when it comes to exogenous chemicals that people are exposed to in workplaces. The lung is constructed for the exchange of oxygen and carbon dioxide between the air and the blood, and the contact surface between the two elements is large (approximately 70 m^2, a badminton court!). The exchange takes place in the functional unit called alveolus, but before reaching this region of the lungs the inspired air passes pharynx, larynx, trachea, and the branching bronchi (one per lung) which end in the alveoli (see Figure 6. 12).

In the alveoli, the distance between air and blood is very small, approximately 1 μm, and this ensures that the diffusion of oxygen and carbon dioxide (as well as most compounds that end up in the alveoli) is very rapid and efficient.

An important variable for absorption in the lungs is the working load of an exposed individual. At rest, an adult inhales approximately 5 l air per minute, which has increased to 20 l/ minute for a person working in a normal manufacturing industry. At extreme workloads, this will of course increase even more. As more air passes the lungs per minute, more of the air contaminants will be absorbed, and this will be further discussed in Section 6.4.3.

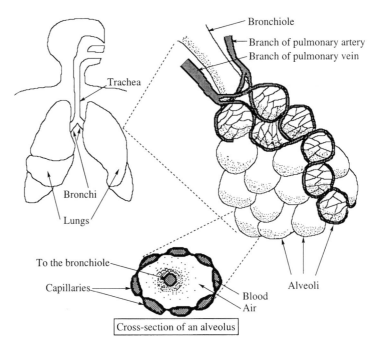

Figure 6.12. The respiratory system with the alveoli.

6.2.3.1 Air contaminants in the form of aerosols

Dust (solid particles), fume (particles formed by combustion in general), smoke (particles formed by combustion of organic material), fog and mist (liquid droplets) and smog (particles and gas formed from car exhausts) contain particles and/or droplets that are sufficiently small to hover in the air, and are called aerosols. The absorption of aerosols in the lungs depends on the weight of the particles or droplets. If they are light enough and, as a consequence, small enough they can follow the inspired air all the way down to the alveoli, while the bigger (and heavier) will not get that deep. The reason for this is that inhaled air passes a number of turns and windings at rather high speed, and the bigger particles are thrown simply by inertial force into the mucus membranes of the nose, the larynx and the bronchi. This protects the lungs from particles, and especially against bacteria that are adsorbed on dust. The mucus is transported by small hairs on the epithelial lining to the pharynx where it is swallowed [this is called the mucociliary escalator (see Figure 6.13)], and the mucus covering the inside of the respiratory system is continuously renewed. Anything that goes down with the mucus to the stomach may of course be absorbed in the intestines, so the mucociliary escalator does not excrete the aerosol from body. However, as a protective device for the lungs it is very important, and chronic exposure to hazardous air contamination (i.e. smoking) may decrease and eventually inhibit its function.

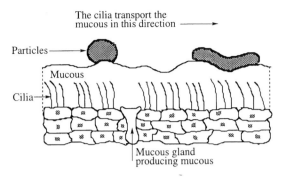

Figure 6.13. Mucociliary escalator.

Very small particles and droplets stay in the air, and although there is no sharp limit it is normally considered that only particles/droplets with a diameter of less than 1 μm are light enough to escape the mucuciliary escalator to a significant extent. Down in the alveoli, such a small particle will through Brownian diffusion (very small particles "diffuse" because they are constantly bombarded by gas molecules) come in contact with and be adsorbed on the moist inner walls. The dissolvable constituents will be absorbed quickly, but anything macroscopic that does not dissolve can be left here for some time because in the alveoli there is no ciliary escalator. Eventually, special cleaner cells of the immune system called macrophages enter the alveolus from the blood, engulf the particle by endocytosis and degrade it both chemically and enzymatically with the contents of their lysosomes. However, particles of some materials, for instance quartz and asbestos, cannot be degraded, and the macrophage will succumb from its efforts. This leads to the stimulation of cell (fibroblast) division in the area, and new layers of connective cells are formed on the inside of the alveolus. As a result, the distance between air and blood increases and the diffusion rate for oxygen decreases. Individuals exposed chronically to quartz dust (working in a quarry for example) or asbestos (used as insulation in older buildings) may develop an irreversible and disabling fibrosis called silicosis and asbestosis, respectively. Asbestos is also carcinogenic, in synergy with smoking, but the mechanism for that effect is not known.

6.2.3.2 Air contaminants in the form of gases and vapours

Chemicals that are present in air as single molecules will not be affected by any inertial forces, but stay in the air until they are condensed on a surface or absorbed by an adjacent solid or liquid. In principle, airborne molecules therefore have no problem to reach the alveoli, but on their way down they also come in contact with and can be absorbed by the mucus membranes. In that sense, the upper respiratory tract can be compared with a scrubber that purifies the air from water soluble and reactive contamination. Hydrophilic molecules therefore do not reach the alveoli as efficiently as lipophilic molecules, although they do get there as well. Highly reactive gases, such as hydrogen chloride (HCl), ammonia (NH_3) and sulphur dioxide (SO_2), are strongly irritating on the mucus membranes in the nose. Respiration is blocked and air containing such gases in significant amounts can not be inhaled in any larger quantities.

Chemicals with intermediate reactivity and lipophilicity, e.g. phosgene, ozone, isocyanates, the halogens, nitric oxide and nitrogen dioxide will be irritating in high concentrations but possible to inhale in lower concentrations. Such compounds can give injuries all the way down, depending on their reactivity and concentration. When they reach the alveoli, their reactivity may harm the cell walls so that the epithelial cells start to leak. The additional liquid in the alveoli will increase the distance between air and blood and hinder the diffusion of oxygen. The condition, which can be compared to internal drowning, is called pulmonary oedema and is believed to be caused by lipid peroxidation of the membranes of the epithelial lung cells (lipid peroxidation is described in Section 10.1.2). Anybody accidentally exposed to such gases should be kept under observation for 48 hours as the pulmonary oedema takes time to develop.

Lipophilic compounds with little or no reactivity reach the alveoli without problems, and are efficiently absorbed by the blood. Examples are the common organic solvents, which are volatile and which we can inhale and absorb the vapours of without any difficulties.

6.3 Bioaccumulation

There are no principal differences between man and other organisms when it comes to the uptake of exogenous chemicals, because there are always biological membranes and cell membranes to traverse (although the uptake of chemicals in the roots of plants is rather complicated). The same rule, that lipophilic compounds are taken up easily, therefore applies for any creature, and organisms will continuously "extract" lipophilic contaminants from the environment. The process when chemicals are absorbed and concentrated by organisms is often called bioconcentration. The bioconcentration factor (BCF), that is the ratio between the concentration of the compound in an organism and the concentration in the surrounding environment, depends largely on the chemical properties of the compound, for example, for non-volatile compounds in the hydrosphere the logP value is strongly correlated with the BCF value. It also depends on the time that the organisms are in contact with the polluted environment, for many compounds it will take days until equilibria are established, and the ability of the organism to convert/excrete the compound will of course also play a role.

Chemicals that are relatively stable in an organism and will not rapidly be converted or excreted, e.g. polyhalogenated aromatic hydrocarbons, may be absorbed by another organism that uses the first as food. In this way chemicals may enter and be concentrated in food chains, a process called biomagnification or ecological magnification. Especially during the period when halogenated hydrocarbons were used in large amounts there were several examples of how the higher organism at the end of a food chain got affected by pollutants as a result of biomagnification. Birds of prey that fed on marine organisms could contain such large amounts of DDT and PCB that their reproduction was inhibited, while birds of prey that fed mainly on plant-eating animals contained less. However, such birds would instead get more of the seed desinfectants used in agriculture (e.g. hexachlorobenzene). Note that bioconcentration and biomagnification do not only apply to lipophilic organic compounds, also inorganic chemicals such as metal ions (see Chapter 12) may be strongly bioconcentrated as they are bound by proteins.

The term bioaccumulation is used for the mixture of bioconcentration and biomagnification, especially in natural environments when it is not possible to separate the two processes. In summary, the accumulation of a chemical in an organism can take place if it is extremely lipophilic and resistant to chemical and/or biological degradation, or if it is complexed and stored out of circulation by the organism. In the following Section we shall see some examples of this.

6.4 Distribution of chemicals in organisms

6.4.1 Chemical properties

Systemic effects of chemicals requires that the compound is absorbed by the blood and, if it is not the blood itself that is the target, distributed so it can reach its target organ. The initial distribution of a compound absorbed into the blood via any of the routes discussed above depends mainly on two things, the ability of the chemical to pass biological membranes and the blood flow in various tissues and organs. As already discussed, compounds that are small and/ or lipophilic will rapidly be distributed to all vascular tissues. However, with time this initial distribution may well change because a number of equilibria will be established. Some organs and tissues, notably the body fat, the nervous system, and the bone marrow, can be considered to constitute a more "lipid" environment, while others, e.g. the blood, the muscles and the eyeballs, are more "watery". In the blood, both endogenous and exogenous chemicals with different polarity may be bound to plasma proteins (e.g. albumin), which may serve as a temporary storage for such compounds. The competition between two compounds for the same binding site on albumin may even lead to a dramatic and dangerous release of one compound as another takes its place.

In general, lipophilic compounds will eventually be distributed to lipophilic tissues like the adipose tissue, but as there are few blood vessels in our fat this will take time. The opposite is of course also true, it takes time for lipophilic compounds to leave the adipose tissues, which in practice function like a long-term storage that in extreme cases can be compared to accumulation. In spite of what is generally believed, even lean and athletic individuals have body fat. Approximately 20 % of the weight of such persons is fat. If body fat is consumed, during extreme slimming or starvation, significant amounts of exogenous lipophilic compounds may suddenly be mobilized (depending on the amount present from the beginning and the rate of fat burning) and this may lead to intoxication.

In addition, many exogenous compounds are distributed in somewhat unexpected ways. The liver and the kidneys, as we will soon come to, are important excretion organs, and they frequently come in contact with compounds that should be expelled. Among other things they take care of many metals, notably toxic heavy metals such as cadmium and lead. If a dose of lead (as Pb^{2+}) is given intravenously, the concentration in the liver is 50 times higher than in the blood 30 minutes after administration. In the liver and the kidneys the heavy metals are complexed by special proteins called metallothioneins, which are small (approximately 6,500 g/mol) and contain large amounts of the amino acid cysteine (25–30 %). In this way the metallothioneins are also thought to function as antioxidants, as metal ions (e.g. Cu^+) that may participate in the generation of reactive oxygen species (Section 7.2.3) are inactivated. Lead is also, together with strontium (Sr), distributed to the skeleton, as both may change places with calcium, and the fluoride ion may take the place of a hydroxyl group in bone. Lead has no known toxic effect in bone, although it will slowly be released as our bones are continuously degraded/rebuilt. Strontium is a radioactive element formed for example during nuclear fission in atom bombs and the distribution of a radioactive element to the skeleton where it will stay for a long time is of course serious. Also fluoride depositions in bone will give chronic effects.

6.4.2 Biological barriers

Although the term "barrier" may be misleading, the passing between the blood and the brain is especially difficult and this resistance is called blood–brain barrier. In the brain, the endothelial cells of the blood capillaries are held together particularly tight, not leaving any intercellular pores, and they are surrounded by glial cells that obstruct the passage. In addition, the interstitial fluid (the fluid between cells and tissues) of the brain has a low protein content, which hampers the diffusion of both lipophilic and hydrophilic chemicals (e.g. metal ions) that prefer to diffuse attached to a protein. Only freely soluble molecules that readily diffuse through the membranes, or are transported by carrier proteins, will pass the blood–brain barrier efficiently. TCDD is an example of a compound that is distributed to all organs but the brain, due to its strong binding to plasma proteins. The blood–brain barrier protects the brain from many chemicals, but it may also act like a cage for compounds that are able to get in and then are transformed to something that cannot get out. We have briefly discussed the absorption of mercury, liquid mercury is poorly absorbed in the intestines but elemental mercury vapour (mercury is a liquid at room temperature and slightly volatile) is readily absorbed in the lungs and distributed throughout the body. When arriving at the brain, dissolved elemental mercury [Hg(0)] is lipophilic enough to diffuse through the blood–brain barrier, but if it is oxidized (which Hg(0) will be everywhere in the body) to divalent mercury ions (Hg^{2+}) inside the brain it cannot get out because of the blood–brain barrier. Chronic exposure to mercury vapour will therefore, besides other effect on the kidneys for example, also give CNS depression. Most organometal compounds, e.g. tetraethyllead and methylmercury (see Chapter 12), readily enter CNS. The blood–brain barrier develops during the first years of a child, and this is a significant difference between infants and adults which explains some of the differences in sensitivity to toxicants (e.g. the neurotoxicity of lead in children).

The placenta (or placental barrier as it also is called) is only present during a special period, when it performs an extremely important task regulating the flow of chemicals between the blood of the mother and the blood of the fetus (which are not mixed, see Figure 6.14). The nutrients are passed through the placenta by active transport while exogenous compounds may pass it just as they pass other biological membranes, and to call placenta a barrier is therefore wrong. The functions of placenta are also discussed in Chapter 10.

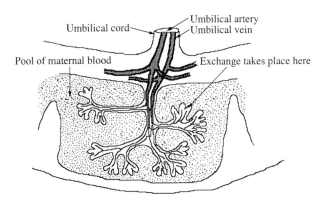

Figure 6.14. Placenta.

6.4.3 Uptake as a function of distribution

The way a chemical is distributed in a body may affect its absorption, especially via the lungs. Volatile compounds can be inhaled and pass to the blood in the alveoli, but they can of course also diffuse in the other direction from the blood to the air. As the balance between uptake and excretion in the lungs depends on the amount of the chemical in the air and in the blood, the most efficient absorption is obtained when the blood during each turn it passes through the body is purified from the chemical by metabolism, distribution to other organs, etc. The distribution can be described by partition coefficients between the different media and organs (or similar substitutes) involved, for example between blood and air, blood and brain, blood and body fat, blood and oil etc. If the blood/air coefficient for a chemical is 10, the brain/blood coefficient is 30 and the fat/blood coefficient is 100, and the person is breathing air containing 1 mg/l (1,000 mg/m^3), the concentrations in the different tissues at equilibrium will be 10 mg/kg blood, 300 mg/kg brain tissue and 1,000 mg/kg body fat. If two compounds with different blood/air coefficients, > 30 for compound A (e.g. styrene) and < 5 for compound B (e.g. 1,1,1-trichloroethane), are compared it is clear that the compound with higher solubility in blood will be absorbed more efficiently. Figure 6.15 shows how much of the two compounds will be absorbed by persons breathing air containing similar (and relatively high) concentrations of the compounds, at different workloads. At rest, the uptake of compound A is approximately 80 % of the inhaled amount, and the proportion taken up is more or less independent of the person's workload (i.e. how much air that is inhaled per minute). For compound B with its smaller blood/air coefficient the uptake at rest is lower, less than 50 % of the inhaled amount, but it decreases even more when the person works harder. This is caused by a fairly rapid saturation of the blood, compound B is not only absorbed from the air but also excreted from the blood back to the air in the lungs.

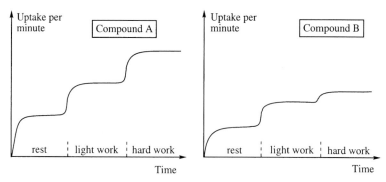

Figure 6.15. The uptake of compounds with large (A) and small (B) blood/air coefficients. See text for explanation.

Saturation can also take place further away in the body. If the coefficients for distribution from the blood to other organs are high, the blood coming back to the lungs will have low concentrations of the compound and new amounts can be absorbed from the respiration air. Toluene, having the intermediate blood/air coefficient of 12 and a high oil/blood coefficient of 1200 (a high oil/blood coefficient indicates that a compound is readily absorbed by tissues containing a lot of lipophilic material), will be absorbed in the lungs to approximately 60 % at rest. The amount present in blood leaving the lungs will efficiently be distributed to other organs and when the blood returns it is almost free from toluene. Even if the workload is increased gradually, 50–60 % of the inhaled amounts are always absorbed (see Figure 6.16, left part). This can be compared with the solvent trichloroethylene, having a similar blood/air coefficient (16) but a much lower coefficient for oil/blood (15). (Remember that blood contains large amounts of proteins that may "dissolve" lipophilic compounds, so the partition between oil and blood cannot be compared with that between oil and water.) It will at rest also be absorbed by the blood in the lungs to 60 %, but its poor solubility in the rest of the body makes it difficult for the

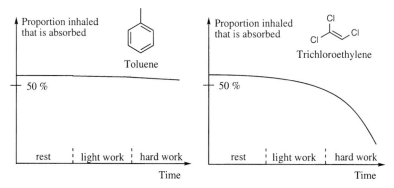

Figure 6.16. The uptake of compounds with similar blood/air but different oil/blood coefficients. See text for explanation.

blood to get rid of the compound. Therefore, the blood coming to the lungs will contain increasing amounts of trichloroethylene as the exposure continues, and the exposure for high concentrations of trichloroethylene may in the end saturate the body completely resulting in the absorption of 0 % of the inhaled amounts (see Figure 6.16, right part).

It should be noted that the experiments discussed in this Section were carried out with healthy volunteers, and that while acute exposure to the high concentrations of the solvents discussed above is not considered harmful, chronic exposure for similar concentrations certainly would be.

6.5 Excretion of chemicals from man

As accumulation of any chemical in one organism in principle is a bad thing, it is important for organisms to be able to convert chemicals. This can be done in several different ways, for example by making use of its nutritional value and metabolizing it to water and carbon dioxide, or by excreting it as it is or as a metabolite.

Just as there are several routes for chemicals to get into the body, there are also a number of ways out. While the three principal ways out, via the lungs, via the liver and via the kidneys will be discussed in some detail, a few minor routes also deserve to be mentioned. Compounds can in limited amounts be excreted by the sweat, in hair, nails and teeth, by salivary and pancreatic secretion, etc. In addition, and more important because this route is coupled with the exposure of an extremely sensitive individual, some chemicals may be excreted via the mother's milk. The milk is characterized by its high content of fat and calcium, and a relative low pH (6.5), and it can therefore be suitable for the elimination of lipophilic compounds, metals that can change place with calcium, and basic amines. It has been shown that the milk is the major excretion route for DDT in breast-feeding women, and if she recently has been exposed to lead this will also be present in the milk. Excretion via the saliva also takes place, but as the saliva is normally swallowed a reabsorption can take place in the intestines and nothing has really changed.

6.5.1 Excretion via the lungs

Volatile compounds that have been absorbed by any route and that are present in the blood will readily diffuse from the blood to the air in the alveoli of the lungs and be excreted by leaving the body with the exhaled air. This is contrary to absorption via the lungs (see Section 6.2.3), and the direction of the net flow of a volatile compound in the lungs depends on the concentration in the air, the concentration in the blood and the blood/air coefficient. So, if the air contains nothing of a volatile compound present in the body it will purify the blood coming to the lungs, and the blood may in its turn extract the compound from the tissues it passes. For very volatile and lipophilic compounds (e.g. diethyl ether) this is the major excretion route, and as it is relatively rapid only small amounts of such compounds will ever be metabolized in the body. The concentration in the exhaled air is directly proportional to the concentration in the blood, notably for less volatile solvents such as ethanol that are only excreted to a small proportion by the lungs, and this is taken advantage of in breath analysers.

6.5.2 Excretion via the liver

The liver is fed by blood coming from the intestines and anything absorbed there will immediately come in contact with the liver cells. These will absorb all kinds of compounds, convert them in various ways and pass them back to the blood or to the bile (see Figure 6.17). After passing through the liver, blood containing the metabolites produced in the liver is collected in veins and pumped to the heart. The bile produced by liver cells is secreted in bile ducts which eventually are emptied in the small intestine. Bile contains mainly amphipathic compounds (bile salts, cholesterol and lecithin) that are responsible for the emulsification of fats in the intestines, but it also serves as a vehicle for the compounds that the liver excretes.

The liver will excrete many metals as ions, e.g. lead, manganese, and also to some extent mercury. Lead is concentrated (approximately 100 times compared to the blood) in the bile by active transport, while mercury is not. Nevertheless, excretion via the bile is still the major route of excretion for mercury, indicating that the half-life of this highly toxic metal in the body is long. In addition, other organic compounds are deposited in the bile, especially acidic and basic compounds but also some neutrals having a high molecular weight. There is no sharp limit for how big they should be, but excretion via the liver is normally the major route for compounds with a molecular weight exceeding 300 g/mol in rats and 500 g/mol in humans. Although this may appear to be a lot, we shall see in Chapter 7 that many smaller organic molecules are metabolized in the liver by conjugation with fairly large endogenous compounds, and the resulting conjugates are considerably heavier than the original compound. An example of an organic compound that is excreted via the liver is the disodium salt of sulphobromophthalein

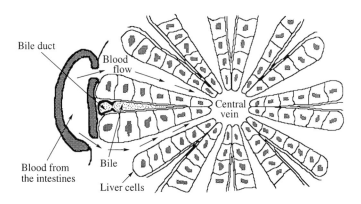

Figure 6.17. The liver.

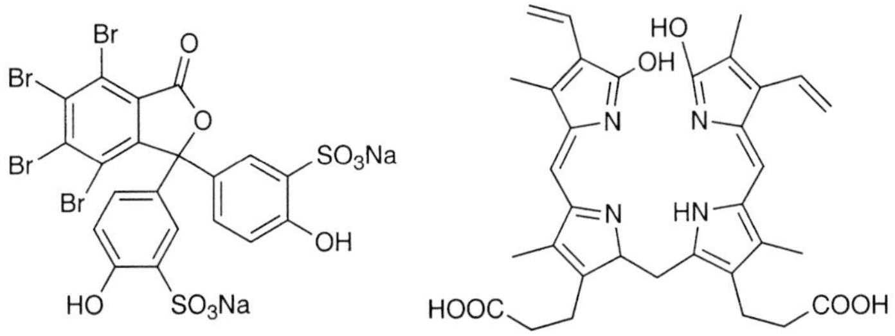

Figure 6.18. Sulphobromophthalein (left) and bilirubin.

(see Figure 6.18), weighing 838 g/mol, which is used to test the function of the liver. It is injected into the blood, and should disappear rapidly if excretion through the liver is functioning normally. Another compound excreted this way is the endogenous compound bilirubin, a coloured breakdown product of haemoglobin, and a person whose liver is malfunctioning is jaundiced because bilirubin is retained in the body.

The exact mechanisms for transporting chemicals that are to be excreted from the liver cells to the bile are poorly understood, but it appears to be carried out by active transport. It can also be noted that this route for excretion has not matured in infants (another example of physiological differences between adults and infants). Excreting organic compounds into the intestines may not appear to be a very good idea, because we have already noted that few organic molecules will escape absorption in the intestines. An example is the highly toxic organometallic compound methyl mercury, well-known as the Minamata-poison (Section 12.7), which may be excreted by the liver, absorbed in the intestines, transported with the blood to the liver where it is excreted. In addition, the bacterial flora has the ability to reverse some of the metabolic conversions carried out in the liver and re-convert the excreted metabolite to an absorbable compound. The process when a excreted compound is reabsorbed in some form is named the enterohepatic circulation, and although it is not a perpetual machine with no losses it still prolongs the time that a exogenous chemical spends in the body. We shall return to the enterohepatic circulation in Chapter 7. Besides being an organ involved in the metabolism and excretion of exogenous compounds the liver is also the target organ in many intoxications, and this will be discussed in Chapter 10.

6.5.3 Excretion via the kidneys

The kidneys are the most important excretion organs. Although the two kidneys only comprise 1 % of the body weight they receive approximately 25 % of the blood coming directly from the heart, and 20 % of this is filtered in the functional unit of the kidneys, the nephron (see Figure 6.19). The kidneys contains approximately a million of these units, each with a glomerulus, a set of fine capillaries with relative large pores that allow fairly big molecules (up to small proteins) to leave the blood, surrounded by a sac (Bowman's capsule) that collects the filtered

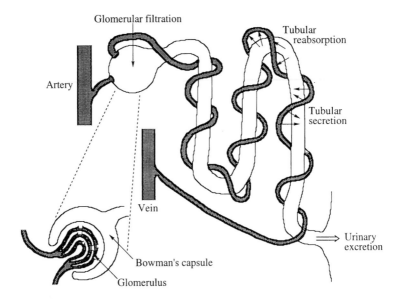

Figure 6.19. The nephron.

fluid. As the blood passes the glomerulus, it loses part of its water together with any dissolved relatively small compound. Albumin and other plasma proteins (with adsorbed lipophilic compounds) are retained. The liquid produced by the glomerular filtration, approximately 200 1/24 hours, is called the primary urine, and it passes through the tubuli which are surrounded by blood vessels. (The glomerulus and the tubule together make up a nephron.) Evidently, a person can not afford to lose 200 1 of water per day, and most of the water as well as nutrients and minerals are reabsorbed back to the blood again (tubular reabsorption). Besides simple diffusion, several specific transport systems are involved to make the loss of valuable and useful chemicals as limited as possible.

What is left in the concentrated urine (approximately 1.5 1 is produced per day) is a solution of anything originally present in the blood but filtered in the glomerulus, unable to diffuse back to the blood and not fitting any of the transport systems. If an exogenous compound that cannot be used and should be excreted is too lipophilic to stay in the urine, it must be further converted in the body to a more hydrophilic form. In addition, there are also transport systems in the tubuli that actively excrete (tubular secretion) certain compounds (organic acids and bases) from the blood to the urine. In the beginning of the era of penicillin as an antibiotic drug it was expensive and its rapid elimination via the kidneys was a problem. Penicillin was therefore administered together with an acid (probenecid, see Figure 6.20) that occupied the transport system normally pumping penicillin from the blood to the urine, which prolonged the time penicillin stayed in the body (a synergistic effect!). Today penicillin is relatively cheap, and higher doses are given instead. Another compound that is excreted in this way is uric acid (see Figure 6.20), present in our urine and associated with gout. If the transport of uric acid is inhibited, the concentration of it goes up and the risk of gout increases.

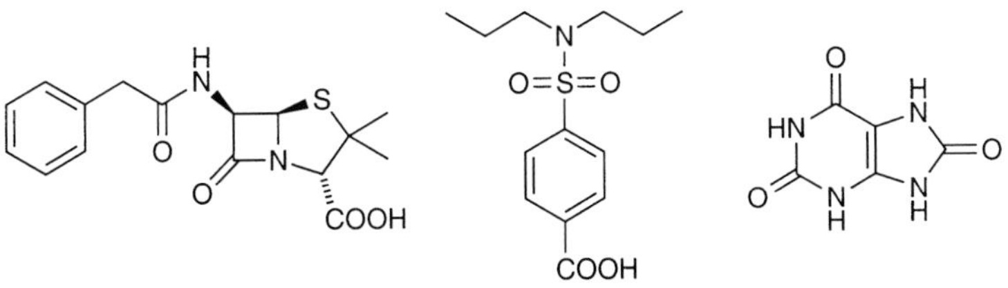

Figure 6.20. Penicillin G (left), probenecid and uric acid (right).

7 Metabolism of exogenous compounds in mammals – the principal conversions

Although the presence of exogenous chemicals in our environment has increased dramatically during the last century, the chemical soup that life on earth has evolved in has always contained numerous compounds that are of little use and toxic to any organism. Consequently, life on earth has always needed ways to handle such chemicals, and it is not surprising that enzymatic systems that exclude/degrade/detoxify/expel exogenous chemicals have evolved. One way of achieving this is by having specific membrane transport proteins that are selective in their choice of compounds to be taken up or excreted (discussed in the preceding Chapter). Such systems only protect us to a certain degree, and we have seen how easily lipophilic exogenous compounds pass biological membranes and enter our bodies. In some rare cases compounds can be converted from being exogenous to become endogenous, and thereby in principle be useful. However, as we shall learn more about in this chapter, the conversions, even if in the end they produce a harmless compound, often generate more toxic intermediates and byproducts. The excretion of compounds is facilitated if they have chemical properties that make them suitable for excretion, e.g. are volatile (excretion with the exhaled air), have a molecular weight over 500 (excretion with the bile) or have high solubility in water (excretion with the urine). If the compounds do not possess the required properties to begin with, they can obtain them by being converted enzymatically by the metabolism. The volatility of a compound may be increased if the metabolism decreases its molecular weight (by breaking bonds) and/or converts functional group. A high molecular weight can be achieved by combining the exogenous compound with an endogenous compound, while increased water solubility can be accomplished by converting chemical functionalities to more polar forms. There are many examples of compounds that are transformed to more volatile forms or to compounds with higher molecular weight, and at least partially excreted by the lungs and the liver, but the kidneys are the main excretion organs and the conversion of exogenous compounds to highly water soluble derivatives is the most important way for the body to get rid of such compounds.

In this chapter the principal enzymes involved and the most important conversions will be discussed, while Chapter 8 will account for how a number of well-known organic compounds are metabolized in mammals. The survey of the principal conversions will concentrate on how one functionality at a time may be changed chemically, while the practical examples in Chapter 8 will demonstrate that it may be difficult to predict the outcome of the metabolism of a compound containing several groups that participate in conversions.

7.1 Overview

The conversion of exogenous compounds is commonly divided into two principal reaction types, the phase I and the phase II reactions. The result of a phase I reaction (or, in some cases, a combination of several phase I reactions) is the introduction of a suitable functional group that increases the water solubility and can be used in a phase II reaction. This is achieved either by the enzymatic introduction of such a functionality into the framework of the molecule, or by exposing a group already present but masked by other groups. The phase II reactions, which are often called conjugations, take advantage of this chemical handle and attach polar hydrophilic groups to it, producing products with considerably higher water solubility. In addition, substitutions are also included among the phase II reactions. Obviously, it is not always necessary to go through both phases, if the phase I reaction produces a compound that is sufficiently water soluble to be excreted with the urine or if a compound already contains suitable functionality for the phase II reactions. An example of a compound that is metabolized by both phase I and phase II reactions is benzene (see Figure 7.1). Benzene is oxidized to phenol (a phase I reaction

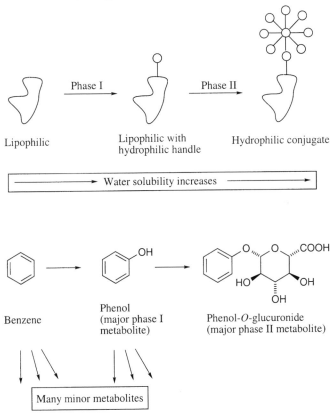

Figure 7.1. The two phases of the metabolism of exogenous compounds, exemplified by the major conversion of benzene.

introducing the hydroxyl group as a chemical handle) which subsequently is conjugated with glucuronic acid. The latter is a carbohydrate derivative that has a carboxylic acid functionality that is ionized at physiological pH, besides several hydroxyl groups, and its size and polarity will facilitate the excretion of any glucuronic acid conjugate via the kidney or the liver.

Benzene is a cheap chemical obtainable in unlimited amounts from oil. It has several useful properties and has been an important industrial solvent as well as an additive in petrol. However, it is also toxic and chronic exposure may lead to the development of leukaemia, and its use is today strictly regulated in most countries. The mechanism by which benzene is carcinogenic is still not completely understood, although neither benzene itself nor its two major metabolites phenol and phenol-*O*-glucuronide are directly involved. Instead it is one or several other metabolites that are responsible for this specific effect, and the example of benzene illustrates an important aspect of the metabolism of exogenous compounds. It is always to be expected that a compound is metabolized by several parallel routes, and a minor route, perhaps only responsible for the conversion of one or a few percent of the original compound, may generate the metabolites that finally turn out to be responsible for the toxic effect. We shall therefore return to benzene in Chapter 8, and identify some minor metabolites that are involved in its toxicity. This is also a common reason for the difference in sensitivity observed between different species (discussed in Chapter 5), as the balance between different metabolic pathways may also be slightly different in closely related species.

It is not unusual that the term "detoxification" is used as a synonym for the metabolism of exogenous compounds, but this is not correct. For example, most carcinogenic compounds require enzymatic conversion (metabolic activation), and it is actually the metabolites that transform the normal cells to tumour cells. Metabolic conversions will always change the toxicity of a compound. The metabolites may be more or less toxic than the original compound but never equally toxic, although in some cases a metabolite can be more toxic in one way and less in another.

Another concept that has also been mentioned is the induction of the enzymes that are required for a certain conversion, for example by the presence of the substrate for that conversion in certain amounts, and induction is very important for the enzymatic systems involved in the conversions of exogenous compounds. Many exogenous compounds are powerful inducers and, as we already have noted, the induction of metabolizing enzymes is often the reason for potentiating and synergistic toxic effects.

7.2 Enzymatic systems involved in oxidations

Enzymes that will metabolize exogenous compounds are present in all tissues in the body, although the liver has the highest metabolic capacity of all organs. However, for a special enzymatic conversion other organs and tissues may have a similar or even higher metabolic capacity, and differences in the metabolic profiles between various organs and tissues may be responsible for systematic toxic effects of compounds that require metabolic activation (which for natural reasons often affect the liver). In the cells, most of the phase I reactions take place in the endoplasmic reticulum, a relatively lipophilic organelle to which lipophilic compounds would be attracted, while most of the phase II reactions are carried out in the cytoplasm. As will be discussed in this section, the phase I reactions are dominated by one group of enzymes, the cytochrome P-450s which in principle are oxidative but also can perform reductions under special circumstances. Several important phase II systems are normally active, and often compete for the same types of substrates. Again, there is reason to remember the existence of the "11th organ system" composed of bacteria in the intestines (Section 6.2.2), with its huge and different metabolic capacity.

7.2.1 Primary and secondary metabolism

The metabolism of all organisms can be divided into a primary and a secondary part, although they are not completely separate. The primary deals with all nutrients, their degradation into suitable units (amino acids, saccharides, nucleotides, etc.) that can be used for the construction of macromolecules and for the production of energy, etc. It takes care of all metabolic conversions necessary to keep a cell (organism) alive, and if the primary metabolism fails or is obstructed, a life-threatening situation results. It is also noteworthy that the primary metabolisms of different species are more or less identical, even organisms as different as man and bacteria share many of the principal primary metabolic pathways. The secondary metabolism is less well understood. It emanates from primary metabolites and converts them to metabolites with other functions, and the secondary metabolites (e.g. alkaloids and terpenoids) produced in various organisms are often completely different from each other. Examples of possible roles that secondary metabolites may play are as substances used for signalling, within an organism as hormones and between organisms as pheromones, or as substances used for defensive reasons, that taste bad or are toxic. The toxic compounds from natural sources that we have discussed (e.g. strychnine and nicotine in Table 5.1) are secondary metabolites, and so are also many antibiotics (e.g. the penicillins), the cardiac glycosides (e.g. digitoxin) and several antitumour agents (e.g. taxol).

Exogenous compounds will be metabolized by enzymes that belong to both the primary and the secondary metabolism. The principal conversions of the primary metabolism are covered by textbooks in biochemistry and will not be discussed here. In this chapter we shall confine ourselves to an overview of other important conversions, mainly belonging to the secondary metabolism, that take place with exogenous compounds in mammals.

7.2.2 Cytochrome P-450

The cytochrome P-450 system (discussed as a system but in reality the enzyme cytochrome P-450 exists as a large number of variants, *vide infra*) catalyzes oxidations mainly but also some reductions, and is the most important metabolic system for the conversion of exogenous compounds. It is not uniquely mammalian enzyme system, having also been found (in identical or similar forms) in many other organisms. In mammals it is found in most tissues although its function does not appear to be the same everywhere, reflecting the role of cytochrome P-450 in the conversion of steroid hormones. The highest concentrations are naturally found in liver cells, but its presence has also been demonstrated in for example the lungs, kidneys, placenta and testes. The cytochrome P-450 system is also called the polysubstrate monooxygenase system or the mixed function monooxygenase system, reflecting the fact that it is not specific for one type of substrate but instead accepts a surprisingly broad range of structures. It is composed of two enzymes, cytochrome P-450 and a reductase (NADPH-cytochrome P-450 reductase) providing the electrons necessary for the reactions in the form of the reduced coenzyme NADPH. Interestingly, there are approximately 20 cytochrome P-450 enzymes for each reductase, placed in a ring around the reductase and the whole complex is embedded in the phospholipid matrix of the endoplasmic reticulum. The cytochrome P-450 system will not work outside this membrane, it is dependent on the membrane for its function and the phospholipids improve the contact between the two enzymes. However, when liver cells are disrupted by homogenisation the endoplasmic reticulum is fragmented into vesicles, diameter 0.1 μm, and these are fairly easily purified from the rest of the cell debris. The vesicles are called microsomes and can be used when the effects of cytochrome P-450 conversions are to be studied *in vitro*. The cytochromes are proteins that readily take part in electron transfers because they contain a haem group with an iron atom that easily goes between the ferric [Fe(III)] and ferrous [Fe(II)] states, and we shall return to cytochromes when we discuss toxic effects on the respiration chain and energy production. Similar haem groups are found in the haemoglobin of the blood and the chlorophyll of plants, although in the latter the iron is exchanged for magnesium. The fact that the cytochrome P-450 system is a monooxygenase reveals that it uses molecular oxygen as an oxidation agent, and that one of the two oxygen atoms forms water. The overall reaction for the oxidation (= hydroxylation in this case) of a substrate RH can be written as shown in Figure 7.2. An example of the physiological role of this system, to prepare various variants of vitamin D, is also showed in Figure 7.2.

The molecular mechanisms by which the cytochrome P-450 system oxidizes various substrates have been intensely studied over the years, but they are still not understood in all details. In this section the general mechanism will be discussed, while the details linked to the oxidation of certain functional groups will be given in section 7.4. Cytochrome P-450 oxidations are initiated by the association of a lipophilic substrate molecule with the cytochrome P-450 enzyme in its oxidized form [Fe(III)] in the endoplasmic reticulum (an lipophilic environment). The complex is then accepting an electron from the reductase via the coenzyme NADPH, reducing the iron to Fe(II). Just as haemoglobin with a similar Fe(II)-containing haem group binds oxygen, the reduced form of cytochrome P-450 will immediately attract a molecule of oxygen (which under normal circumstances is present in all tissues) to the iron atom. Now the stage is set for the acceptance of a second electron from the reductase, two protons are picked up from the surrounding water and a molecule of water is split off from the very unstable complex. The second oxygen of O_2 has now combined with the iron to the final oxidation reagent, a complex containing what could be a perferryl group with the iron in an extremely oxidized state [Fe(V)]. How this final form looks is not understood, and the oxidation states of iron and oxygen in the final reactive enzyme could also be Fe(III)-O$^+$ or Fe(IV)-O·. However, it is known that other perferryl compounds prepared synthetically are very efficient oxidizing agents and perform

$$R\text{-}H + O_2 \xrightarrow{\quad NADPH + H^+ \qquad NADP^+ \quad} R\text{-}OH + H_2O$$

A general oxidation performed by cytochrome P-450

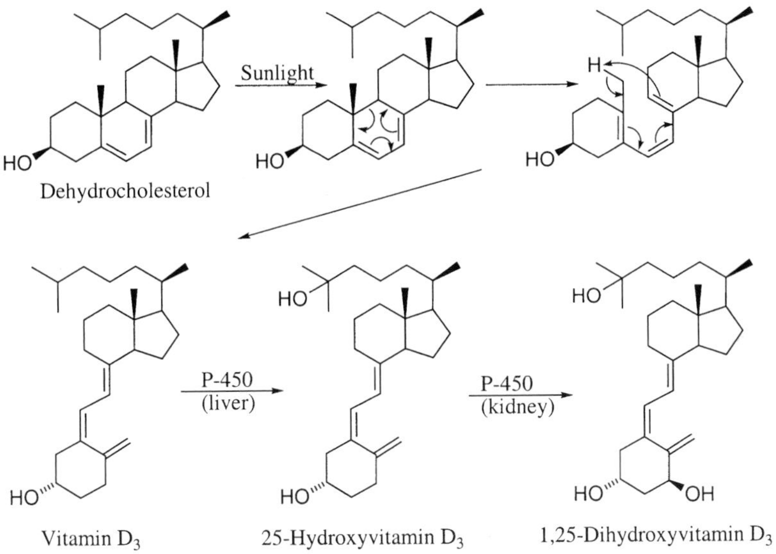

Dehydrocholesterol

Sunlight

Vitamin D$_3$ 25-Hydroxyvitamin D$_3$ 1,25-Dihydroxyvitamin D$_3$

P-450 (liver)   P-450 (kidney)

Figure 7.2. The oxidation of vitamin D$_3$ by cytochrome P-450.

similar oxidations as cytochrome P-450, because Fe(V) is eager to accept electrons and be reduced to something more normal [Fe(II) or Fe(III)]. The substrate is now oxidized, for example by the insertion of an oxygen in a carbon–hydrogen bond so that hydroxylation has taken place as shown in Figure 7.3, whereafter the substrate can leave the enzyme and cytochrome P-450 returns to its original state.

Just as haemoglobin is blocked by carbon monoxide (CO) (discussed in Chapter 10), the reduced form of cytochrome P-450 is also sensitive to this toxic gas. This effect can be utilized when cytochrome P-450 is studied *in vitro*, but it will not be relevant *in vivo* because of the more serious effect of CO on the blood. In certain situations, e.g. during anaerobic conditions, cytochrome P-450 can make use of another molecule than molecular oxygen as the oxidation agent. That compound, whatever it is, will be reduced, and this is one way that metabolic reductions of exogenous compounds takes place. Although the oxidation cycle of cytochrome P-450 appears to be secure, several complexes are not very stable and especially that formed after oxygen has added to the reduced haem group (in the lower right-hand corner of Figure 7.3) has a tendency to get into trouble.

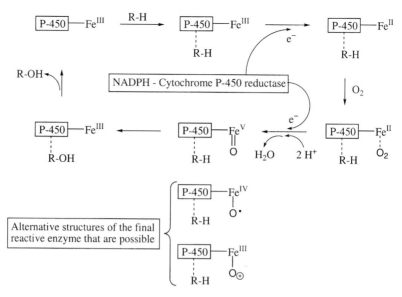

Figure 7.3. The oxidation cycle of cytochrome P-450.

7.2.3 Generation of reactive oxygen species

Instead of accepting an electron from the reductase, an electron may be transferred internally from the iron (which will be oxidized) to the O_2 (see Figure 7.4). The negatively charged molecular oxygen possessing an additional electron (called superoxide, superoxide anion or superoxide radical) will leave the complex, the cytochrome P-450 with its substrate will re-enter the oxidation circle, and the effect of this extra loop is that it generates superoxide which as an anion radical may participate in radical reactions.

The generation of superoxide during electron transfer to molecular oxygen puts the finger on a problem that any life-form that uses oxygen for the production of energy has to cope with. Molecular oxygen itself has a certain reactivity, in its ground state it is a biradical with two unpaired electrons which is fairly stable (because the two unpaired electrons have parallel spin in the ground state of O_2) but that readily participate in and promote radical reactions. Oxygen would therefore be expected to give synergistic effects with anything that generated radicals in the body, for example the exposure to ionizing radiation or to any exogenous compounds that are metabolized to radicals. This is believed to be a major cause of the toxic effects observed in mammals exposed to high concentrations or high pressures of oxygen, as some radicals always are generated in our bodies by, for example, the normal background radiation. In its ground state, molecular oxygen can only function as an oxidation agent by accepting one electron at the time, and the second electron transforms superoxide anion to the dianion of hydrogen peroxide which during physiological pH is protonated. The hydroperoxyl radical, the corresponding acid of superoxide anion, has the pKa 4.8 and is consequently ionized at pH 7.4. Hydrogen peroxide

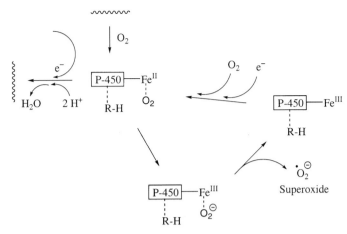

Figure 7.4. Reactive oxygen species are generated during cytochrome P-450 oxidations.

is a well-known oxidation agent, used to bleach hair, as a rocket propellant and a disinfectant, and although it is not limited to one-electron reductions like molecular oxygen it is in some situations reduced to the hydroxyl radical on the reduction path to water (see Figure 7.5). The hydroxyl radical is one of the most reactive compounds known, it will immediately initiate radical chain reactions and its and other radicals involvement in various toxic effects will be discussed in Chapter 10.

Also other oxidizing enzymatic systems will leak superoxide, as will oxyhaemoglobin (the oxygenated form of haemoglobin), and in addition we will encounter several examples of how superoxide is generated by the autoxidation of exo- and endogenous compounds with molecular oxygen. It is believed that several percent of the oxygen consumed in the mitochondria (the cell organelle where energy is produced and most oxygen is consumed) is reduced to superoxide and hydrogen peroxide, and this continuous generation of reactive oxygen species, as they are called, is a threat to any organism. It has been counteracted by the evolution of several protective enzymatic systems which are present in tissues where oxygen is being reduced and take care of the reactive oxygen species. Examples are superoxide dismutase (SOD), catalase (*vide*

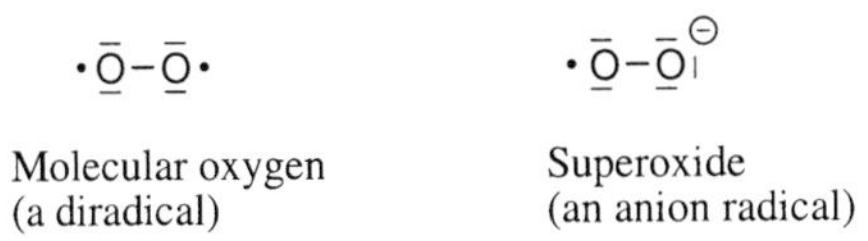

$$\cdot\bar{O}-\bar{O}\cdot \qquad\qquad \cdot\bar{O}-\bar{O}^{\ominus}$$

Molecular oxygen   Superoxide
(a diradical)   (an anion radical)

$$O_2 \xrightarrow{e^-} \cdot O_2^{\ominus} \xrightarrow{e^-,\, 2\,H^+} H_2O_2 \xrightarrow{e^-,\, H^+} \cdot OH + H_2O \xrightarrow{e^-,\, H^+} 2\,H_2O$$

Figure 7.5. The reduction of O_2 to H_2O via reactive oxygen species.

Superoxide dismutase (SOD): $2 \overset{\cdot}{O_2}^{\ominus} + 2 H^{\oplus} \longrightarrow H_2O_2 + O_2$

Catalase: $2 H_2O_2 \longrightarrow 2 H_2O + O_2$

Glutathione peroxidase : $H_2O_2 + G\text{-}SH \longrightarrow 2 H_2O + G\text{-}S\text{-}S\text{-}G$

Glutathione disulphide reductase

Figure 7.6. The enzymatic disarmament of superoxide and hydrogen peroxide.

infra), and glutathione peroxidase, and the reactions they catalyze are shown in Figure 7.6. The dismutation of two molecules of superoxide may also take place spontaneously, but is much more efficient when SOD is present. The reduction of hydrogen peroxide by glutathione peroxidase is coupled with the oxidation of the thiol groups of two molecules of glutathione (GSH) to glutathione disulphide (GSSG) (two thiol groups are oxidized to a disulphide). To regenerate glutathione, which is an important endogenous compound in several ways, the disulphide is reduced back to the thiols by glutathione disulphide reductase with the coenzyme NADPH.

Glutathione peroxidase depends on both vitamin E and selenium as cofactors for its function. Shortage of these two essential compounds, e.g. because of malnutrition, will weaken the defence against reactive oxygen species and increase the risk of toxic effects. Investigations have show that there is a correlation between insufficient intake of vitamin E and selenium and an increased incidence of tumours, and it is suspected that at least part of this is due to their role in hydrogen peroxide reduction. Another possible disturbance of this defence, which has a limited capacity, is that a sudden generation of large amounts of reactive oxygen species will saturate the enzymes and increase the possibility for these compounds to take part in other reactions. Examples of some deleterious reactions are shown in Figure 7.7, the reaction between superoxide and hydrogen peroxide and between hydrogen peroxide and Fe(II), generating the extremely reactive hydroxyl radical. Although most of the 4 g of iron that an adult contains is

$$\overset{\cdot}{O_2}^{\ominus} + H_2O_2 + H^{\oplus} \longrightarrow \overset{\cdot}{O}H + H_2O + O_2$$

The Haber Weiss reaction

$$H_2O_2 + Fe^{2+} + H^{\oplus} \longrightarrow \overset{\cdot}{O}H + H_2O + Fe^{3+}$$

The Fenton reaction

Figure 7.7. Left unattended, superoxide and hydrogen peroxide may generate hydroxyl radicals.

not available for the latter reaction, the small part that is makes it biologically relevant. Iron, as all other essential compounds, is toxic in high concentrations, and the effects observed in iron overload suggest that they are caused by increased formation of hydroxyl radicals.

Interestingly, reactive oxygen species are also put to good use by our immune system. They are deliberately produced by phagocytes, and used to degrade foreign material. In addition, it has been shown that in certain plants lipid peroxidation generates reactive oxygen species as an efficient defence against parasites.

The number 450 in the name cytochrome P-450 comes from its absorption of visible light with an absorption maximum at 450 nm in its reduced form with carbon monoxide bound to the Fe(II) ion, while the letter P stands for pigment. However, it is not one single enzyme but rather a range of isoenzymes that all have slightly different properties, and the exact composition of a cytochrome P-450 subset varies with species, organ, sex, age, health, stress, etc. They all absorb light with approximately the same wavelength, and the exact number is sometimes used to identify individual enzymes (e.g. cytochrome P-448, etc). It is not understood why so many isoenzymes are produced, if they are intended for example to carry out slightly different reactions, and in this text we will simply call them all cytochrome P-450. An important aspect of the cytochrome P-450 system is its ability to be induced, so that the number of enzymes increases in situations when the need for cytochrome P-450 conversions is high. The presence of exogenous compounds is a common cause for induction, and different compounds will induce different isoenzymes (or sets of isoenzymes). Potent inducers of cytochrome P-450 are for example polyhalogenated aromatic and alicyclic hydrocarbons, e.g. TCDD, PCBs, PBBs, and the pesticides aldrin, dieldrin, lindane as well as the chlorinated phenoxy acids.

7.2.4 Other oxidative enzymes

Besides cytochrome P-450, there is a number of other enzymes involved in phase I oxidations. These are less general compared to the cytochrome P-450 system with which they in most cases compete, and they appear to be intended for more specific conversions. Only a brief overview is given in this subsection, further details about the conversions can be found in Section 7.3.

The microsomal FAD-containing monooxygenase is, like the cytochrome P-450 system, localized in the endoplasmic reticulum and depends on NADPH and molecular oxygen for its action, and it is obvious that it is not easy to distinguish the two. It oxidizes mainly heteroatoms, nitrogen (primary, secondary and tertiary amines), sulphur (thiols, thioethers and thiocarbamates) and phosphor (phosphines and phosphonates). Many of these compound classes are also believed to be substrates for cytochrome P-450, especially before the existence of the microsomal FAD-containing monooxygenase had been established, and the balance between the two has not yet been determined. In contrast to cytochrome P-450, it does not appear to be inducable, instead it has been suggested that its activity is regulated by hormones.

The dioxygenases, which for example are important in the primary metabolism of mammals for the degradation of amino acids, will just as the monooxygenases use molecular oxygen for their oxidations. The difference between the two is that the dioxygenases insert both oxygen

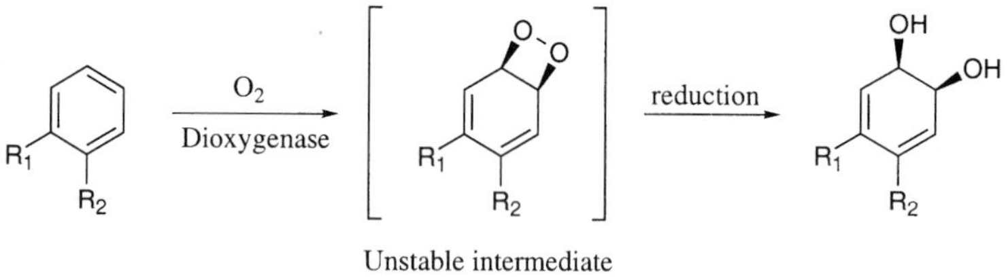

Figure 7.8. Oxidation of an aromatic hydrocarbon with a dioxygenase.

atoms, as indicated in Figure 7.8. Dioxygenases are also present in bacteria and, as we shall discuss in Chapter 9, they are very important for the degradation of organic compounds by microorganisms.

Amine oxidases (monoamine oxidases) typically oxidize primary amines to aldehydes, but also secondary amines to ketones (compare with the transaminases of the primary metabolism that oxidize α-amino acids to α-keto acids). They are not located in the endoplasmic reticulum but instead in the mitochondria or dissolved in the cytoplasm of the cell. They are present in different tissues, and in CNS for example they are involved in the conversion of neurotransmittors.

Peroxidases are enzymes that couple the reduction of hydrogen peroxide or a hydroperoxide to the oxidation of a substrate. Metabolic activation by peroxidases has received attention recently, and it is believed that they play a significant role in the generation of several types of electrophilic compounds. Several examples of this will be given in the next section. An interesting example is prostaglandin synthase, which actually generates its own hydroperoxide. It will oxidize arachidonic acid (an essential fatty acid) to prostaglandin G, which subsequently is reduced by the same enzyme to prostaglandin H_2 (see Figure 7.9). During the second step, exogenous compounds can be cooxidized, and a number of examples of this with various types of substrates have been reported.

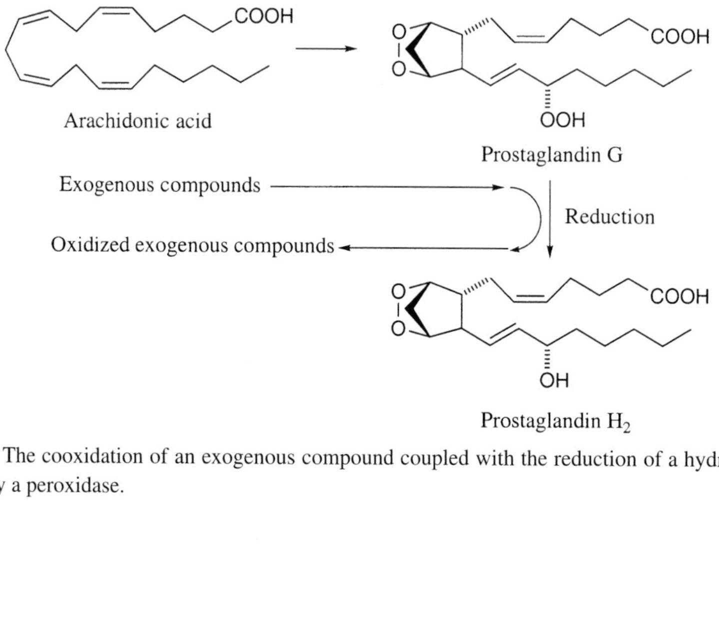

Figure 7.9. The cooxidation of an exogenous compound coupled with the reduction of a hydroperoxide, catalyzed by a peroxidase.

Another example is catalase which uses hydrogen peroxide produced by for example SOD (*vide supra*) to oxidize various substrates, e.g. phenols, formic acid, formaldehyde and alcohols according to the general formula $RH_2 + H_2O_2 \rightarrow R + 2H_2O$. Actually, a substantial part of the ethanol consumed by humans is converted to acetaldehyde by catalase. The conversion shown in Figure 7.6 is only carried out when excess hydrogen peroxide accumulates in the cell.

Alcohol dehydrogenases, which exist in several similar forms, will oxidize primary alcohols (to aldehydes) and secondary alcohols (to ketones), although some amounts of alcohols will also be oxidized by cytochrome P-450 and catalase/hydrogen peroxide. Tertiary alcohols will normally not be oxidized at the alcohol function. Aldehydes are considerably more toxic than primary alcohols, and also more lipophilic, so this oxidation is clearly a metabolic activation. The alcohol dehydrogenases are soluble enzymes found especially in the liver, the kidney and the lungs. It should be noted that these are reversible reactions, the aldehydes and ketones may be reduced back to alcohols again by the same enzyme (but with the reduced form of the coenzyme instead of the oxidized, *vide infra*) and the equilibrium depends on weather other conversions or any excretion that take place with the different forms.

Aldehyde dehydrogenases oxidize both aliphatic and aromatic aldehydes to carboxylic acids, a reaction that shows all the features (i.e. less toxic and more hydrophilic products are formed) of a metabolic detoxification. The reaction is irreversible and the carboxylic acid is a good substrate for the phase II reactions, which in practice will be a strong driving force for pushing the equilibrium between primary alcohols and aldehydes to the right. Aldehyde dehydrogenases exist both in a general form (as many isoenzymes) and one specific for formaldehyde. Other enzymes that oxidize aldehydes have also been identified, including the molybdenum containing flavoprotein aldehyde oxidase and xanthine oxidase.

Acyl coenzyme A dehydrogenase takes part in the β-oxidation pathway which is responsible for the conversion of fatty acids to acetyl units that subsequently are oxidized in the citric acid cycle, a part of the primary metabolism. It converts the coenzyme A thioester of a straight-chain saturated carboxylic acid to an α,β-unsaturated thioester, and exogenous carboxylic acids (possibly formed by the oxidation of the corresponding alcohol or aldehyde) may of course also be activated as a thioester and oxidized.

7.3 Phase I conversions

Although the classification of certain metabolic conversions of a exogenous compound as a phase I or a phase II reaction is not always unambiguous, phase I reactions can be considered to be either oxidations, reductions or hydrolyses.

7.3.1 Oxidations

7.3.1.1 Epoxidations of carbon–carbon multiple bonds

Carbon–carbon multiple bonds are readily epoxidized by cytochrome P-450, and as the reaction increases the electrophilicity of the compound this is a metabolic activation from a toxicological point of view. However, although most of the exogenous compounds that we are exposed to contain a carbon–carbon multiple bond, this is not as serious as it may appear, and this for two main reasons. Firstly, alkenes and aromatic hydrocarbons also undergo other conversions which do not generate epoxides. Secondly, the epoxides formed in a phase I oxidation are in most cases either chemically unstable and will spontaneously form other products, or are good substrates for phase II enzymes that will hydrolyze the epoxide to a relatively harmless 1,2-diol (*vide infra*). Assuming that the mechanism for cytochrome P-450 oxidations presented in Figure 7.3 with the perferryl complex as the true oxidizing agent is correct, the epoxidation of an alkene can be imagined as either a one- or a two-electron transfer (see Figure 7.10). The intermediate complex formed may then either cyclize to an epoxide, rearrange via a 1,2-alkyl (-hydride if R_4 is a hydrogen) shift to a carbonyl compound, or react with a nucleophilic atom (e.g. a nitrogen of the haem group) of the enzyme and form a covalent bond to it (the two latter possibilities are only shown for the two-electron transfer route in Figure 7.10). The reaction

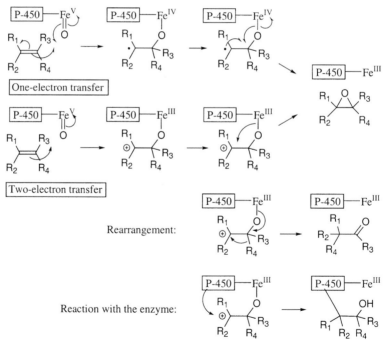

Figure 7.10. Possible mechanisms for the oxidation of alkenes.

with the enzyme, which has been shown to take place with an intermediate and not with the epoxide end product, will inactivate the cytochrome P-450. This has been called a suicide mechanism, as the enzyme activates a substrate to a form that also destroys the enzyme.

Aromatic compounds may also be epoxidized, by an electrophilic attack of the perferryl complex (or via the corresponding radical). If the cation is formed as an intermediate, as shown in Figure 7.11, it may yield the corresponding products (via routes a and b in Figure 7.11) as were obtained when alkenes are epoxidized (see Figure 7.10). However, epoxides of aromatic compounds are unstable in general and will spontaneously rearrange to the corresponding ketone as indicated in Figure 7.11, and the ketone (formed directly by route a or via the epoxide by route b) would immediately be transformed to the corresponding phenol (as a result of the keto-enol equilibrium). A hydroxylated aromate (a phenol) is therefore to be expected to be the initial product after cytochrome

> Furan is used as a solvent and as an intermediate in chemical synthesis. It is known to be toxic to the liver and kidneys and exposure to furan has been associated with liver tumours. The furan ring, which is aromatic, is a common functionality in natural products and we contact furans in many different ways.

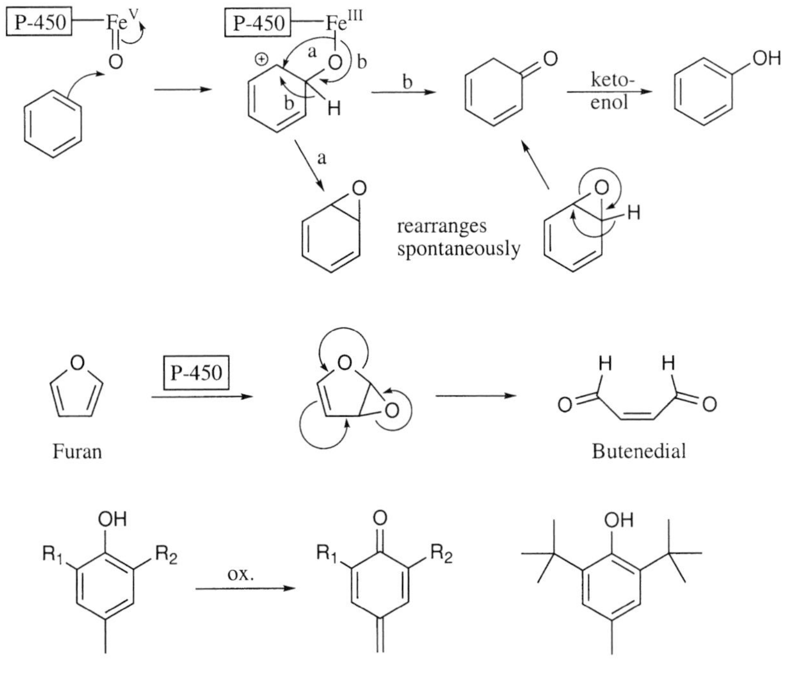

P-450 oxidation of aromatic compounds, if the aromatic nucleus is attacked. Heterocyclic aromates may also be epoxidized, as the example with furan in Figure 7.11 indicates. The

Figure 7.11. Oxidation of aromatics.

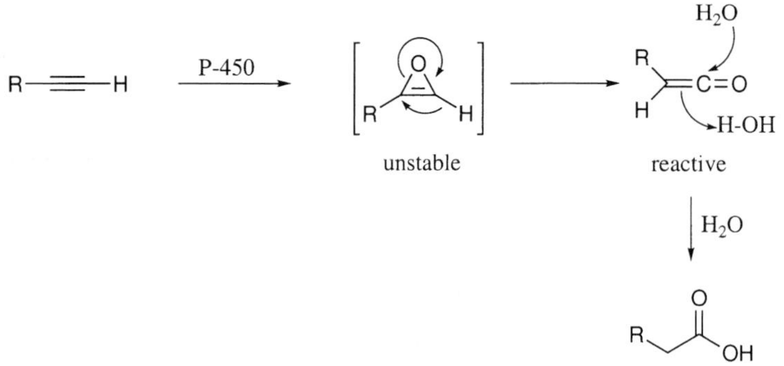

Figure 7.12. Oxidation of alkynes.

transient epoxide formed will rearrange, in the case of a furan with no substituents in positions 2 and 5 to an unsaturated 1,4–dialdehyde which is reactive.

Frequently observed when aromates containing suitable substituents (e.g. phenols with a *para*-hydroxy or -methyl group) are oxidized is that reactive benzoquinone derivatives are formed. In the example shown in Figure 7.11, a *p*-hydroxytoluene is oxidized to a quinone methide which has electrophilic properties. BHT, butylated hydroxytoluene (3,5-di-*t*-butyl-4-hydroxytoluene) is a synthetic antioxidant widely used as an additive in foods, cosmetics and drugs, and has been shown to damage the lung tissue of experimental animals fed large doses. The toxic effect of BHT is due to its conversion to a reactive quinone methide.

It is believed that alkynes also are epoxidized, although they have not been so thoroughly investigated as the use of alkynes (except for acetylene gas) is not wide-spread. The product formed initially from the epoxidation of a terminal alkyne is too unstable to be isolated, and rearranges to the ketene. This in turn is reactive and may add water to form a carboxylic acid, and experiments with labelled compounds have supported the mechanism shown in Figure 7.12. Cytochrome P-450 is destroyed to some extent also during the oxidation of alkynes, and it is reasonable to believe that this is due to the formation of a reactive ketene (that of course may react with another nucleophile instead of water) at the surface of the enzyme.

7.3.1.2 Hydroxylations of saturated carbon

Hydroxylations will insert an oxygen atom into the bond between any saturated atom and a hydrogen. It is possible that also unsaturated atoms are hydroxylated although the normal pathway for oxidizing unsaturated compounds appears to be via the epoxide as discussed above. However, as the epoxide of especially aromatic compounds will rearrange to phenols, this oxidation could formally be regarded as a hydroxylation. Hydroxylations of carbon–hydrogen bonds are the most common, resulting in alcohols (for example the endogenous hydroxylations of steroids), although nitrogen–hydrogen hydroxylations (producing hydroxylamines) are important because of the toxicity of the products. In principle it is possible to hydroxylate any X–H

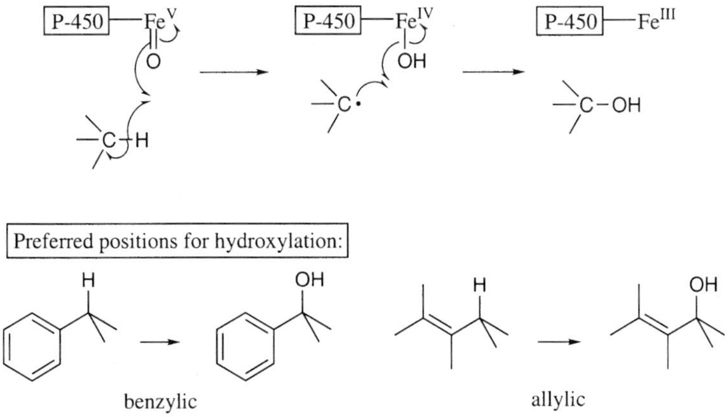

Figure 7.13. Hydroxylation of hydrocarbons.

bond, although in practice it is the position that for one or another reason is activated that are hydroxylated at a reasonable rate. Important positions that relatively easily are hydroxylated are the allylic and the benzylic positions, i.e. a saturated carbon with at least one hydrogen adjacent to a double bond or an aromatic system (see Figure 7.13). The reason for this is obvious when the most likely mechanism for the hydroxylation is examined, as it involves the formation of a relatively stable allylic or benzylic radical after the abstraction of a hydrogen atom from the substrate as the rate-determining step (or the corresponding allylic or benzylic cation if the oxidation proceeds with a two-electron transfer). This will then collapse to form the hydroxylated substrate and the regenerated cytochrome P-450 is ready for the next oxidation cycle.

Another position that is activated for hydroxylation is the C–H α to heteroatoms, i.e. carbons with at least one hydrogen and a heteroatom bound to it. In the case the heteroatom is N, O or S, this hydroxylation is often called α-hydroxylation or oxidative dealkylation. In Figure 7.14 a reasonable mechanism for the α-hydroxylation of an amine is shown, the heteroatom is thought to take part in the initial step by being oxidized and the α-hydrogen is picked up by the perferryl complex. The unpaired electron on the heteroatom together with one of the electrons of the C-H bond forms a π-bond between the heteroatom and the α-carbon, whereafter the hydroxyl group is transferred from the iron to the carbon and the electrons in the double bond moves back to the heteroatom so that it recovers its lone pair. As has been discussed in Chapter 5, the product formed after α-hydroxylation, with two heteroatoms bound to the same saturated carbon, is in principl unstable in contact with water. The initially formed N-hydroxymethyl derivative formed in Figure 7.14 will therefore spontaneously generate the corresponding amine and formaldehyde. The overall result of the hydroxylation is therefore that the bond between the nitrogen and carbon that is oxidized is cleaved, explaining why this hydroxylation also is called oxidative dealkylation.

The same hydroxylation may also take place if the heteroatom is a halogen, chlorine, bromine or iodine, although this reaction has been given a different name; oxidative

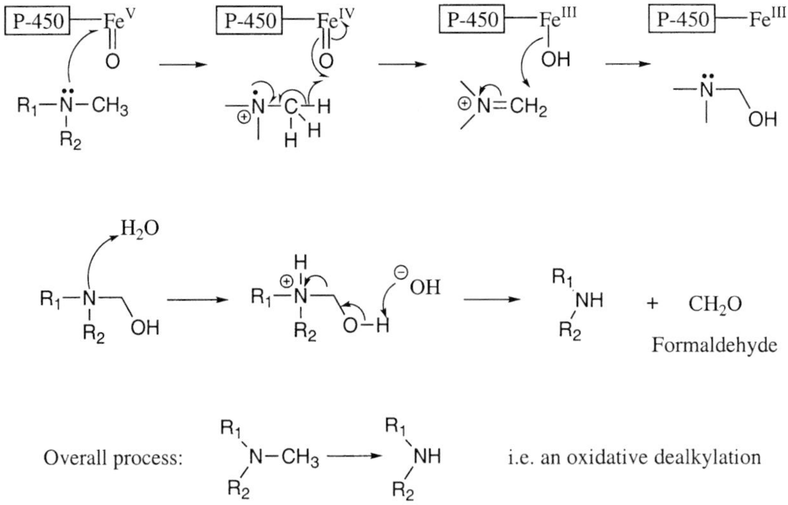

Figure 7.14. α-Hydroxylation, leading to oxidative dealkylation.

dehydrohalogenation. As the name implies, HCl, HBr or HI is eliminated after oxidation, the hydrogen comes from the hydroxyl group added by the hydroxylation and the double bond formed by the elimination is a C=O bond. A primary halide would yield an aldehyde, a secondary a ketone, while a tertiary halide is not able to undergo oxidative dehydrohalogenation as it lacks α-hydrogens. A compound with two halogens attached to the same carbon is hydroxylated faster, and the result of the dehydrohalogenation is a carboxylic acid halide which normally is quite reactive. Chloramphenicol (see Figure 7.15), a natural antibiotic used to treat infections, has been shown to deactivate cytochrome P-450, presumably by reacting irreversibly as the corresponding carboxylic acid chloride derivative with the enzyme. A compound with one hydrogen and three halogens on the same carbon may also be hydroxylated, and chloroform will generate phosgene (the dichloride of carbonic acid) which is highly toxic and at least partly responsible for the toxicity of chloroform.

Saturated hydrocarbons that completely lack functional groups are not readily oxidized in mammals. The smaller members are often volatile enough to be excreted by the lungs, but the larger members are not possible to excrete as they are. As we already have indicated, the mechanism for hydroxylation of saturated carbon goes via the carbon radical, and the more a molecule can stabilize an unpaired electron the more likely is it that it is formed. The more branched a hydrocarbon radical is, the more stable is it and the more likely is it that this carbon is hydroxylated. For unbranched (linear) alkanes it has been noted that the favoured (but not sole) positions for hydroxylation is the carbon adjacent to the terminal methyl group (called ω-1 hydroxylation, as the terminal carbon is called the ω-carbon). We shall return to this topic in Section 8.1.

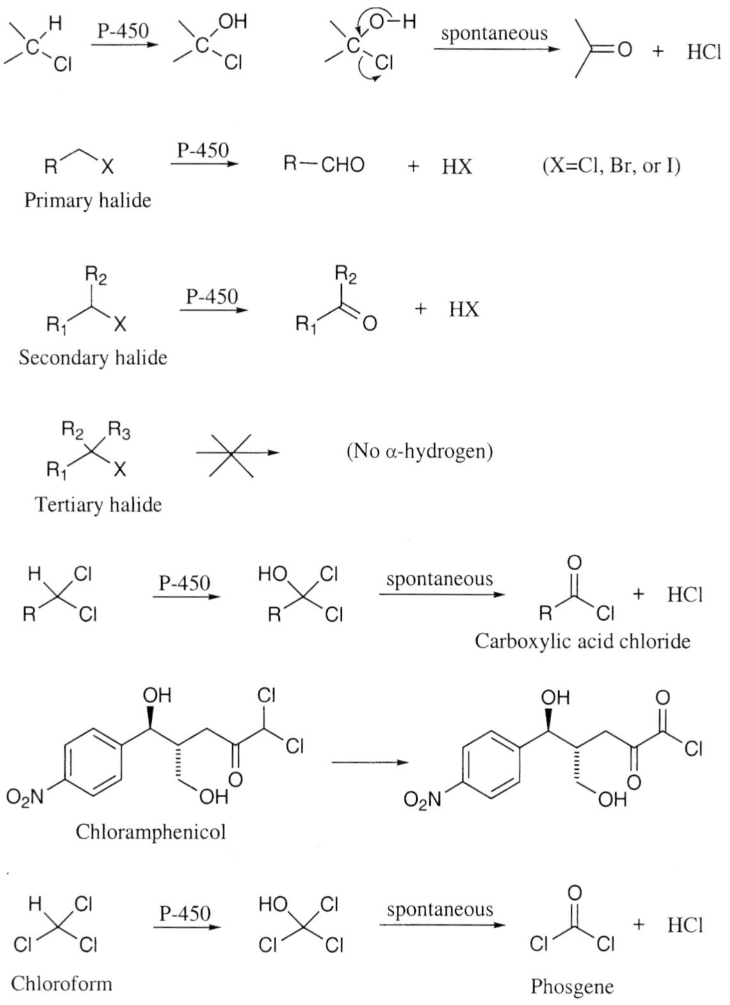

Figure 7.15. Oxidation of halides.

7.3.1.3 Hydroxylation of amino groups

Also nitrogen in activated positions may be hydroxylated, especially if α-hydroxylation (*vide supra*) is not possible. Aromatic amines are a class of compounds that has received a lot of attention over the years, because they are toxic and because they have been used in several industrial chemical processes. In an aromatic amine the carbon α to the amino group has no hydrogen and consequently cannot be oxidized by α-hydroxylation, instead oxidation may take

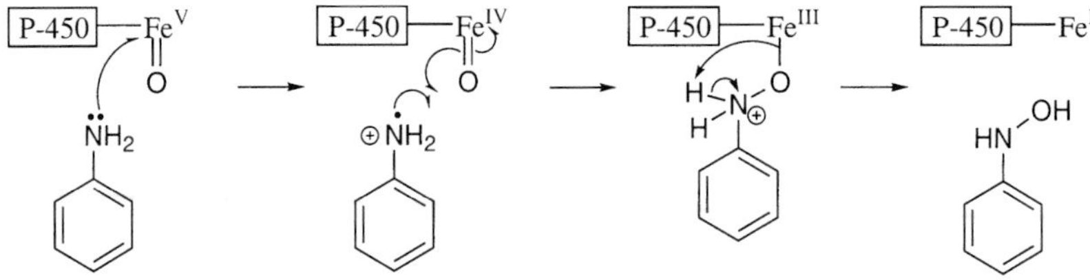

Figure 7.16. *N*-Hydroxylation of an aromatic amine.

place in the aromatic system or in the amino group. As the nitrogen is benzylic and thereby activated, nitrogen hydroxylation takes place fairly easily. Hydroxylation of an amino group can be carried out by both cytochrome P-450 and by the microsomal FAD-containing monooxygenase, and peroxidases have also been shown to be involved in the oxidation of aromatic amines in some tissues (e.g. in the epithelial cells of the urinary bladder). A possible mechanism for the hydroxylation of an aromatic amine to the corresponding hydroxylamine by cytochrome P-450 is shown in Figure 7.16.

7.3.1.4 Oxidation of single bond to double bond

Although this is less common, several examples are known. One obvious possibility is that carboxylic acids with two CH_2 groups next to the carboxylic acid functionality are attacked by the β-oxidation of the primary metabolism. In this, an activated (as the coenzyme A thioester) fatty acid is oxidized by the enzyme fatty acyl CoA dehydrogenase with the coenzyme FAD, to become unsaturated. As mentioned, this is intended to oxidize fatty acids to acetic acid units (acetyl CoA) that can be used for the production of energy in the mitochondria. The oxidation is carried out according to Figure 7.17, via the oxidation of the α,β-single bond to a double bond.

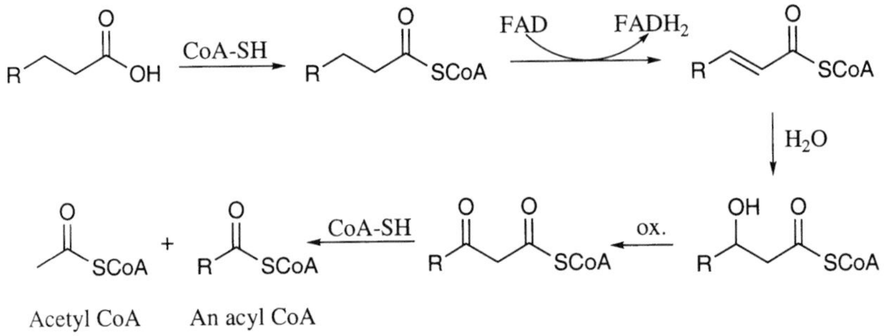

Figure 7.17. The oxidation of a single bond to a double bond during β-oxidation.

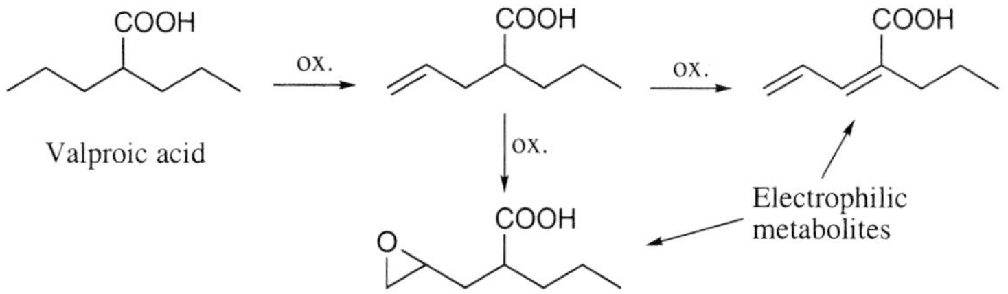

Figure 7.18. The metabolic activation of valproic acid.

Exogenous carboxylic acids may also undergo β-oxidation. However, the reaction will normally not stop after the first oxidation but proceeds via the β-hydroxyl and the β-keto derivatives to the new carboxylic acid lacking two carbons (see Figure 7.17). In addition, the oxidation of C–C to C=C bonds (presumably by dehydrogenases) have also been observed in other situations, and one case where the oxidation appears to be an important activation step is with valproic acid (an anti-epileptic drug). The oxidized derivatives of valproic acid shown in Figure 7.18 are responsible for the liver toxicity of the drug.

Amines (primary, secondary or tertiary, as long as they have an α-hydrogen) have been shown to be oxidized by amine oxidases to imines (see Figure 7.19). The imines are not stable in contact with water and will be hydrolyzed to the free amine and an aldehyde or a ketone (depending on the nature of R_1 and R_2 in Figure 7.19). The hydrolysis proceeds via the α-hydroxylamine, formed after the addition of water to the imine double bond. Consequently, α-oxidation of amines by cytochrome P-450 and oxidation of amines to imines by amine oxidases both yield the same end products.

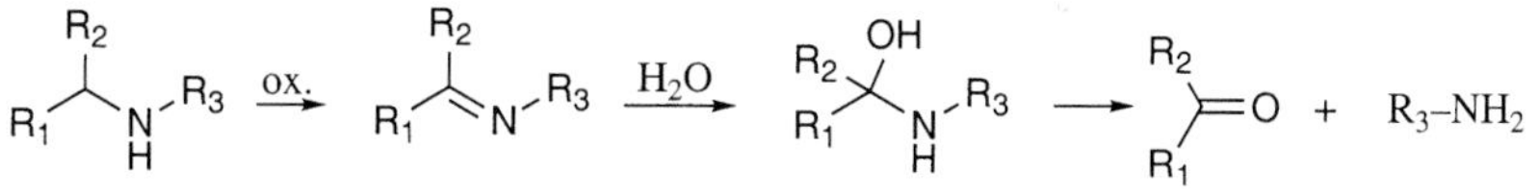

Figure 7.19. The oxidation of an amine to an imine, which can be hydrolyzed to a ketone.

7.3.1.5 Oxidation of heteroatoms

Different heteroatoms, e.g. N, S, P, Se, and I, may be oxidized by cytochrome P-450, by the microsomal FAD-containing monooxygenase and by amine oxidases. The hydroxylation of nitrogen has already been discussed, in addition nitrogen (in tertiary amines) can also be oxidized to the corresponding aminoxide (e.g. nicotine to nicotine *N*-oxide, see Figure 7.20) which has high water solubility and can be excreted without any conjugation. Sulphur can be oxidized in several steps, and some examples are given in Figure 7.20.

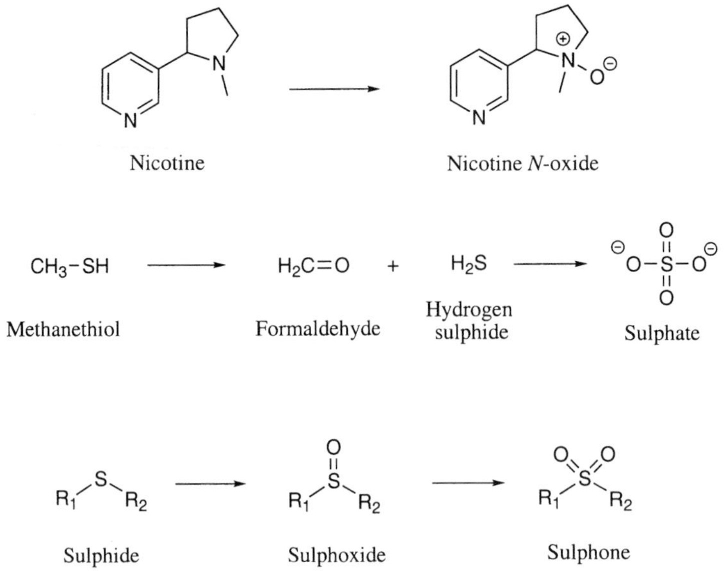

Nicotine   Nicotine *N*-oxide

CH_3-SH $\longrightarrow$ $H_2C=O$ + H_2S $\longrightarrow$

Methanethiol   Formaldehyde   Hydrogen   Sulphate
                                     sulphide

Sulphide   Sulphoxide   Sulphone

Figure 7.20. Oxidation of nitrogen in tertiary amines, and sulphur.

7.3.1.6 Oxidation of alcohols and aldehydes

The oxidation of alcohols to aldehydes or ketones are reversible reactions, catalysed by alcohol dehydrogenases that use NAD^+ as a coenzyme, and the equilibrium depends on the further destinies of the oxidized/reduced forms. On the contrary, the oxidation of aldehydes to carboxylic acids is irreversible. Aldehydes, formed from primary alcohols, are in most cases oxidized to carboxylic acids so efficiently that only negligible amounts are reduced back to the alcohol. For secondary alcohols a true equilibrium can often be observed. The reduced form has a good handle (the hydroxyl group) for conjugations, which facilitates its excretion by the kidneys, while ketones are more volatile compared to alcohols and may, at least the smaller ketones, be excreted by the lungs. In addition, ketones are in some instances further oxidized on the carbon α to the keto function, acetone is for example oxidized to pyruvic acid which enters the citric acid cycle and is converted to carbon dioxide and water.

7.3.2 Reductions

Aldehydes and ketones can be reduced to alcohols by the same enzymes that oxidize alcohols but with (the reduced form of) the coenzyme NADPH (see Figure 7.22). The oxidation of alcohols is favoured by the fact that the $NAD^+/NADH$ ratio in mammalian cells normally is high, while reductions are favoured by a low $NADP^+/NADPH$ ratio. Aldehydes are, as discussed above, irreversibly oxidized to carboxylic acids and are therefore less likely to be reduced.

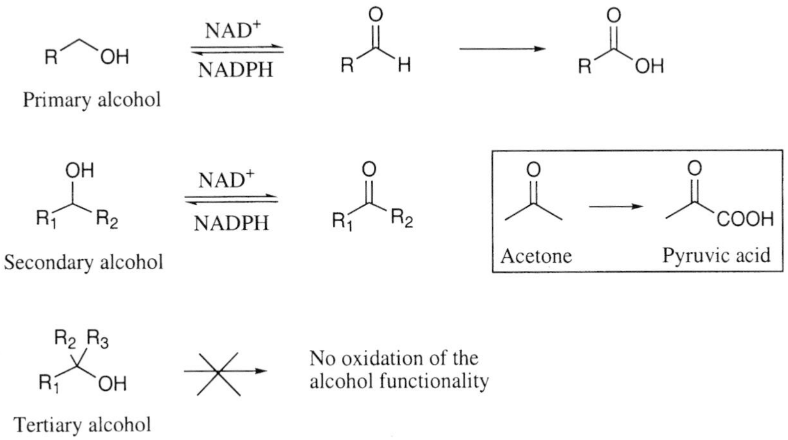

Figure 7.21. Oxidation of alcohols.

Cytochrome P-450 will to a small extent also carry out reductions of exogenous compounds, in spite of its character as an oxidase. The substrate will then replace oxygen as the electron sink and be reduced, and the substrate and molecular oxygen will compete for the electrons supplied by the NADPH-cytochrome P-450 reductase. Low oxygen concentration will promote the reduction of such a substrate while high concentrations will inhibit it. Reduction is of course most important under anaerobic conditions, which we primarily find in the distant parts of the intestines, and here bacteria will carry out reductions of exogenous compounds with enzymes that show similarities with cytochrome P-450. Nitro compounds, for example, will be reduced to the corresponding nitroso compound, which in turn may be further reduced to the hydroxylamine and the amine (see Figure 7.23). Also azo compounds may be reduced, initially to hydrazines but also after cleavage of the nitrogen–nitrogen bond to two amines, and also other oxidized heteroatoms may be reduced. Halogenated organic compounds may be reduced via reductive

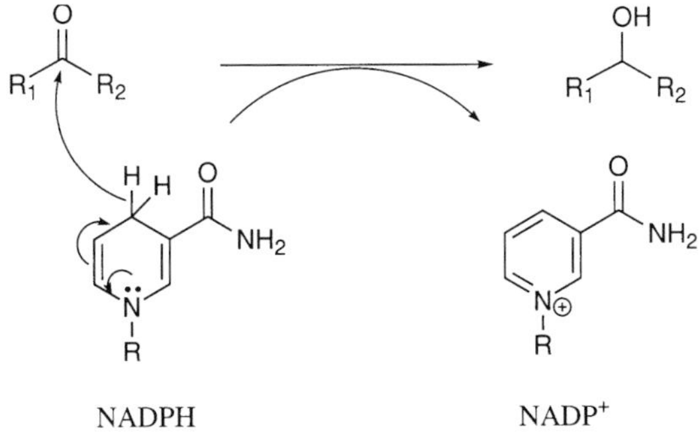

Figure 7.22. Reduction of a ketone with the coenzyme NADPH.

Heteroatom reduction:

Reductive dehalogenation:

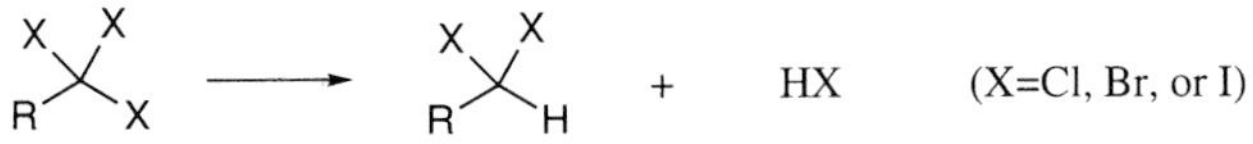

Figure 7.23. Examples of reductions that may take place in mammals.

dehalogenation (further discussed in Chapter 9), by which, in principle, a halide ion is replaced by a hydride ion. The banned insecticide DDT is metabolized by this route (discussed in Chapter 12), although very slowly, and we shall encounter some other examples in Chapter 8.

7.3.3 Hydrolyses

Oxidations and reductions are chemically more complicated reaction, involving the transfer of electrons. They are carried out with the help of coenzymes (e.g. NADPH in cytochrome P-450 oxidations) which have been prepared by the primary metabolism at the expense of chemical energy. Hydrolyses do not involve the transfer of electrons, and are reactions between a substrate and a molecule of water that are energetically favourable but hampered by an activation energy that makes them slow at body temperature. The enzymes that catalyze hydrolyses take this energetic threshold down and speed up the reaction considerably.

7.3.3.1 Hydrolysis of epoxides

trans means that two substituents are situated on opposite sides of a ring or a double bond, while *cis* indicates that they are on the same side. For double bonds, the notations *E* (entgegen, apart in German) and *Z* (zusammen, together in German) are also used:

trans-2-Butene, or *E*-2-butene

cis-2-Butene, or *Z*-2-butene

Some epoxides are used as industrial chemicals, e.g. ethylene oxide which is used for the production of certain polymers and for the sterilization of medical equipment, and the handling of such chemicals is very hazardous. In addition, we have seen that epoxides are formed inside our bodies as a result of phase I conversions, and this is of course a serious threat to our health. However, as we already have indicated the most epoxides formed from aromatic compounds are unstable and generally rearrange to phenols within seconds (although

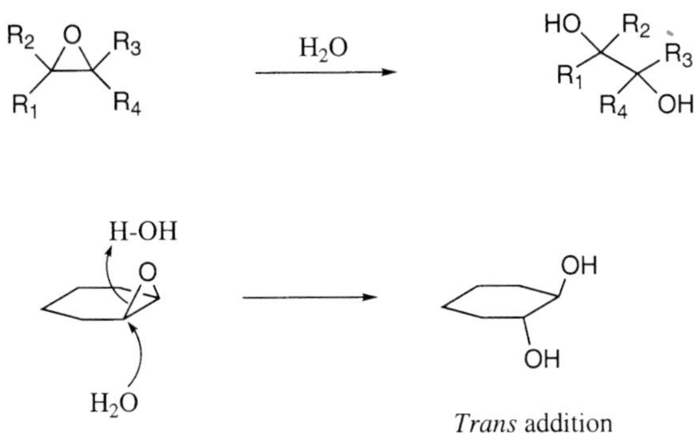

Figure 7.24. The hydrolysis of epoxide by epoxide hydrolase.

some are more stable). In addition, the majority of the reasonably stable but still reactive epoxides that are formed as a result of phase I conversions are detoxified by hydrolysis. The efficient weapon against the epoxides is the enzyme epoxide hydrolase, present in the endoplasmic reticulum of all tissues in the body although in quite different amounts, and it will simply hydrolyze an epoxide to a diol. As always, there are many different kinds of epoxide hydrolases, with different substrate specificities, and some appear to be present also in the cytoplasm.

Cyclic diols formed by this procedure are always *trans*, as shown in Figure 7.24, indicating that the mechanism for the hydrolysis is that the water molecule that opens the epoxide is attacking from the opposite side of the epoxide ring (see Figure 7.24). The reaction with water as a nucleophile also takes place spontaneously, but without the enzyme it is much slower and much less selective. That is to say that without the enzyme epoxide hydrolase the epoxide would preferentially react with better nucleophiles, for example thiol groups in proteins and nitrogens in the nucleotides of the DNA. The enzyme picks up a molecule of water and a molecule of the epoxide and brings them together, and facilitates their reaction by creating both conditions acidic (to protonate the epoxide oxygen) and basic (to make the water oxygen more nucleophilic). The same reaction can be carried out in a chemical laboratory, by adding acid or base as catalyst but never the two at the same time, and it is difficult to imitate the mild and efficient conditions from which enzymatic catalysis benefit.

For some compounds, e.g. styrene which is used for the manufacture of polyfoam, it was suspected that the vinylic double bond is epoxidized by cytochrome P-450, and synthetically prepared styrene oxide was shown to be genotoxic *in vitro*. However, it was not possible to obtain any conclusive results from *in vivo* experiment with respect to the carcinogenicity of styrene, although chronic exposure to styrene is well-known to damage the CNS. The reason for the discrepancy of the genotoxicity is the existence of epoxide hydrolase. Styrene oxide is indeed formed *in vivo*, but it is so efficiently hydrolyzed by epoxide hydrolase (see Figure 7.25) that it has problems reaching the DNA to a significant degree. The diol obtained after the hydrolysis of styrene oxide is oxidized by alcohol dehydrogenase to both the ketone and the aldehyde/carboxylic acid, and mandelic acid is the major urine excretion product of styrene.

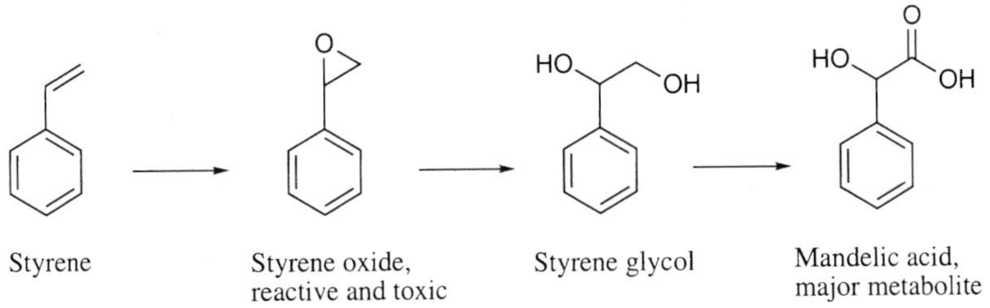

| Styrene | Styrene oxide, reactive and toxic | Styrene glycol | Mandelic acid, major metabolite |

Figure 7.25. The metabolism of styrene.

A dangerous situation is at hand when an epoxide formed in a phase I oxidation for some reason is a bad substrate for the epoxide hydrolase. Such epoxides are not hydrolyzed, at least not efficiently, and they will be able to diffuse through the cell and react with critical targets. An example of this is benz[a]pyrene, a chemical found in trace amounts in the exhaust of any combustion with a non-perfect balance between fuel and oxygen (car exhaust, cigarette smoke, etc.). Benz[a]pyrene is genotoxic and carcinogenic, and believed to be responsible for a significant proportion of the lung cancers caused by cigarette smoking. As a fairly large and lipophilic molecule it has no chance to be excreted without first being metabolized, and there are not much choice for the metabolism but to attack the aromatic system itself. Several different positions are oxidized, but the most important for the genotoxicity of benz[a]pyrene is the C-7/C-8 epoxidation as shown in Figure 7.26. (The [a] in benz[a]pyrene indicates to which side of pyrene, composed of the four central rings, an additional benzene ring has been added.) The epoxide thus formed is a relatively stable, good substrate for epoxide hydrolase that will convert it to the *trans*-diol shown in Figure 7.26. However, the diol will be epoxidized a second time, generating the diolepoxide shown in Figure 7.26 that also is relatively stable but for some reason not a suitable substrate for the epoxide hydrolase. This epoxide will be able to diffuse into the cell nucleus and react with the DNA.

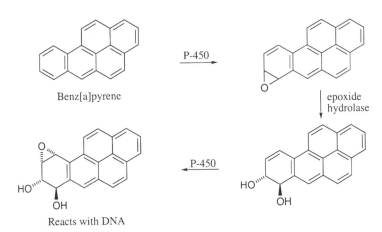

Figure 7.26. Metabolic activation of benz[a]pyrene.

The epoxide hydrolase present in the microsomes has been shown to be inducable, although not by one specific compound but instead by a general need for "epoxide hydrolysis". Some compounds, e.g. cyclohexeneoxide formed after the epoxidation of cyclohexene, has been shown to inhibit the enzyme, possibly by reacting with it. It is interesting in this context to imagine the possibility that also endogenous compounds containing double bonds, of which there are many, can be epoxidized. No systematic investigation has been carried out, but cholesterol has a carbon–carbon double bond that in principle could participate in such a conversion. The epoxide of cholesterol (prepared synthetically) is genotoxic to some simpler organisms but not to mammalian cells, due to its efficient hydrolysis. This could be an indication that evolution has adapted our metabolism to the generation of reactive metabolites from endogenous compounds.

7.3.3.2 Hydrolysis of esters and amides

The need for our cells to hydrolyze ester and amide bonds is huge, and it is no surprise that we have a number of enzymes that carry out such hydrolyses. There are many examples of important tasks for esterases (hydrolyzing esters) and amidases (hydrolyzing amides), for example to hydrolyze fats to glycerol and free fatty acids, peptides to free amino acids, and some neurotransmittors to inactive compounds. Many of the esterases and amidases present in our cells are known to be non-specific, and several esterases will even hydrolyze thioesters (and even some amides!). It is reasonable that exogenous esters and amides in most cases should be readily hydrolyzed. In addition, the products of the hydrolysis of an ester or an amide would be considerably more hydrophilic and good substrates for phase II reactions, as a ester yields a carboxylic acid and an alcohol whole an amide yields a carboxylic acid and an amine (see Figure 7.27).

The hydrolysis of an ester is normally much more rapid than that of an amide, partly because the amide bond is stronger (because the lone pair on the nitrogen will more efficiently be donated to the carbonyl group). However, this depends largely on the substituents on the oxygen/nitrogen. The esterases and amidases are both soluble and membrane bound, and besides hydrolyzing thioesters and some amides the esterases in the endoplasmic reticulum may also attack ureas, phosphoric esters and thiophosphoric esters. Lately, it has been shown that esters also may be metabolized by α-hydroxylation by cytochrome P-450.

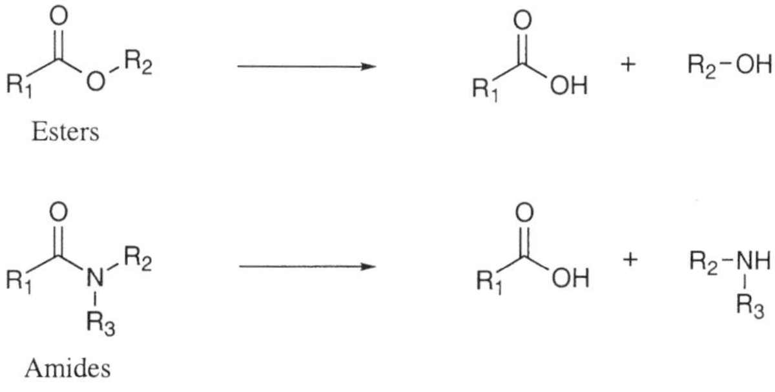

Figure 7.27. Hydrolysis of esters and amides.

7.4 Phase II conversions (conjugations)

The general purpose of phase II reactions in the metabolism of exogenous compounds can be said to be to increase the water solubility and to facilitate excretion, although some other important results are achieved in addition. It involves the reaction of a suitable compound, which may or may not have undergone phase I conversions, with different endogenous compounds which are said to be conjugated to the substrate. The reactions carried out are condensations (generating one molecule of water as a by-product), additions and substitutions. Most conjugations are biosynthetic reactions and will consequently cost energy to perform. In general, this is achieved by activating the substrates prior to the reaction.

7.4.1 Conjugations with sulphate

The attachment of a sulphate group to a compound that should be excreted from the body is of course an attractive possibility, as this is a negatively charged group that will increase the water solubility of any compound considerably. The conversion of various compounds to their sulphate ester is an important phase II reaction, and the conjugates are excreted by the kidneys. The reaction takes place in the cytoplasm and is catalyzed by enzymes called sulphotransferases. They utilize a coenzyme with a complicated chemical structure (its name is abbreviated PAPS), which in principle is nothing but an activated sulphate group (see Figure 7.28). The origin of the sulphur atom in this sulphate group is originally cysteine, an essential amino acid that is used

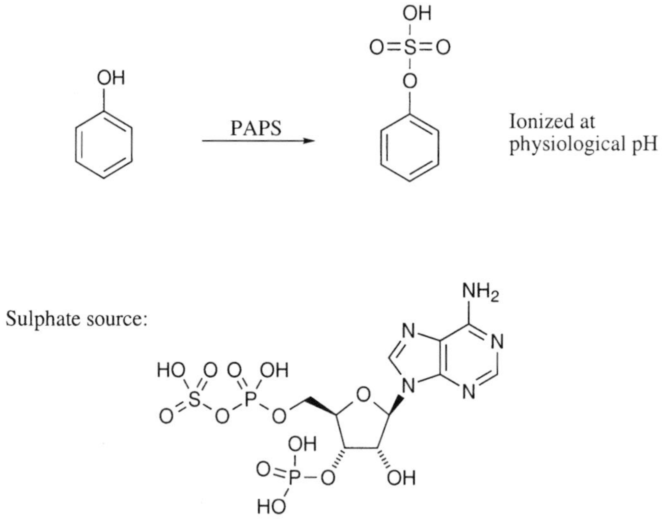

3'-Phosphoadenosine-5'-phosphosulfate (PAPS)

Figure 7.28. Conjugation with sulphate.

for many purposes in the body. However, the stock of PAPS is limited, because cysteine is valuable for the body, and the endurance of this rapid and very efficient phase II reaction is unfortunately not very good. For some potentially toxic substrates, conversion to the sulphate ester is competing with other phase II reactions, and in some cases it can be observed that low doses are harmless (because of the efficiency of the conjugation with sulphate) while higher doses are toxic (because the stock of PAPS is depleted and other, less efficient, phase II reactions take over and the time spent in the body is prolonged). The substrates for sulphate conjugations are primarily alcohols and phenols, both exogenous and endogenous, but also aromatic amines and aromatic hydroxylamines may use this phase II reaction.

Most sulphates formed are stable conjugates, in contrast to dialkylsulphates and alkylsulphonates, the monoesters of sulphuric acid are not very electrophilic. However, when attached to strongly activated positions the sulphate ion may be substituted by a nucleophile. Sulphate esters of aromatic hydroxylamines especially have received a lot of attention. The sulphate group in such esters is not only benzylic but also attached to a heteroatom (the N–O bond is relatively weak). The benzylic position is also encountered in sulphates of benzyl alcohols. Examples are the sulphate groups of the hydroxylamine formed from 2-acetylaminofluorene (2-AAF), a carcinogenic chemical previously used in the dye industry, and safrole, a natural compound present in several spices (e.g. sassafras and black pepper) and associated with a weak carcinogenic activity. As indicated in Figure 7.29, both compounds react with DNA and are carcinogenic because the sulphate groups are poor but sufficiently good leaving groups in nucleophilic substitutions.

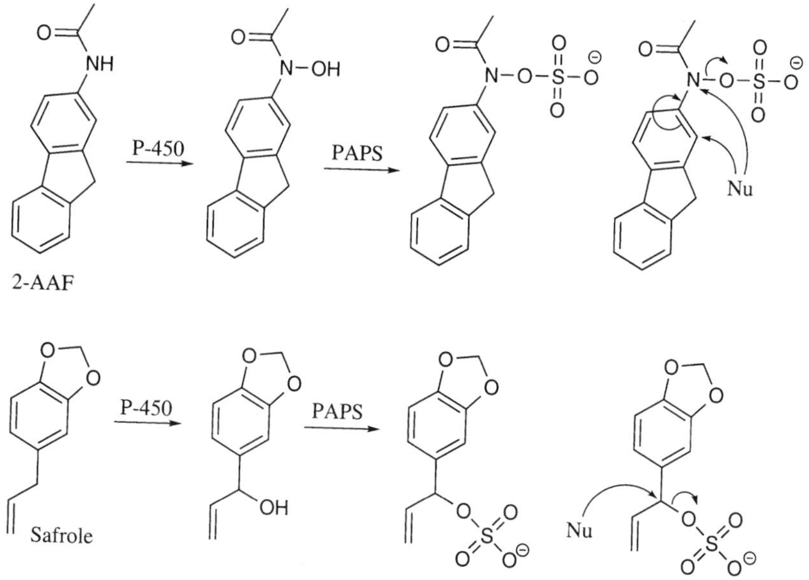

Figure 7.29. The sulphate group as a leaving group.

7.4.2 Conjugations with glucuronic acid

Glucuronic acid is derived from glucose, which by itself is water soluble. In addition to the hydroxyl groups glucuronic acid also has a carboxylic acid functionality that is ionized at physiological pH and that also may serve as a handle for specific excretion membrane transport systems. Unlike the other phase II reactions, the conjugation with glucuronic acid takes place in the endoplasmic reticulum, not in the cytoplasm, mainly in the liver cells. The enzyme that catalyzes the conjugation is called UDP-glucuronosyl transferase, of which several forms (some that are inducable) have been identified. The enzyme uses UDP-glucuronic acid as a coenzyme, and again we notice (Figure 7.30) that although the coenzyme appears to be complicated it is simply an activated form of glucuronic acid. A special feature with the UDP-glucuronosyl transferases is that they are non-specific, and will conjugate almost anything with glucuronic acid. Examples of functional groups that are acceptable are alcohols, phenols, carboxylic acids, hydroxylamines, aromatic amines and sulphonamides.

Due to the water solubility of the conjugates they may be excreted by the kidneys, but also by the liver as many of them have quite high molecular weights. Glucuronic acid weighs 194 g/mol itself, and if the substrate is of the same size, excretion with the bile becomes significant. The glycoside bond between the substrate and glucuronic acid is relatively stable under normal condition, but can be hydrolyzed by certain enzymes and by acidic conditions. Enzymes called glucuronidases will efficiently hydrolyze *O*- and *S*-glycosides, and although they are not present to any large extent in our cells they are common in the bacteria of the intestines. It is therefore to be expected that glucuronic acid derivatives that are excreted by the liver will eventually be

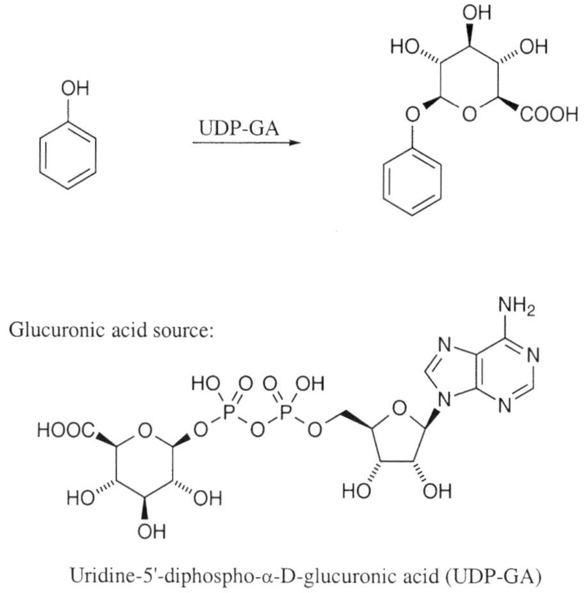

Glucuronic acid source:

Uridine-5'-diphospho-α-D-glucuronic acid (UDP-GA)

Figure 7.30. Conjugation with glucuronic acid.

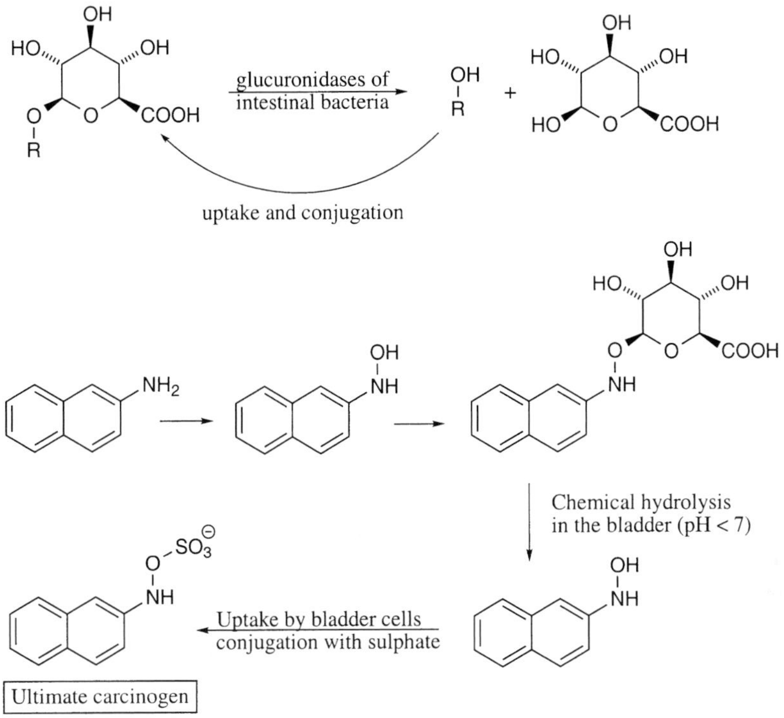

Figure 7.31. The hydrolysis of conjugates with glucuronic acid.

hydrolyzed again (see Figure 7.31). If it as such is reasonably lipophilic it will be reabsorbed from the intestines, transported to the liver, conjugated with glucuronic acid once more, and be excreted with the bile. Even if it is not possible that this process, called the enterohepatic recirculation, is 100 % efficient because of other metabolic conversions and other excretion routes, it will nevertheless increase the time that a toxic chemical spends in the body. Chemical hydrolysis catalyzed by acid (no enzymes involved) may take place in the urine bladder, as urine often has a lower pH compared to the rest of the body fluids. This has been suggested as the reason why the bladder is the target organ for aromatic amines. If they are excreted as the glucoronic acid conjugate of the corresponding hydroxylamine and hydrolyzed in the bladder, the bladder cells may absorb the hydroxylamines and possible sulphate them to their electrophilic and genotoxic form (see Figure 7.31). In such a case one can say that the glucuronic acid masks the reactive functionality and delivers a toxic compound to its target.

7.4.3 Conjugations with amino acids

Carboxylic acids may be conjugated with the amino group of an amino acid, to form an amide. In most cases this is carried out with the simplest amino acid, glycine, although in some

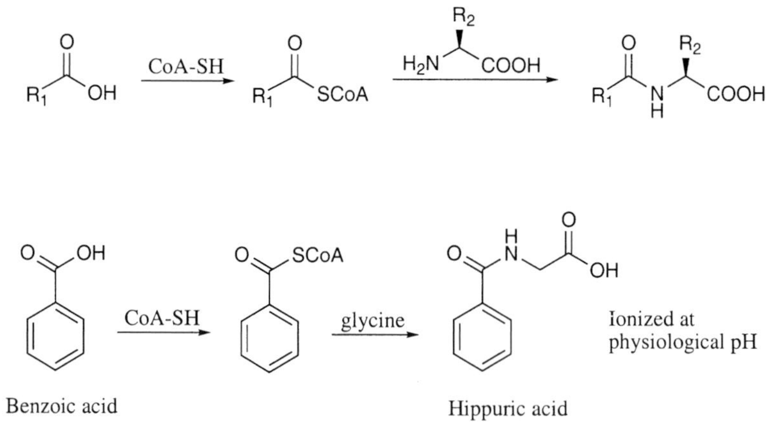

Figure 7.32. Conjugation with an amino acid.

cases others are used (e.g. glutamic acid and serine). The result of the conjugation is that the compound receives an extra amide function, which increases the water solubility and serves as a handle for the specific excretion systems. The conjugation is carried out in two steps, first the substrate is activated as a coenzyme A thioester by the enzyme coenzyme A-ligase, whereafter acyl transferases will catalyze the coupling of the activated acid and the amino acid. In Figure 7.32 the conjugation of benzoic acid with glycine is illustrated, and as the end product (hippuric acid) was first isolated from the urine of a horse (Latin for horse is hippos) this phase II reaction is called "hippuric acid synthesis".

7.4.4 Conjugations with glutathione

In several ways, the conjugation of exogenous compounds with glutathione is the most important phase II reaction, because it gives the body a direct possibility to take care of electrophilic compounds. Glutathione is a tripeptide composed of glycine, cysteine and glutamic acid, and its nucleophilicity is of course due to the presence of a thiol group in the cysteine residue (see Figure 7.33). Cysteine itself is actually a better nucleophile than glutathione, because the sulphur has a slightly more ionic character in cysteine than in glutathione, and it is not evident why our cells go through the extra trouble of adding two other amino acids on each side of cysteine if it was only to be used in conjugations (which it is not!). The reaction between a electrophilic substrate and glutathione is catalyzed by enzymes, glutathione-S-transferases, and the trick is that a molecule of glutathione is a much more reactive nucleophile when it is in its position in the enzyme compared to when it is free. This is accomplished by enzymatic stabilization of the anion with a negative charge on the sulphur, and we have already seen how dramatic this will change the nucleophilicity. Most of the conjugations with glutathione takes place in the cytoplasm, although some are carried out in the endoplasmic reticulum. In the cells of the liver, the

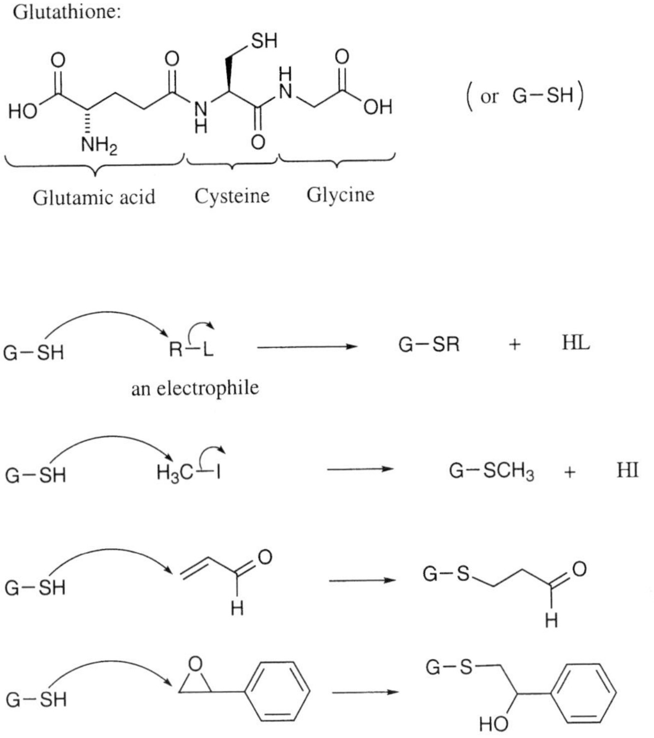

Glutathione:

Glutamic acid   Cysteine   Glycine

an electrophile

Figure 7.33. Conjugation of electrophiles with glutathione.

concentration of glutathione is extremely high, 1–5 mM, and there are plenty of glutathione-*S*-transferases as well.

Note that the bond between cysteine and glutamic acid is not a normal peptide bond, which makes glutathione less prone to degradation by peptidases (enzymes cleaving peptide bonds), although other enzymes (γ-glutamyl transferases) will be able to cleave this bond. After a conjugation with glutathione, glycine and glutamic acid may be hydrolyzed from the conjugate, and the remaining cysteine with the former electrophile attached to the sulphur atom can be further metabolized by two main routes. Either the glycine adduct is acetylated (*vide infra*) to a mercapturic acid derivative (this phase II reaction is sometimes called "mercapturic acid synthesis"), or the bond between the sulphur and the rest of the cysteine is cleaved by a β-lyase whereafter the resulting thiol may be methylated (*vide infra*) (see Figure 7.34). If an exogenous compound is conjugated with glutathione, indicating that it is electrophilic or is converted to electrophilic metabolites by the metabolism, one can expect to find mercapturic acids and thioethers in the urine (this is the case for example in the urine of cigarette smokers) and these can be detected by analytical procedures.

In almost all cases the product formed is less toxic than the original electrophile, but the reverse can of course happen. For example, acrolein has in spite of its general reactivity been

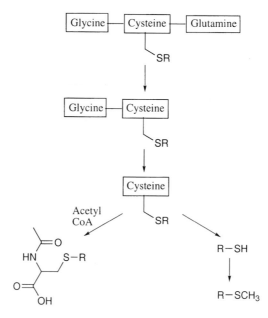

Mercapturic acid derivative

Figure 7.34. Further processing of glutathione conjugates.

found to be specifically toxic to the kidneys, and the toxic principle is believed to be formed from its mercapturic acid derivative. In the kidneys this can be oxidized to the sulphoxide, which is more prone to β-elimination and will regenerate acrolein (see Figure 7.35). 1,2-dichloroethane may as an electrophile be conjugated with glutathione, and the product will have some structural features in common with mustard gas.

Another important role for glutathione is to assist during the conversion of hydrogen peroxide to water, as a coenzyme of glutathione peroxidase. In addition it is believed to protect various membrane proteins from oxidation. Any situation causing the concentration of glutathione to drop would therefore be a potential threat, especially if it is accompanied by an exposure for electrophilic or pre-electrophilic chemicals. Extreme diets may result in the intake of insufficient amounts of cysteine leading to a decrease in the glutathione levels in the liver. As a consequence, the toxicity of a number of chemicals (e.g. chloroform, carbon tetrachloride, acetaminophen and thioacetamide) is aggravated. In order to strengthen the defence against reactive compounds, especially after intoxications, it is possible to administer cysteine, normally as acetylcysteine, to the victim. Acetylcysteine is able to react as a nucleophile with the reactive compound, fill the supplies of cysteine and glutathione, and also facilitate the production of PAPS. The protection offered by glutathione may at least for some types of chemicals result in that no effects are observed up to a threshold concentration, above which glutathione is no longer able to take care of things. Besides all these protective functions, glutathione also acts as

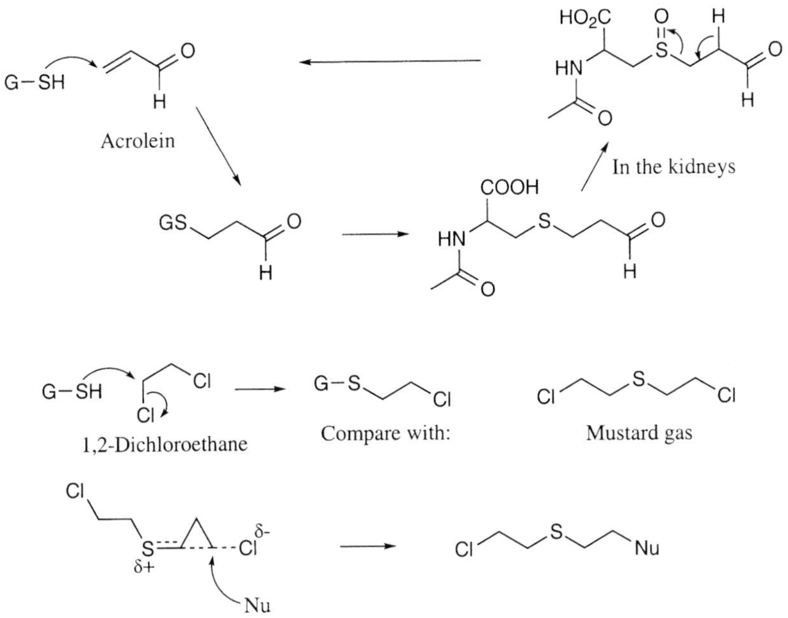

Figure 7.35. The conjugation of acrolein and 1,2-dichloroethane with glutathione.

a reserve for cysteine, which when needed can easily be produced by the hydrolysis of gluta-thione.

7.4.5 Other conjugations

In addition to the major phase II reactions discussed above, various acylations and methylations of heteroatoms also take place with exogenous compounds, of which the acetylation of nitrogen to form acetamides and the methylation of sulphur to form thioethers are especially interesting. However, acylations and methylations differ from the others as they decrease instead of increase the water solubility, and they are therefore of minor importance when it comes to converting exogenous compounds to a excretable form. The coenzyme for acetylations is acetyl-coenzyme A, which also plays a prominent role in energy production. The enzyme catalyzing the reaction is called *N*-acetyl transferase, and important substrates are aromatic amines and sulphonamides, and substituted hydrazines. The amide formed may of course be a substrate for amidases (*vide supra*). Acetylation is interesting because is has been found to be involved in the carcinogenicity of aromatic amines, since acetylated as well as free aromatic amines can be *N*-hydroxylated by cytochrome P-450. An aromatic *N*-hydroxy-*N*-acetylamine may be converted by a special enzyme, an acyl transferase, to the corresponding aromatic *N*-acetoxyamine which is an efficient electrophile because acetate is a very good leaving group (see Figure 7.36).

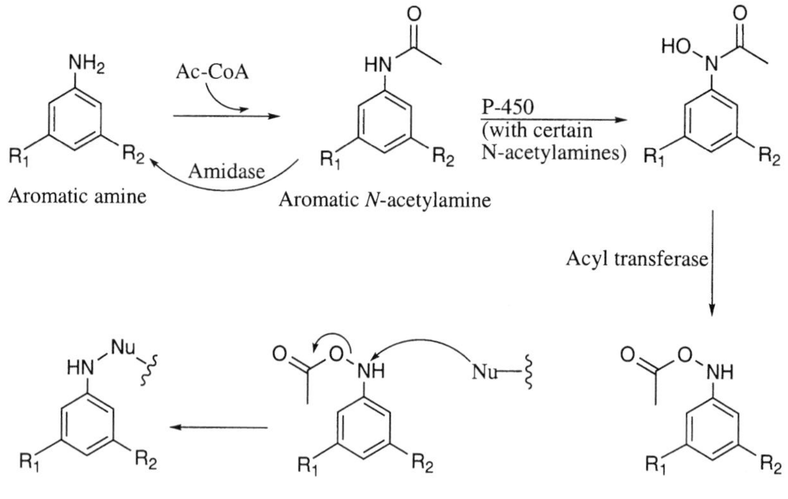

Figure 7.36. A possible way to activate aromatic amines.

Methylation of heteroatoms is an important biochemical reaction that is used for example for the regulation of DNA and RNA. The coenzyme used in this reaction, S-adenosylmethionine (SAM for short), is unusual as an electrophilic endogenous compound. Interestingly, it has been shown that SAM will react with and methylate DNA *in vitro*, although it is not known if this reaction takes place to any significant degree *in vivo*, where in principle it is only used together with enzymes (methyl transferases). The methylations of exogenous compounds primarily takes place with thiols produced after glutathione conjugation (*vide supra*) as indicated in Figure 7.37.

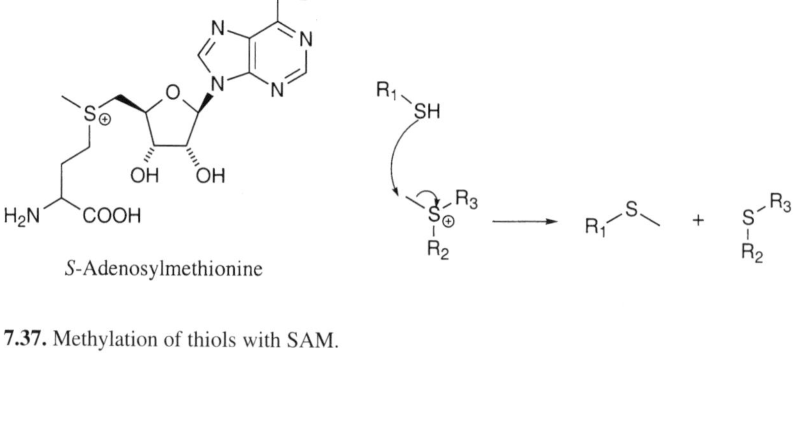

Figure 7.37. Methylation of thiols with SAM.

7.5 Summary of the major phase I and phase II reactions

Oxidation	epoxidation of	alkenes
		alkynes
		aromatic systems
	hydroxylation of	allylic and benzylic C-H
		C-H α to heteroatoms
		allylic and benzylic N-H
		ω-1 position in hydrocarbons
	of single to double bonds	
	of heteroatoms	
	of alcohols/aldehydes	
Reduction	of aldehydes/ketones	
	of oxidized heteroatoms	
	reductive dehalogenation	
Hydrolysis	of epoxides	
	of esters	
	of amides	
Conjugation	with sulphate	
	with glucuronic acid	
	with amino acids	
	with glutathione	

8 Metabolism of exogenous compounds in mammals

Chapter 8 will give a number of examples of how the metabolism of exogenous compounds in mammals is carried out in practice, which is not necessarily exactly the same as the theoretical reactions discussed in Chapter 7. The most significant difference is that most "real" compounds contain more than one chemical functionality that can be attacked by metabolic reactions, and it is not apparent in which order different conversions take place. The conclusions of Chapter 8 should be that although we cannot predict in detail which metabolites are formed from the metabolism of a certain compound, or their relative amounts, the general rules discussed in Chapter 7 still apply. With those in mind we can make "intelligent guesses" about what to expect and by anticipating the worst we can identify compounds that may be converted to toxic metabolites.

8.1 Metabolism of hydrocarbons

Hydrocarbons are among the most common chemicals that we contact, daily as fuel but also in the working environment as for example solvents and as pollutants, for example formed during the combustion of various materials. This section describes the metabolism of some saturated, unsaturated and aromatic hydrocarbons, and also some other compounds which are metabolized by conversions involving carbon–hydrogen or carbon–carbon bonds.

Among the alkanes, the smallest (methane to pentane) are very volatile and are not likely to stay long enough in the body to be metabolized at all, unless we are continuously exposed to high concentrations. With hexane (boiling point 69 °C) the volatility starts to become limited, hexane is a cheap and excellent solvent for various purposes and has been used in enormous quantities by the chemical industry. However, it has been noted that chronic exposure to hexane may affect the peripheral nervous system, which is noted as the loss of sensation in the hands and the feet, and in addition it may damage the testicular tissues. The cause of these injuries is a metabolite of hexane, formed from the cytochrome P-450 mediated hydroxylation of both ω-1 carbons according to Figure 8.1.

The toxic metabolite is 2,5-hexanedione, which is believed to react irreversibly with primary amino groups of the amino acid lysine in proteins in the peripheral nerve cells. The product is a N-substituted 2,5-dimethylpyrrole derivative (for organic chemists this reaction is known as the Paal–Knorr pyrrole synthesis), which is aromatic and thereby stabilized, and alkylpyrroles are susceptible to autoxidative dimerisation which also is believed to be an important step in the neurotoxicity of 2,5-hexanedione (see Figure 8.1). This has lead to the exchange of hexane in many industrial processes to for example heptane, which does not give the same toxic effects.

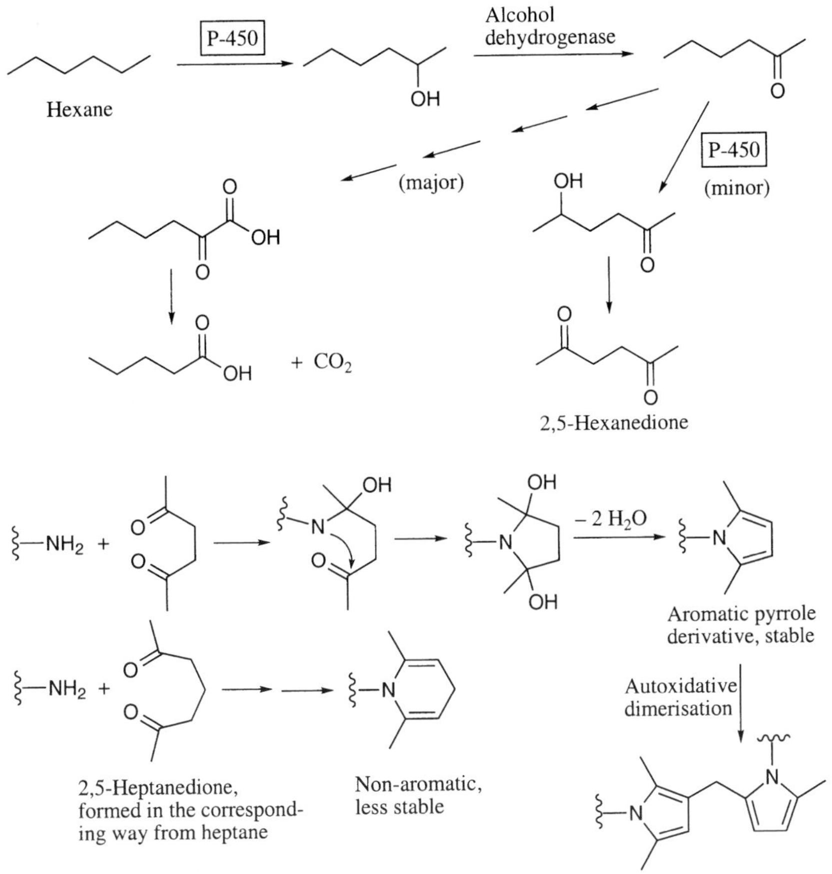

Figure 8.1. The conversion of hexane.

Heptane is oxidized in the corresponding way, but the 2,6-heptanedione formed can not form an aromatic (and stabilized) end product after reaction with an amine, something that makes this reaction reversible.

Alkanes are hydroxylated in various positions, although the ω-1 position is considered to be slightly promoted by being substituted but not too sterically crowded. However, hydroxylations are much easier in positions that are activated, e.g. by the presence of a aromatic system or a double bond next to the carbon to be hydroxylated. An example is the natural product safrole, which has been shown to be a weak carcinogen and also hepatotoxic in animal experiments. Safrole contains several chemical functionalities that can be (and are!) attacked by the metabolism, but the metabolite responsible for the carcinogenic activity is believed to be formed by hydroxylation of the benzylic position (which at the same time is allylic, i.e. doubly activated for hydroxylation) and conjugation of the alcohol formed with sulphate (see Figure 8.2).

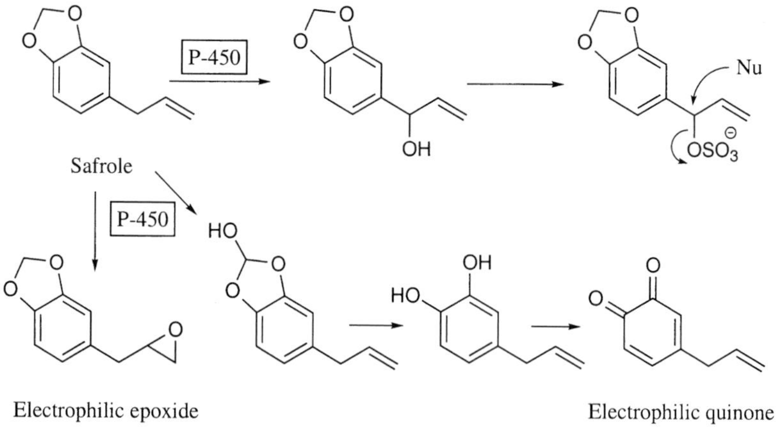

Figure 8.2. The bioactivation of safrole.

Safrole, an important component of many essential oils and the main component (approximately 75 %) of sassafras oil (*Sassafras officinale*), with weak insecticidal and antimicrobial activity. It is used for various purposes in the perfume industry, but is not allowed as an additive in food.

The sulphate ester formed is a poor electrophile, but will to some extent be able to react for example with DNA. As suggested in Figure 8.2, safrole could also be activated to other electrophilic metabolites that could contribute to the toxicity of the sulphate ester, but their importance is not known.

Hydroxylation is the most common conversion that saturated hydrocarbons are subject to, but carbon–carbon single bonds can also be oxidized to double bonds. We have already encountered the example of valproic acid, and another is urethane, or ethyl carbamate. Urethane has been used for many applications, but has been found to be carcinogenic and should be avoided. It may react with genetic material, e.g. the DNA base adenosine according to Figure 8.3, after oxidation to the alkene and epoxidation of the double bond by cytochrome P-450.

Besides the oxidation of the carbon–carbon single bond, urethane will be oxidized by hydroxylation of both the nitrogen and the methyl group, although these reactions are not considered important for toxicity.

Alkenes can be both epoxidized and hydroxylated, preferably in an allylic position, by cytochrome P-450, and although it is not evident which route will be chosen, epoxidation seems to more common. Cyclohexene is epoxidized and the epoxide is hydrolyzed to the diol by epoxide hydrolase (note the formation of the *trans*-diol). Small amounts of the epoxide will also react with glutathione (see Figure 8.4). In 4-ethenylcyclohexene (4-vinylcyclohexene), with a double bond in both a ring and a chain, it is almost exclusively the ring that is epoxidized (see Figure 8.4). On the other hand, in styrene, where the choice stands between epoxidation of the vinyl group and the aromatic ring, it is the side chain that is converted to give styrene oxide (see Figure 7.25 for a complete picture of the metabolism of styrene).

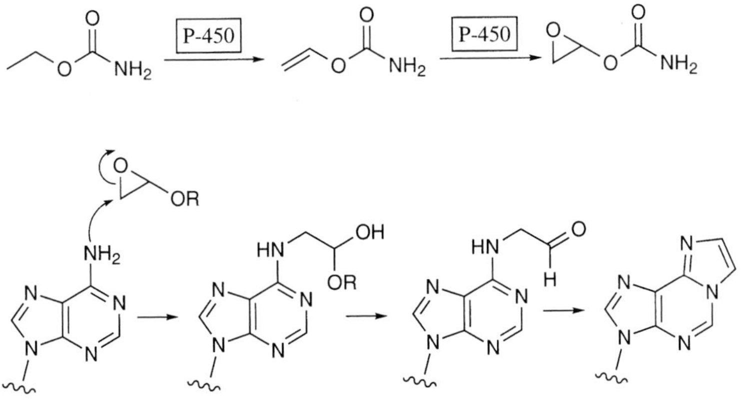
P-450
P-450
O
NH2
OR
OH
HN
H
N

Figure 8.3. Bioactivation of the carcinogenic compound urethane to an epoxide, and its reaction with adenosine.

Aromatic compounds can be epoxidized in the ring, resulting in unstable epoxides that readily rearrange to phenols. The simplest aromatic compound, benzene, is a relatively cheap organic solvent that has been used in large quantities, until it was realized that exposure to benzene for years increased the risk for leukaemia. Benzene is still used in industry, although the handling is surrounded by restrictions, and it is still a component of gasoline which means that most people actually get in contact with the compound in small amounts. There are consequently a number of reasons for investigating how benzene gives rise to its carcinogenic effect,

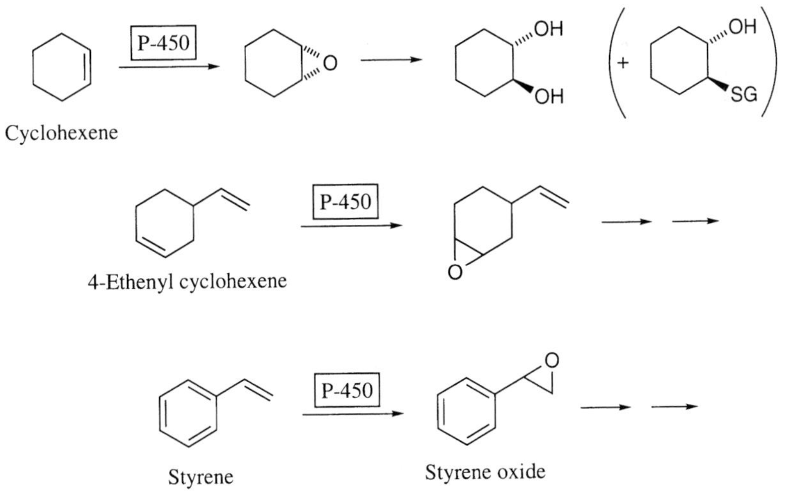
P-450
OH
OH
SG
Cyclohexene
P-450
4-Ethenyl cyclohexene
P-450
Styrene
Styrene oxide

Figure 8.4. Epoxidation of various alkenes.

but in spite of the many man-years of research in this matter we still do not have a clear picture of its metabolic fate. It is obvious that most is converted to phenol, and the fact that some amounts of the corresponding 1,2-dihydro-1,2-diol and the glutathione conjugate are formed suggests that the phenol formation passes via the epoxide. However, phenol itself does not give the same effects as benzene (phenol is also a carcinogen but acts by a different mechanism). This difference could be due to the fact that phenol formed from the oxidative metabolism of benzene is directly present in the endoplasmic reticulum and can more efficiently be further converted to the final carcinogenic metabolite, or that this is formed by another metabolic route, not via phenol.

The epoxide was long considered to be responsible for the toxicity of benzene. It survives long enough to be hydrolyzed by epoxide hydrolase and conjugated with glutathione, but these conversions only take place with small amounts (a few percent) of the epoxide. The diol formed can spontaneously eliminate water and be transformed to phenol, it can be oxidized by alcohol dehydrogenase to the keto alcohol that spontaneously turns into the corresponding enolic phenol, or it can be further oxidized to open the ring. The detection of both Z,Z-muconic acid and

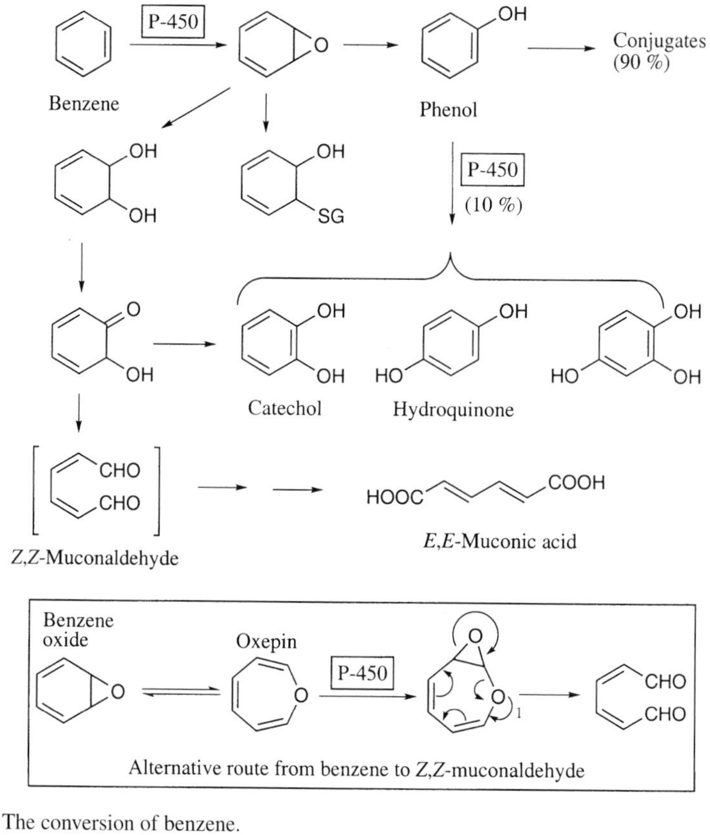

Figure 8.5. The conversion of benzene.

the more stable *E,E*-muconic acid (with both carbon–carbon double bonds in the *trans* configuration, see Figure 8.5) as metabolites of benzene indicates that the unsaturated dialdehyde Z,Z-muconaldehyde is formed after oxidative opening of the ring (either via dihydrocatechol or via another route, and an alternative metabolic route to Z,Z-mucoaldehyde via benzene oxide and it's isomer oxepin is shown in Figure 8.5). Although the dialdehyde itself has not been isolated, and the evidence for its formation is only circumstantial, it is nevertheless likely that this very reactive compound is formed and responsible for at least part of the toxicity of benzene. As described previously, the epoxide of benzene spontaneously yields phenol, and the half-life of the epoxide in water at 37 °C and pH 7 is only approximately 2 minutes. Most of the phenol formed is conjugated with sulphate and glucuronic acid, but a portion is further oxidized by epoxidation/rearrangement to form catechol and hydroquinone (the *ortho*- and *para*-positions of phenol are actually activated for further oxidation by the electron-donating hydroxyl group already present), and even to trihydroxybenzenes. Catechol and hydroquinone can be oxidized to the corresponding quinones, which are electrophilic and toxic, and hydroquinone is actually autoxidized by the molecular oxygen present in the cells (no enzymes required). Besides the formation of quinone, which with its unsaturated keto functionality is electrophilic and mutagenic and can react with DNA bases, this autoxidation also generates superoxide which adds to the toxic effect (see Figure 8.6).

However, the oxidation of hydroquinone is faster in the presence of peroxidases, which normally have higher activity in tissues where the monooxygenase activity is low, and this may be a piece in the puzzle to explain the specific carcinogenic effect of benzene on the blood cells.

Toluene has replaced benzene as a solvent, it has similar chemical properties but it is considerably less toxic. The reason for this difference is that the methyl group in toluene is more readily oxidized than the aromatic ring, and that this metabolic path leads to less toxic metabolites (see Figure 8.7). Benzylic hydroxylation leads eventually to hippuric acid, after conjugation of the carboxylic acid with glycine, which is efficiently eliminated with the urine.

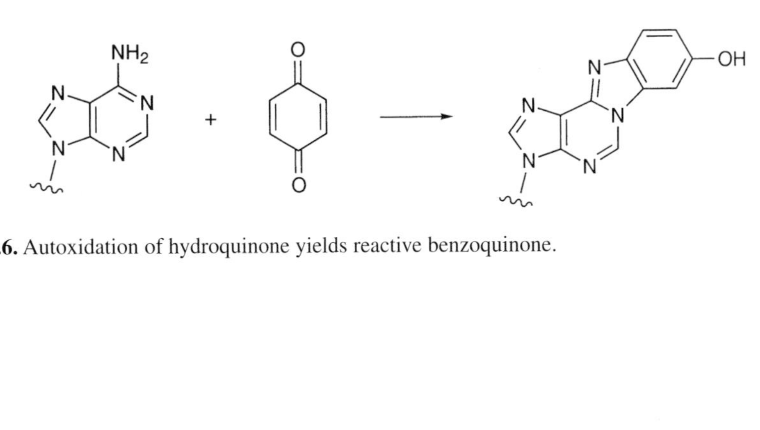

Hydroquinone   Semiquinone   1,4–Benzoquinone

Figure 8.6. Autoxidation of hydroquinone yields reactive benzoquinone.

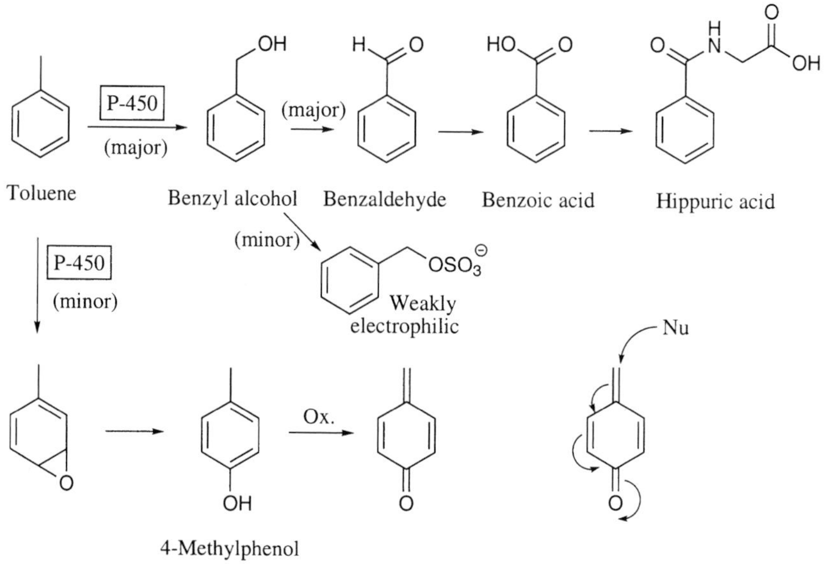

Figure 8.7. The conversion of toluene.

However, epoxidation of the aromatic ring of toluene takes place to some extent, generating for example 4-methylphenol, and this minor route is believed to be responsible for the toxic effects that toluene will give after chronic exposure. 4-Methylphenol, as well as any 4-alkylphenol, will be oxidized by peroxidases to a quinone methide that is electrophilic and harmful (see Figure 8.7). In addition, it has been shown that small amounts of the benzyl alcohol formed is conjugated with sulphate, yielding a slightly electrophilic product having a sulphate group in a benzylic position (similar to the toxic metabolite of safrole, *vide supra*). The hazards associated with toluene and the xylenes (dimethylbenzenes) have therefore been re-evaluated during the last decade.

Naphthalene has been used as moth-proofer, the classic moth-balls are pure naphthalene, it is not very efficient, it does not kill the moths but it works. The reason is that naphthalene is slightly volatile, it will slowly evaporates from its solid state (it melts at 80 °C) at room temperature and the vapour will cover everything in a closed space (e.g. clothes in a wardrobe) with a thin layer of naphthalene. Moths do simply not like the taste of naphthalene, and will not eat such clothes.

Naphthalene is perhaps best known for its use as a moth-proofer, but it is also an important chemical in industry. As it is less volatile compared to benzene persons that use the compound will not be exposed in the same way, even if naphthalene is metabolized in approximately the same way as benzene (see Figure 8.8). Naphthalene affects the blood in relatively small amounts (a LD_{low} value of 100 mg/kg for a child has been reported), and chronic exposure gives effects on the liver and lungs. It has been reported to give allergic reactions in the skin and it is suspected to be a weak carcinogen.

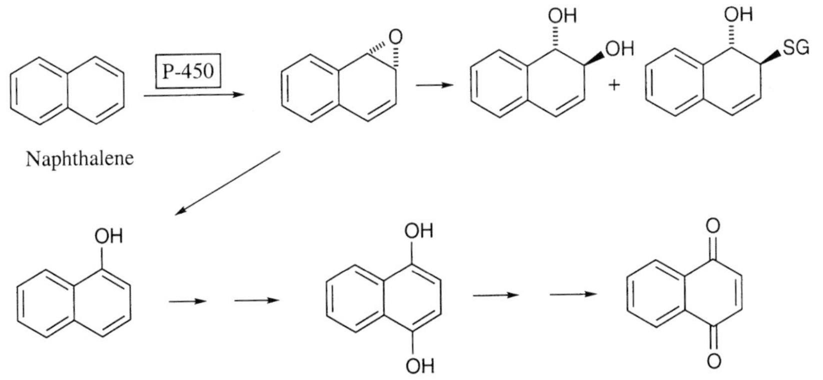

Figure 8.8. Metabolism of naphthalene.

Polyaromatic hydrocarbons will also be epoxidized, the epoxides may be hydrolyzed, conjugated or spontaneously rearrange to phenols which are conjugated and eventually may be oxidized to quinones. Differences between the rates of the various reactions are observed, which will make the polyaromatic hydrocarbons differently toxic. We have already discussed benz[a]pyrene (Section 7.3.3) which is one of the most toxic members of this class (a potent carcinogen), and the reason for this is the relative stability of the epoxide formed which is a poor substrate for epoxide hydrolase and therefore is given time to react with DNA.

8.2 Metabolism of compounds containing nitrogen

The use of simple, aliphatic amines in industry is increasing, and some toxic effects have been noticed (besides the general irritating effect that all amines have because of their basicity). Amines in which the amino group is attached to a saturated carbon having at least one hydrogen bound to it, are oxidized by cytochrome P-450 or by amine oxidases, to aldehydes. When cytochrome P-450 is responsible for the oxidation it proceeds via an α-hydroxylation and a dealkylation as described previously, while the amine oxidases will oxidize the amine to an imine which spontaneously will be hydrolyzed to the aldehyde. Methylamine (aminomethane), for example, is oxidized by an amine oxidase to the imine that is hydrolyzed to formaldehyde (see Figure 8.9). Allylamine (3-aminopropene), a highly irritating and toxic chemical used for example in the pharmaceutical industry, gives a specific and strong toxic effect on the muscle cells of the blood vessels of the heart, even after a single exposure. The effect resembles that seen in arteriosclerosis. Allylamine is oxidized by an amine oxidase which is present in relatively large amounts in the cells of the damaged heart tissue, to acrolein (see Figure 8.9) which is responsible for the effect.

N-Alkylformamides, for example *N*-methylformamide and *N,N*-dimethylformamide (see Figure 8.10), are important solvents used both in laboratories and in industrial processes. They

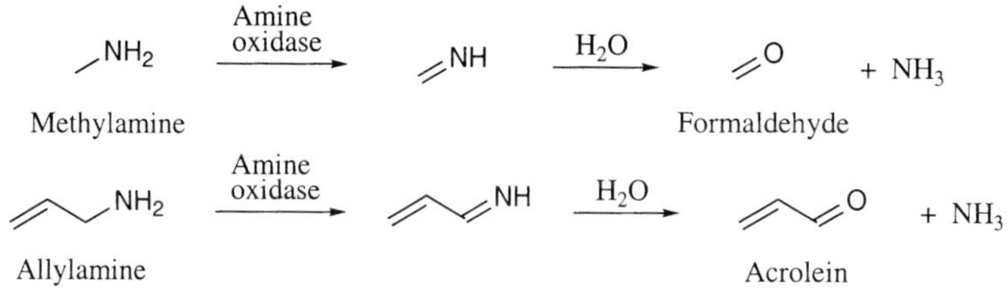

Figure 8.9. The oxidation of amines to aldehydes by amine oxidases.

have been found to be toxic to the liver and in some cases also to be teratogenic. *N,N*-Dimethylformamide undergoes oxidative dealkylation (α-hydroxylation) catalyzed by cytochrome P-450 in the liver to *N*-methylformamide (and formaldehyde). Further oxidation by cytochrome P-450, at the formyl carbon, yields a reactive intermediate that is believed to be protonated methyl isocyanate (see Figure 8.10) that rapidly will react with nucleophiles [e.g. glutathione to form *S*-(*N*-methylcarbamoyl)glutathione].

The four letters in BASF, the German company that started in 1865 and today is one of the most important chemical companies operating world-wide, stand for Badische Anilin- und Sodafabrik. Aniline, the most simple aromatic amine, was in the beginning isolated from coal-tar and used for the production of pigments.

The aromatic amines have attracted considerably more attention, because they are toxic to the blood in general and in many cases carcinogenic, especially to the bladder. The aromatic amides, in which the amino function has been transformed to an amide by condensation with a carboxylic acid, are normally discussed together with their corresponding amines as they are oxidized in the same way in our bodies and they easily are converted to amines (by amidases). They were among the first chemicals to be used by a growing chemical industry in the 19th century, for example for the production of pigments and polymers. Aromatic amines can be metabolized in different way, and Figure 8.11 summarizes the most important routes leading to toxic metabolites.

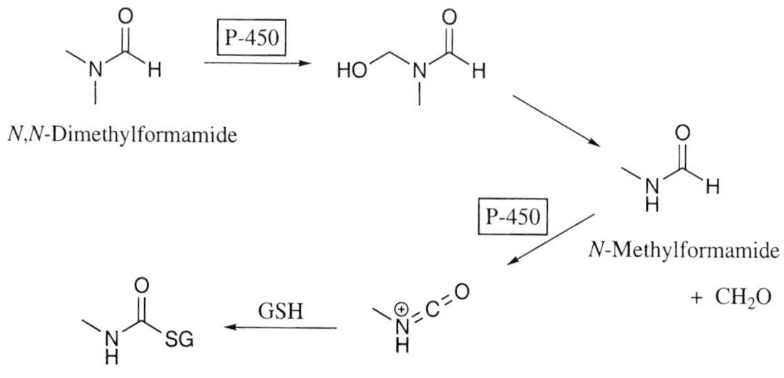

Figure 8.10. The oxidation of *N*-alkylformamides yields an electrophile.

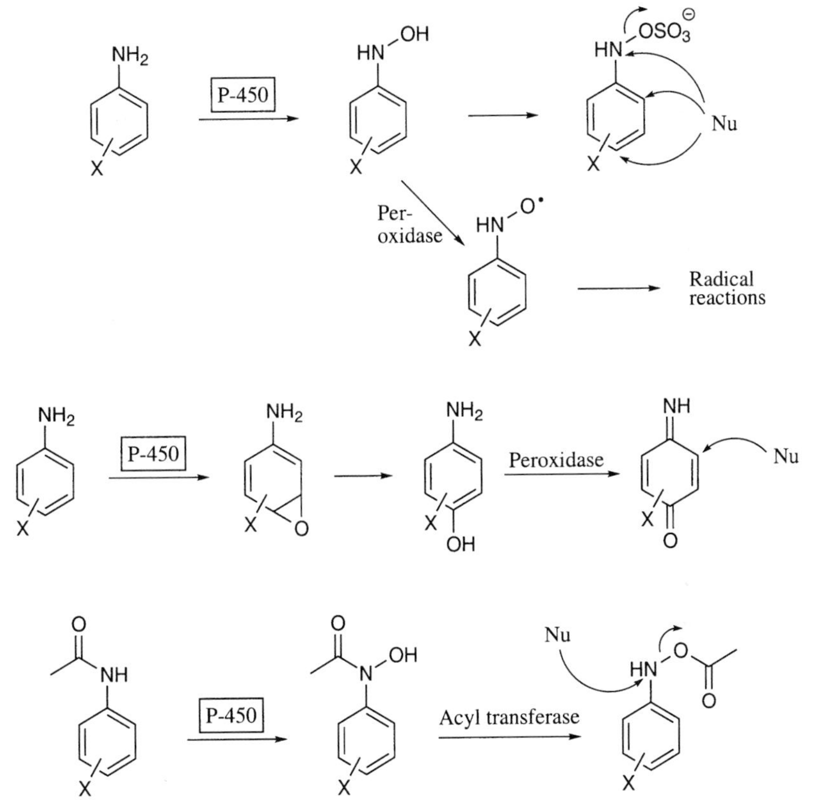

Figure 8.11. Principal metabolic activations of aromatic amines. (X denotes any substituent in any position.)

Hydroxylation of the nitrogen followed by conjugation with sulphate creates a reasonably good electrophile that may be attacked by nucleophiles either directly on the nitrogen but also in conjugated positions in the aromatic ring. The hydroxylamine is actually electrophilic as it is, because the nitrogen–oxygen bond is relatively weak, but conjugation increases its reactivity. In addition, the hydroxylamine may be oxidized by peroxidases to an oxygen radical, that will initiate radical reactions in the cells. Finally, the epoxidation of the aromate, a less likely conversion that still will take place, will produce aminophenols that will be oxidized by peroxidases to reactive benzoquinone imines. Aromatic amides will also be N-hydroxylated, and a special enzyme (an acyl transferase) can move the acetyl group of the hydroxyamide from the nitrogen to the oxygen resulting in an aromatic N-acetoxyamine. This is very electrophilic, because an acetoxy group in this position is an excellent leaving group in nucleophilic substitutions.

Aniline itself and acetanilide (the N-acetyl derivative of aniline) have not been proven to be carcinogenic, although a suspicion for a weak activity remains, but o-toluidine (aniline with a methyl group in the *ortho* position) is clearly carcinogenic. Considerably more potent are for example 2-aminonaphthalene and *N*-acetyl-2-aminofluorene (both used in large quantities in the early days of the chemical industry), for which it has been possible to clearly establish that they cause tumours in exposed workers. Interestingly, the analogues having the amino (or acetylamino) function in position 1 instead of 2 are only very weak carcinogens, while both isomers of the hydroxylamines (or *N*-acetyl-*N*-hydroxylamines) prepared synthetically and assayed in animal experiments are equally potent (see Figure 8.12).

The observed difference is caused by an apparent sensitivity for steric hindrance of the hydroxylation of the nitrogen atom by cytochrome P-450. The 1-hydroxylamine (and 1-*N*-acetyl-*N*-hydroxylamine) are therefore formed much slower, but once formed they are equally potent. Besides *N*-hydroxylation, both 1- and 2-aminonaphthalene are also oxidized in the ring to yield 1-amino-2-hydroxynaphthalene and 2-amino-1-hydroxynaphthalene, which after peroxidase oxidation to the corresponding quinone imines will add to the toxic effects. In addition, conjugation will take place in parallel to oxidation, both of the original amine and the oxidation products, the conjugates are normally less toxic but we have seen that the reverse also may be true.

In this context it should be noted that the toxic effects of aromates containing nitro groups resemble those of the corresponding aromatic amines, simply because the nitro group can be reduced to the hydroxylamine (via the nitroso derivative). As noted when the carcinogenic ac-

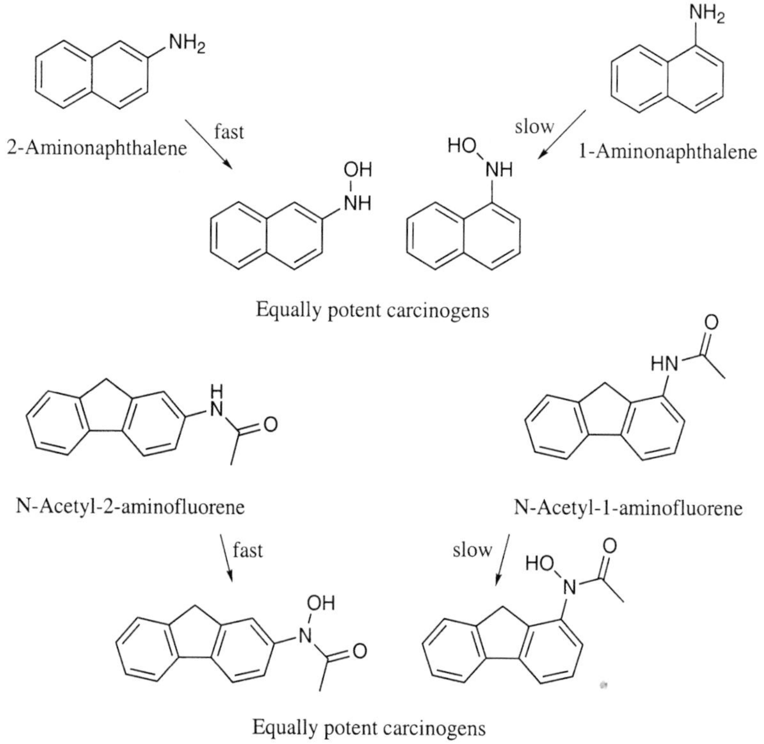

Figure 8.12. The hydroxylation rate differs between various aromatic amines (amides).

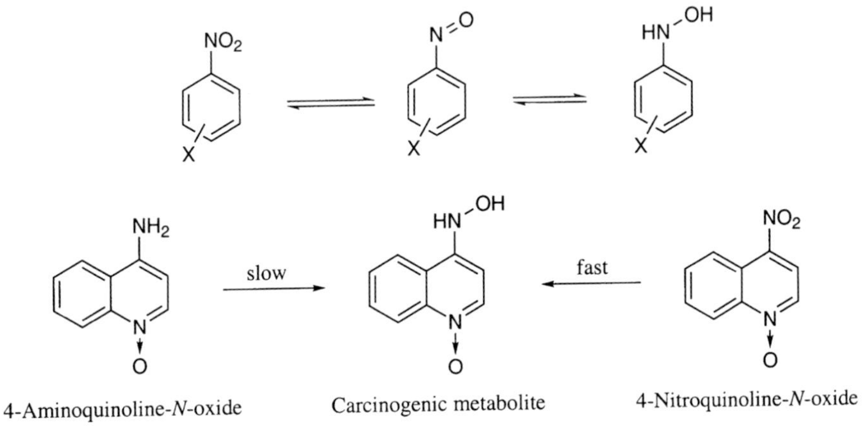

4-Aminoquinoline-*N*-oxide   Carcinogenic metabolite   4-Nitroquinoline-*N*-oxide

Figure 8.13. Aromatic nitro compounds can be reduced to the corresponding hydroxylamines and are often toxic as the aromatic amines.

tivity of 4-aminoquinoline-*N*-oxide (which is inactive) was compared to that of the corresponding nitro derivative (which is a potent carcinogen) (see Figure 8.13), there are no steric factors impeding the reduction of a nitro group to a hydroxylamine as there are for oxidating an amino group to a hydroxylamine.

An interesting aromatic amide that is used in large amounts as a medicament (an analgesic and antipyretic) and available without prescription is acetaminophen, or paracetamol (*N*-acetyl-4-aminophenol). It is considered to be safe, but it has been noticed that misuse (typically several grams per day for a long period) can result in liver damage. In a normal situation the half-life of acetaminophen in our bodies is rather short, because it is efficiently conjugated at the phenolic function with sulphate and glucuronic acid. However, as already noted, the amounts of PAPS available for sulphate conjugation are limited, and if too much of a compound like acetaminophen is taken daily the PAPS stock will be depleted. In such cases only conjugation with glucuronic acid remains, and as it is considerably slower the possibility of *N*-hydroxylation of acetaminophen by cytochrome P-450 increases. The *N*-acetyl-*N*-hydroxyl derivative can eliminate water and form a highly (electronically) conjugated *N*-acetylquinone imine that is electrophilic (see Figure 8.14) and believed to be responsible for the liver toxicity of acetaminophen (by affecting the enzyme glutamate dehydrogenase which inhibits mitochondrial respiration). The direct oxidation of acetaminophen to the reactive quinone imine has also been observed, apparently by a peroxidase but it has been suggested that cytochrome P-450 is responsible for the one-electron abstraction from the phenolic oxygen that initiates this conversion.

Besides the amines that we contact in for example the working environment, there are also a number of endogenous amines that are not necessarily completely harmless. The DNA base adenine has attracted especial attention, and studies have shown that even adenine can be oxidized by cytochrome P-450 to the corresponding *N*-hydroxyadenine which turns out to have

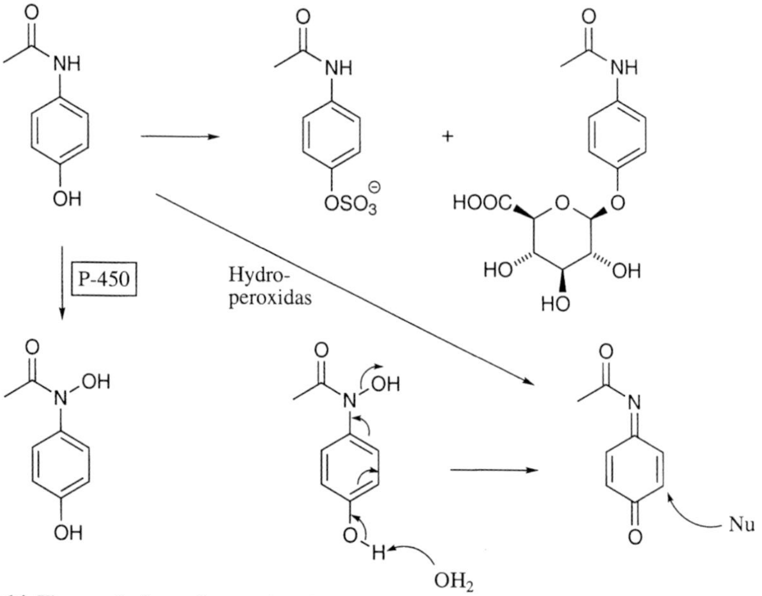

Figure 8.14. The metabolism of acetaminophen.

genotoxic properties *in vitro*. However, it has so far not been possible to determine if this activation takes place also *in vivo*, and in that case if it has any biological significance.

The formation of secondary *N*-nitrosoamines from secondary amines and nitric acid in the stomach has already been discussed in Chapter 6. In addition they are formed in lubricating oils in which sodium nitrite is used as an antibacterial additive, during the vulcanization of rubber if nitrite is a component, and in some foods (e.g. some beers). Natural tobacco contains both amines and nitrite, and the tobacco used for smoking contains several *N*-nitrosoamines (a cigarette smoker inhales approximately 1 µg per cigarette). Some secondary *N*-nitrosoamines, e.g. dimethyl-*N*-nitrosoamine, are potent carcinogens after metabolic activation via a cytochrome P-450 hydroxylation of the carbon α to the *N*-nitrosoamine group (see Figure 8.15), and they give tumours in various tissues.

Dimethyl-*N*-nitrosoamine is converted to diazomethane, a highly toxic and volatile chemical (that in addition is highly explosive), which in its protonated form is a powerful electrophile and will methylate nucleophiles. The hydroxylation of secondary *N*-nitrosoamines is most efficient with methyl groups, where there is as little steric hindrance as possible, and asymmetric compounds will preferentially yield the electrophilic diazonium ion of the bigger alkyl group. Another possibility for *N*-nitrosoamines is a reduction to nitric oxide and the corresponding secondary amine, and for *N*-nitroso-*N*-methylaniline this is actually the most important route (see Figure 8.15). However, in spite of the limited amounts of the electrophilic benzene diazonium ion formed from *N*-nitroso-*N*-methylaniline, it is still carcinogenic and causes cancer of the oesophagus.

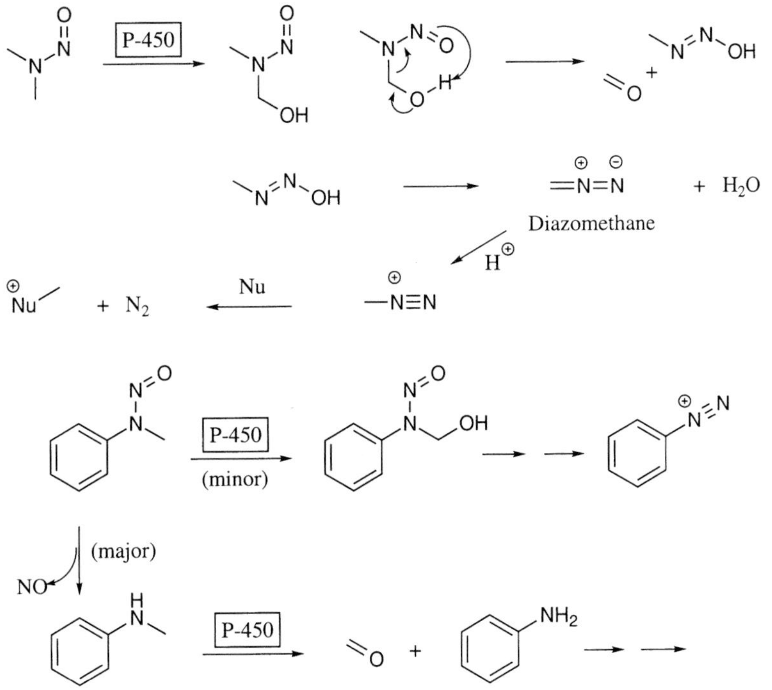

Figure 8.15. The metabolic activation of secondary N-nitrosoamines.

The cyanide functionality in nitriles is peculiar in that the carbon atom is not really part of the carbon skeleton of the molecule, instead the cyanide group can be regarded as a substituent similar to a hydroxyl group or a halogen and is even a (poor) leaving group in nucleophilic substitutions. Nitriles may consequently be subjected to oxidative dealkylation via an α-hydroxylation, yielding an α-hydroxynitrile that spontaneously will eliminate hydrogen cyanide. As this product is a highly toxic compound, it is at least in theory possible that nitriles may be hazardous because of the hydrogen cyanide they generate as a metabolite. However, as will be discussed in detail in Chapter 10, hydrogen cyanide is not very long-lived in our bodies, and it is difficult for the cytochrome P-450 α-hydroxylation to generate life-threatening concentrations of this compound. Acetonitrile is a much used solvent that many working in chemical laboratories are exposed to. It has been shown to be oxidized by cytochrome P-450, although the intermediate hydroxyacetonitrile appears to produce hydrogen cyanide after oxidation by catalase and the expected metabolite formaldehyde (or formic acid) has not been detected (see Figure 8.16). Because of its high water solubility, a substantial part is actually excreted directly with the urine without being metabolised. In bacteria, acetonitrile may be converted by addition of water to the triple bond, to acetamide, but this metabolite is not formed in mammals. Acrylonitrile, used for the preparation of various polymers, is an electrophile because the double bond is activated by the electron–withdrawing cyanide group, and may consequently react with glu-

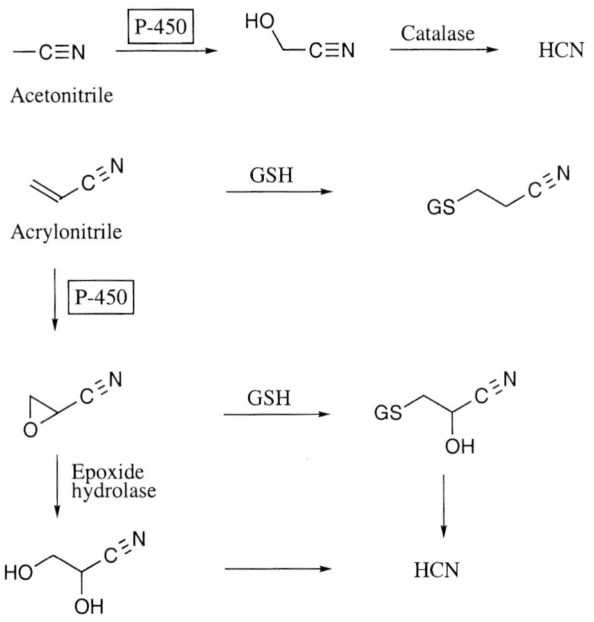

Figure 8.16. The metabolism of acetonitrile and acrylonitrile.

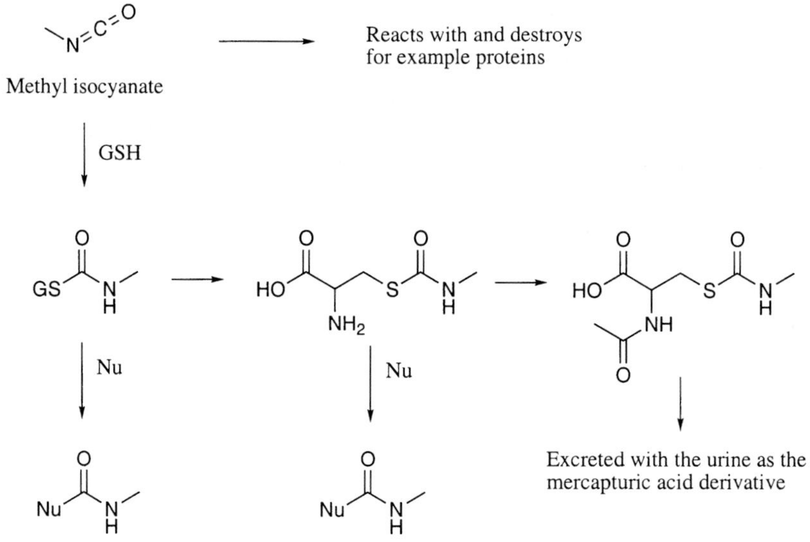

Figure 8.17. The metabolism of methyl isocyanate.

tathione (or other nucleophiles) directly. However, the double bond may also be epoxidized by cytochrome P-450, and after hydrolysis/conjugation of the epoxide a labile α-hydroxynitrile generating hydrogen cyanide is formed (see Figure 8.16)

Methyl isocyanate, the chemical that killed more than 2,000 people after an accident in a pesticide manufacturing plant in India 1984 (Bhopal), is a highly reactive compound that in high concentrations will destroy the respiratory tract. In lower amounts it can be taken up by the body and react with glutathione in a reversible way, and the systemic effects observed some time (up to a year) after the accident in Bhopal are believed to be caused by this masked form of methyl isocyanate that is distributed through the body and later hydrolyzed back to methyl isocyanate in the target organ (see Figure 8.17).

8.3 Metabolism of compounds containing halogen

The principal oxidations and conjugations of halogenated organic compounds have been discussed in Chapter 7. Many chlorinated compounds were used as solvents in huge quantities, but have been exchanged as it turned out that they were in fact hazardous. Carbon tetrachloride (tetrachloromethane) is an example, it is a very good solvent indeed but rather soon turned out to be carcinogenic and toxic to the liver. It is to some extent metabolized by reductive dehalogenation (Section 7.3.2) to the trichloromethyl radical, that will be able to initiate radical reactions and is responsible for the toxic effects observed. The dimer of this radical, hexachloroethane, is also observed as a metabolite, proving that the radical actually is formed (see Figure 8.18). Carbon tetrachloride was replaced by chloroform (trichloromethane), which is less toxic but after having been used for some time also turned out to have similar effects on the liver. It appears to be mainly oxidized to the electrophile phosgene, although a reduction generating radicals has also been suggested. Chloroform was in turn replaced by methylene chloride (dichloromethane). This solvent is less toxic compared to carbon tetrachloride and chloroform, but there are indications that it is genotoxic and possibly carcinogenic. Methylene chloride may react slowly as a weak electrophile with for example DNA, and the conjugate formed with glutathione, *S*-(chloromethyl)glutathione (see Figure 8.18), would be an excellent electrophile. Besides their toxicity, some chlorinated solvents have the ability to affect the ozone layer in the stratosphere, and measures are today taken to decrease the use of this class of solvents. Methylene chloride can possibly be replaced by methyl *tert*-butyl ether (MTBE), a volatile compound that is used in large amounts as an octane booster in gasoline. MTBE lacks chlorine altogether, it is considered to be quite safe from a toxicological point of view but has different properties as a solvent compared to methylene chloride.

Methylene chloride is also metabolized by oxidative dehydrohalogenation, generating the acid chloride of formic acid (see Figure 8.18) which should be a highly reactive and toxic product. However, it is so unstable that it is simply transformed to carbon monoxide and hydrogen chloride the moment it is formed. This is probably the reason why methylene chloride is less toxic to the liver compared to carbon tetrachloride and chloroform, although large doses of the compound will generate toxic (and possibly fatal) amounts of carbon monoxide in the body.

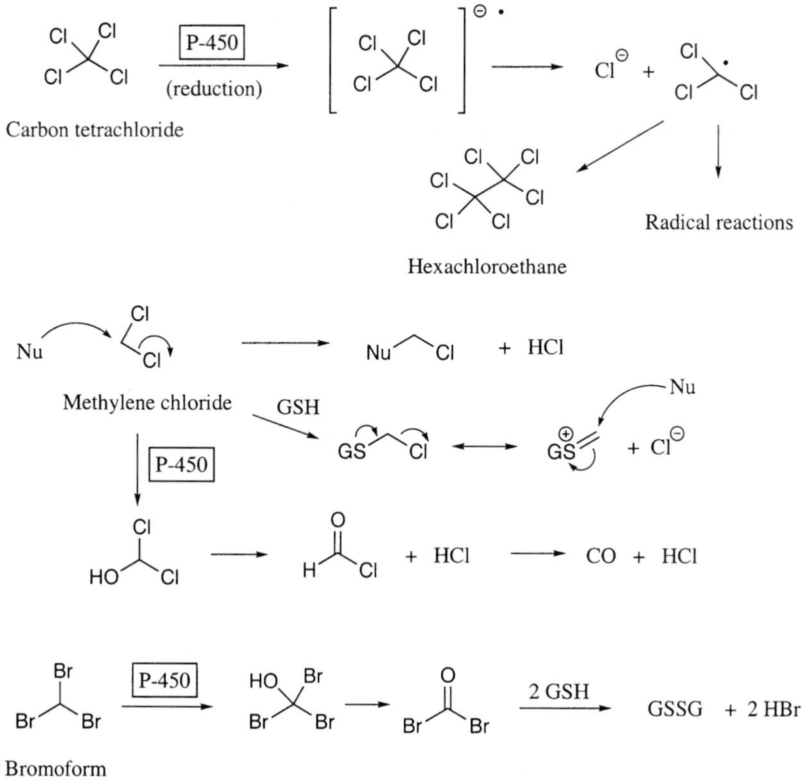

Figure 8.18. Metabolism of carbon tetrachloride, methylene chloride and bromoform.

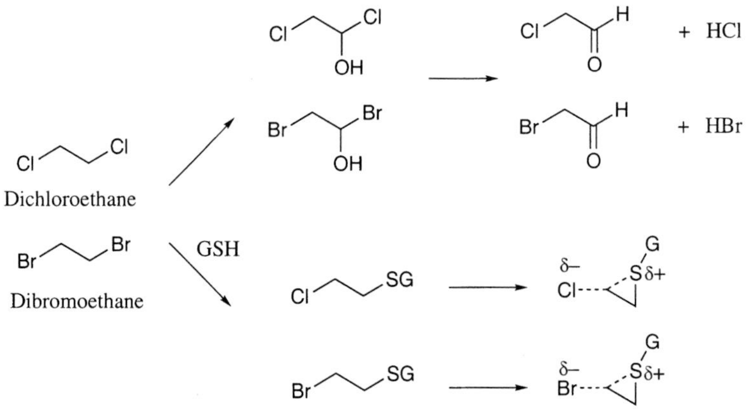

Figure 8.19. Dichloroethane and dibromoethane are electrophilic as they are, after oxidative metabolism, and after conjugation with glutathione.

1,2-Dichloroethane, used as a lead scavenger in leaded petrol, an industrial solvent and a grain fumigant, and 1,2-dibromoethane, also a lead scavenger and a pesticide, are toxic to the liver and kidneys. Both may react as primary halides with nucleophiles, in addition they are hydroxylated to form the highly reactive metabolites α-chloroacetaldehyde and α-bromoacetaldehyde, and they may be directly conjugated with glutathione to S-(2-chloroethyl)-glutathione and S-(2-bromoethyl)-glutathione (see Figure 8.19). The two latter, either as they are or as the corresponding cysteine derivatives formed after hydrolysis of the tripeptide, may react in a similar way to mustard gas, forming an extremely reactive electrophile by an intramolecular attack of the sulphur on the carbon with the halogen atom. The toxic effects are caused by a combination of these reactive forms.

A similar halogenated alkane is 1,2-dibromo-3-chloropropane, used as a soil fumigant and a nematicide, which is nephrotoxic and genotoxic. Besides the reactive metabolites corresponding to those shown in Figure 8.20, 1,2-dibromo-3-chloropropane has also been shown to form reactive acrolein derivatives, epoxides (formed by ring closure of an alcohol substituted with an bromine or a chlorine on the α-carbon) and 1-bromo-3-chloro-2-propanone (see Figure 8.20).

Halothane, used as an inhalation anaesthetic, has low toxicity but has in some cases been shown to provoke an allergic response and may also be toxic to liver cells, and there has been some concern especially for persons exposed to halothane daily in their work. It may be oxidized by cytochrome P-450 to form trifluoroacetic acid chloride (or bromide) via oxidative dehydrohalogenation (see Figure 8.21), which will react with for example proteins in the liver cells. It has also been indicated that halothane can be reduced by cytochrome P-450, to form reactive radicals.

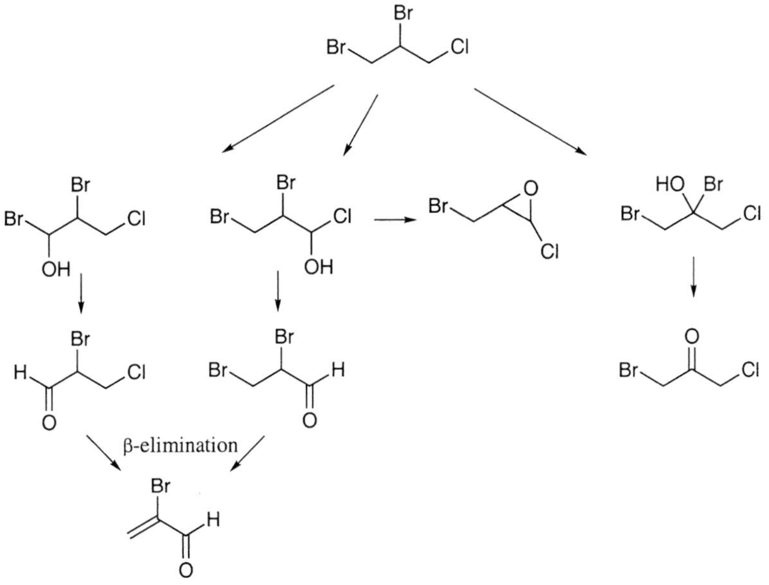

Figure 8.20. Additional reactive metabolites formed from 1,2-dibromo-3-chloropropane.

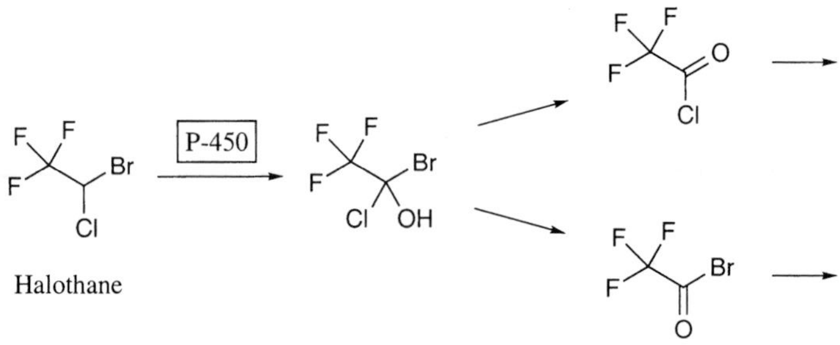

Figure 8.21. Halothane may be converted to reactive acid halides.

Vinyl chloride (chloroethene, chloroethylene) is used in large quantities as the monomer for production of polyvinyl chloride plastic (PVC) and in organic synthesis. It became famous in the end of the 1960s, up to this point it had been regarded as an almost harmless solvent that could be used without any special precautions. However, it could be shown that heavily exposed workers are more likely to get cancer compared to non-exposed workers. Vinyl chloride causes a very unusual form of liver cancer, which is the reason why it was possible to establish the relationship between the exposure and the effect. Its potency as a carcinogen is quite low, the effect would not have been detected if the tumour was of a common type, and only relatively few individuals were ever affected. As a consequence of these findings, its use was severely restricted. It is metabolized by epoxidation, and the epoxide is long-lived enough to be hydrolyzed as well as conjugated with glutathione (see Figure 8.22). However, the epoxide has a tendency to rearrange spontaneously to form the electrophile α-chloroacetaldehyde which can be either reduced to the alcohol or oxidized to the carboxylic acid. The toxic effect is assumed to be caused by the reactivity of the epoxide and α-chloroacetaldehyde.

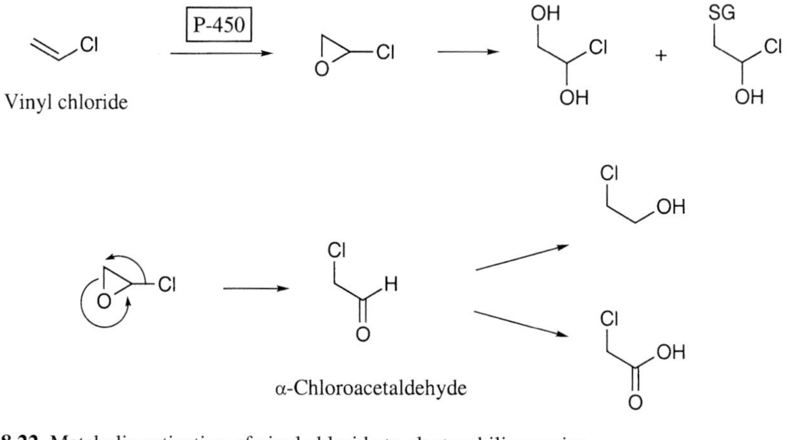

Figure 8.22. Metabolic activation of vinyl chloride to electrophilic species.

Trichloroethylene, a solvent that also has been used for general anaesthesia, is considered less toxic compared to vinyl chloride. It will also undergo epoxidation by cytochrome P-450, but the epoxide is so unstable that it will more or less instantly be rearranged (in a variant of the pinacol rearrangement!) to trichloroacetaldehyde (see Figure 8.23). This reaction is so fast that no epoxide hydrolase product can be observed, and no conjugate formed from the reaction with glutathione. Trichloroacetaldehyde, or chloral, is a well-known and popular sedative and hypnotic which is still used. Chloral can be oxidized to the corresponding acid (trichloroacetic acid) or reduced to 2,2,2-trichloroethanol which is conjugated with glucuronic acid and excreted. Trichloroethylene can also react with cysteine and glutathione to form various *S*-(dichlorovinyl) derivatives that are especially toxic to the kidneys because these ionic compounds are concentrated there. This reaction is quite slow and probably of little importance *in vivo* but has been shown to be important when trichloroethylene is used for the extraction of fat from soybean oil, as the corresponding *S*-(dichlorovinyl)-cysteine formed during the extraction is toxic to animals

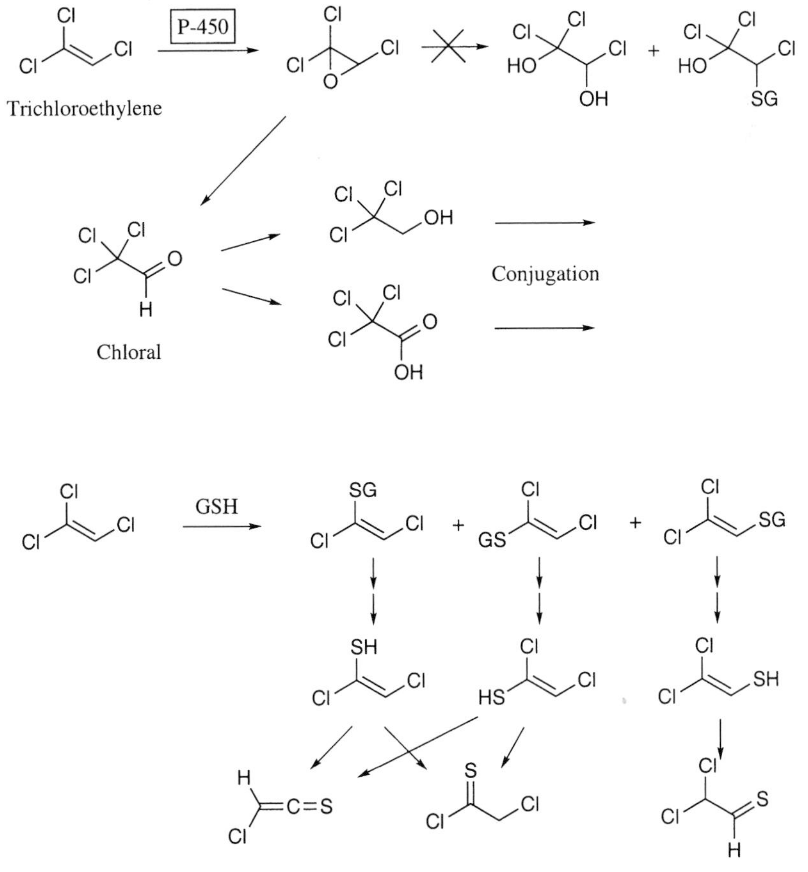

Figure 8.23. Conversions of trichloroethylene and its glutathione derivative.

that are fed the remainings after the extraction. (Chemically the reaction is not a nucleophilic substitution but rather an addition of the thiol to the double bond followed by an elimination.) The enethiol ether is bioactivated via the *S*-(dichlorovinyl)-cysteine by the enzyme β-lyase, which cleaves the bond between the sulphur atom and the rest of the cysteine. The enethiols are transformed to the corresponding thiocarbonyl compounds, or by elimination of hydrochloric acid to a chlorothioketene (see Figure 8.23), which are all reactive compounds (compared with the corresponding oxygen derivatives).

8.4 Metabolism of compounds containing sulphur

Dimethyl sulphoxide (DMSO) is an unusual solvent that readily dissolves both polar and nonpolar compounds. It is characterized by being absorbed relatively easily through the skin and is known to rapidly induce a distinct taste (of onions) in the mouth if applied or spilled on the skin. This is due to the reduction of small amounts of dimethyl sulphoxide to dimethyl sulphide (see Figure 8.24), a compound that has an extremely strong (and characteristic!) smell even in very low concentrations, in our bodies. Most DMSO is oxidized to dimethyl sulphone. Both dimethyl sulphoxide and dimethyl sulphone are water soluble and can be eliminated with the urine.

Dimethyl sulphoxide　　　　　　Dimethyl sulphide　　Dimethyl sulphone

Figure 8.24. Dimethyl sulphoxide is both reduced and oxidized.

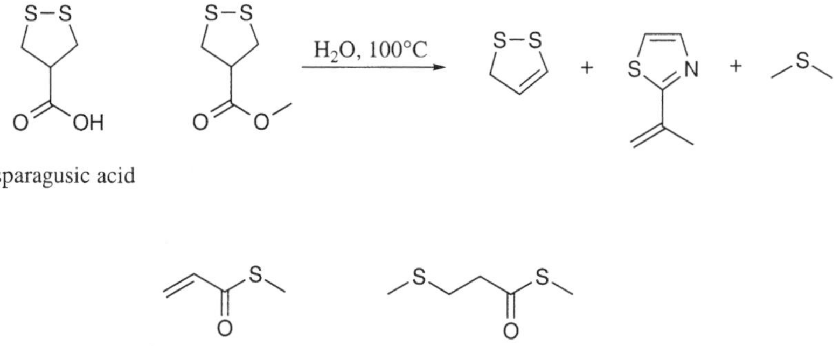

Asparagusic acid

Metabolites excreted in the urine

Figure 8.25. Transformations and conversion of asparagusic acid.

A final example of a compound that has little importance for the environment but that some may have noticed the existence of is asparagusic acid, present together with its methyl ester in one of the most delicious comestibles available; asparagus. After boiling 1,2-dithia-3-cyclopentene, dimethyl sulphide and 2-acetylthiazol are the major components in the aroma, while the thioesters shown in Figure 8.25 are the metabolites responsible for the characteristic odour of the urine after the delight of an asparagus meal.

9 Conversions and transformations in the environment

In the environment (= outside the body of humans) chemicals are subjected to conversions/transformations (degradation) by reactions with other chemicals present in the same medium, by photochemical reactions initiated by sunlight, and by enzymatic conversions in organisms. Such conversions/transformations will change the structure and thereby the toxicity of a chemical, and it is consequently of interest to understand what may be formed under various circumstances. Organic compounds will eventually be mineralized, meaning that they are completely degraded to inorganic chemicals (CO_2, H_2O, etc), although this may take a long time and pass via more hazardous intermediates. Decisive for the fate of compounds in the environment is their distribution in the different media, for example the probability for volatile compounds to undergo photochemical reactions is of course large compared to nonvolatile compounds. Some compounds, even if they contain chemical functionalities that appear to be easily converted/transformed, have so limited water solubility and volatility that they are simply sorbed to macroscopic surfaces and withdrawn from circulation. Such compounds may survive for very long times in various compartments, for example in sediments, and will not pose any hazard until they for various reasons are mobilized again.

9.1 Enzymatic conversions

Biotic conversion of chemicals in the environment is especially important for the degradation of organic compounds in general and pollutants in particular, because they include reactions that hardly would be accomplished, at least not in a comparable time scale, in abiotic ways. Enzymes may facilitate a chemical reaction that is feasible but slow without enzymatic catalysis by lowering the activation energy, or by the input of energy (from cofactors produced by the primary metabolism) that promotes the reaction, and we have already seen several examples of conversions carried out by mammalian enzyme systems in Chapter 7. As was indicated in Chapter 4 by far the most common organisms are the microorganisms, and they will consequently be responsible for most of the metabolic conversions of organic compounds in the environment (some examples of conversions of toxic metal ion are given in Chapter 12). There are many different types of microorganisms (e.g. some algae, bacteria, fungi and protozoans) possessing different metabolizing capacities, and microorganisms often have an incredible ability to adapt to new conditions (e.g. the presence of a new compound in their environment). However, although several of the basic metabolic conversions are similar to those of mammals (Chapter 7), there are many more and it is not possible to summarize the microbial metabolism of organic compounds in this chapter. In addition, it has turned out that some microorganisms carry out a sequence of metabolic steps in cooperation, and that the rate of the overall conversion depends on the presence of all partners at the same place. A major factor that influences the metabolic fate in different organisms is the amount of molecular oxygen present. In mammals, which normally have relatively high concentrations of molecular oxygen, oxidative conversions dominate, while organisms living in a environment containing only little molecular oxygen may have a completely different metabolic profile. Many organic compounds, e.g. hydrocarbons, are more efficiently metabolized by oxidation, but halogenated compounds are often converted more rapidly under reducing conditions. The metabolites that eventually are formed will be substrates for other metabolic conversions as well as for chemical/photochemical transformations, which further complicates the picture.

The ability of some microorganisms to perform unexpected metabolic conversions at a high rate has proven to be useful for the purge of contaminated soil and water. The uncontrolled handling of hazardous chemicals during the 20th century left a large number of industrial and military sites where the soil was heavily contaminated, and to avoid the leaking of such contaminants into the subsoil water and to make such areas exploitable again the soil has to be decontaminated. This can be achieved in several (expensive) ways, and lately biodegradation with microorganisms has become popular. By screening large numbers of microbial species it is often possible to find at least one that can be used for the degradation of a certain organic compound. The conversion of organic compounds by microorganisms can be divided into 4 principal classes:

1. The compound is relatively non-toxic to the microorganism and readily metabolized to components of the primary metabolism. In this case the microorganism may use the compound as nutrient, multiply and mineralize the compound completely.
2. The toxicity of the compound is limited and even if it is not mineralized it is metabolized to new compounds that are as toxic or less toxic compared with the original compound. The products may then be further metabolized by other organisms or transformed by chemical/photochemical reactions.
3. The toxicity of the compound is limited but the metabolites formed are more toxic compared with the original compound. This poses a threat not only to the microorganism itself, which will have problems to multiply, but also to other organisms. The products may be further metabolized by other organisms or transformed by chemical/photochemical reactions. The conversion rate will be low, and as a biodegradation this means low efficiency.
4. The compound is toxic to microorganisms, hampering its multiplication, and probably also to other organisms. Only low concentrations of the compound in the soil/water will be tolerated. If the compound is toxic because it is chemically reactive, a chemical or physical method for degrading the pollutant may be preferable.

In addition, the mass transport from the sorbed state (in soil) to the microorganism is also a critical factor, as is the concentration of oxygen. Completely different metabolites are produced under aerobic compared to anaerobic conditions, and in the case when biodegradation should be performed in a controlled way there are many parameters to optimize. It is important to remember that biodegradation is not simply a question of getting rid of an organic contaminant in soil or water, it is necessary to be aware of the possibility that even more hazardous compounds are formed. Most attempts with biodegradation have been made with aromatic compounds, because many members of this class are hazardous and they are used in very large amounts and released into the environment. Consequently, most of our knowledge about biodegradation of organic compounds concerns aromatic compounds.

9.1.1 Biodegradation of saturated hydrocarbons

Hydrocarbons in crude oil from natural oil leaks or from spills resulting from oil-production and oil refinement are the most common pollutants in both aquatic and terrestrial ecosystems. Even though hydrocarbons, lacking functional groups and unsaturations, are not easily converted by the metabolic reactions, many microorganisms will nevertheless metabolize hydrocarbons so efficiently that they are of practical use for biodegradation of oil discharges. We have seen in Chapters 7 and 8 that two metabolic pathways for the degradation of saturated hydrocarbons are available in mammals, hydroxylation, the insertion of an oxygen between a carbon and a hydrogen, and oxidation of a carbon–carbon single bond to a double bond. Microorganisms will primarily use hydroxylation, catalyzed by cytochrome P-450 monooxygenases or related

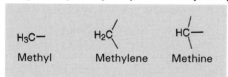

enzyme systems, and the most important conversions are summarized in Figure 9.1. Hydroxylation may occur either in a terminal position, i.e. of a methyl group, or in a chain or a ring, i.e. of a methylene or a methine group.

The hydroxylation of a straight-chained hydrocarbon in either of its methyl groups will produce a primary alcohol that will be further oxidized (alcohol dehydrogenases and aldehyde dehydrogenases) via the aldehyde to the carboxylic acid, which will be degraded by the β-oxidation of the primary metabolism. It is not unusual that both ends of an unbranched alkane are oxidized simultaneously, forming diacids that also are further oxidized by β-oxidation.

Methylene groups that are hydroxylated to secondary alcohols are further oxidized to ketones by alcohol dehydrogenases. Monooxygenases may then incorporate an additional oxygen next to the keto function, in a conversion that is analogous to the chemical transformation called the Baeyer–Villiger reaction, to produce an ester. The ester can be hydrolyzed, and the resulting carboxylic acid and alcohol will be further oxidized. Tertiary alcohols, if formed by the hydroxylation of methine groups, are not oxidized further.

9.1.2 Biodegradation of benzene

In general, different organisms will attack the aromatic hydrocarbons in different ways, and while for example plants and fungi predominantly use monooxygenases and phase II reactions bacteria will use dioxygenases which produce metabolites that are easier to mineralize. The major biodegradation routes for benzene proceed via catechol, formed after the oxidation of benzene by first a dioxygenase followed by a dehydrogenase. The intermediate product ben-

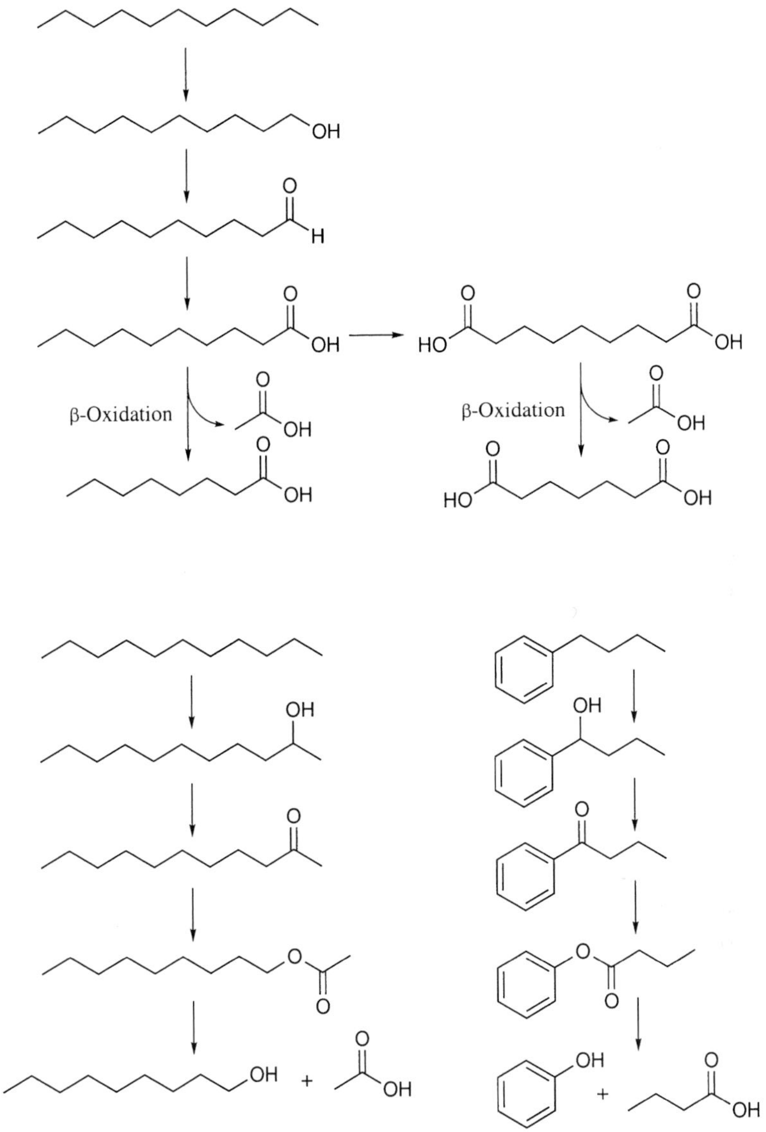

Figure 9.1. Major biodegradation pathways for saturated hydrocarbons by microorganisms.

zene dihydrodiol (note that the two hydroxyl groups are *cis*!) is not too stable and may eliminate a molecule of water to form phenol (and thereby regain its aromaticity). Nevertheless, catechol appears to be the major product and is further metabolized by two principal pathways (see Figure 9.2). Additional oxidation of the diol moiety by the enzyme catechol 1,2-dioxygenase will via ring fission produce muconic acid, via the so called *ortho*-cleavage route. Muconic acid

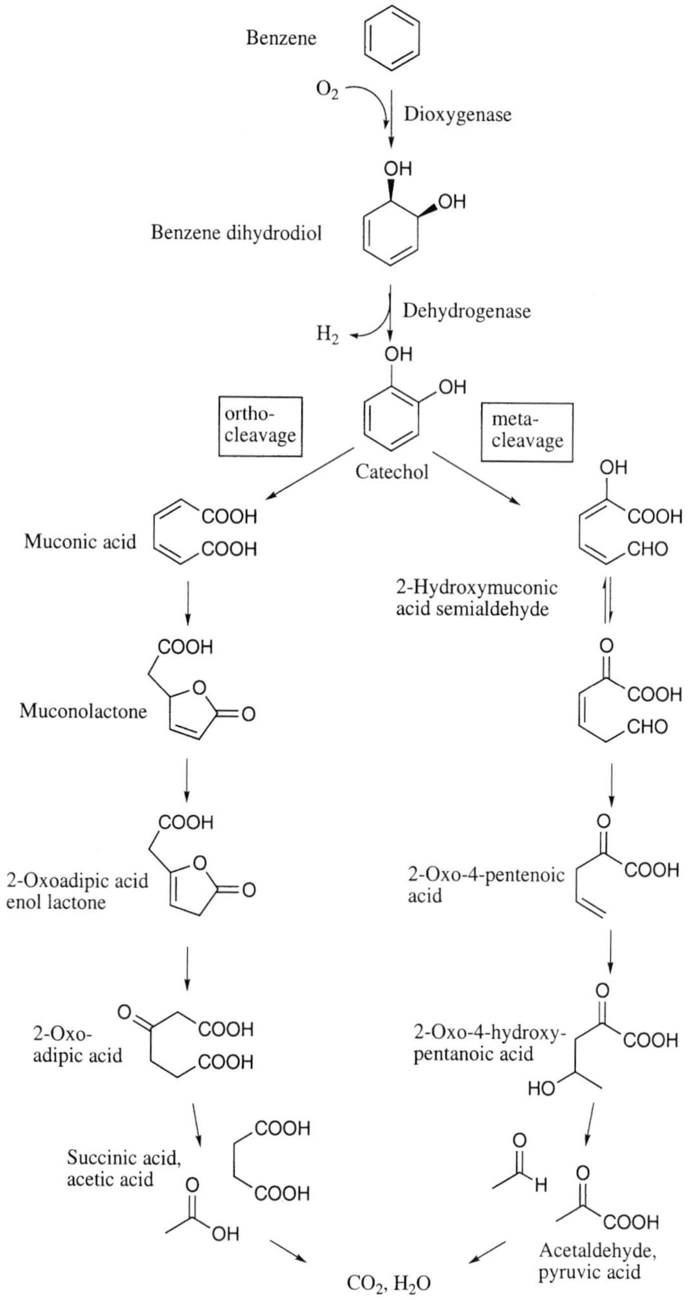

Figure 9.2. Major biodegradation pathways for benzene by bacteria.

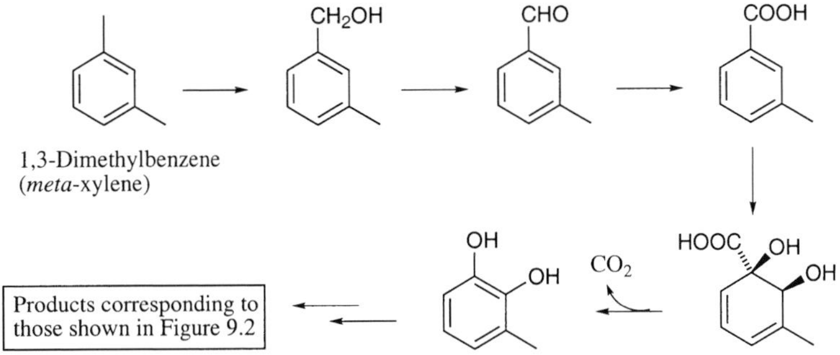

Figure 9.3. Bacterial degradation of 1,3-dimethylbenzene (*meta*-xylene).

will spontaneously cyclize to muconolactone which will be isomerized to the enol lactone of 2-oxoadipic acid shown in Figure 9.2. This will yield acetic acid and succinic acid which are completely degradable. The other pathway, called the *meta*-cleavage route, involves further oxidation of the carbon–carbon bond adjacent to the diol moiety by catechol 2,3-dioxygenase and produces 2-hydroxymuconic acid semialdehyde as its initial product. Its keto form is oxidized and decarboxylated to 2-oxo-4-pentenoic acid, which after the addition of a molecule of water to the carbon–carbon double bond (to 2-oxo-4-hydroxypentanoic acid) is converted to pyruvic acid and acetaldehyde.

9.1.3 Biodegradation of alkylbenzenes

Substituted benzenes are metabolized by the same routes, although variations are of course observed. Alkylbenzenes, e.g. toluene, are also oxidized in the alkyl group, and a methyl group after oxidation to a carboxylic acid functionality may be split off as carbon dioxide. Dimethylbenzenes (xylenes), for example, are oxidized to methylbenzoic acids, which are decarboxylated during the subsequent oxidation to methylcatechols (see Figure 9.3 for 1,3-dimethylbenzene).

9.1.4 Biodegradation of fused-ring aromatic compounds

Some organisms are also able to degrade fused-ring aromatic compounds, compounds with two or more benzene rings joined side by side. Of special interest is of course the polycyclic aromatic hydrocarbons, PAHs, because some of them are highly toxic and carcinogenic. Figure 9.4 shows how phenanthrene is believed to be metabolized by certain bacteria, and it is similar to the conversions discussed above.

However, biodegradation will generate all kinds of metabolites, including toxic ones. Figure 9.5 shows an example of the products identified from the degradation of pyrene by an *Aspergil-*

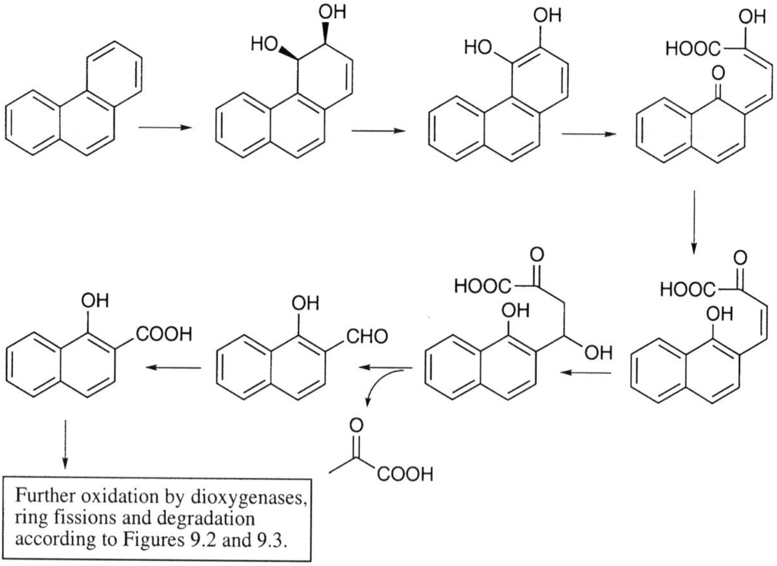

Figure 9.4. Biodegradation of phenanthrene by bacteria.

lus species, and we can observe that this metabolism is more similar to that of mammals. The formation of reactive and potentially hazardous quinones is always a possibility, and we have seen that oxidation of some hydroquinone to the corresponding benzoquinones may take place spontaneously under aerobic conditions.

9.1.5 Biodegradation of halogenated aromatic compounds

As indicated in the previous section, substituted benzenes may be metabolized by microorganisms in the same way as benzene itself, via the *ortho*- or *meta*-cleavage route. In addition, halogenated aromatic compounds may also undergo oxidative dehalogenation (see Figure 9.6), when the halogen is lost during oxygenation of the ring, a sequence that also generates catechols. Another possibility is hydrolytic halogenation, when a halogen simply is replaced by a hydroxyl group without any oxidation or reduction taking place. Finally, under anaerobic conditions halogenated aromatic compounds may undergo reductive dehalogenation, resulting in the replacement of a halogen with a hydrogen.

Again, one should not be surprised to find completely different products. Figure 9.7 shows what a fungus managed to make of 3,4-dichloroaniline, a microbial product from the biodegradation of several chlorinated phenylcarbamate herbicides. Instead of degrading 3,4-dichloroaniline it was condensed to dimers, trimers and tetramers, and it is likely that such products can be incorporated into for example the humic substances.

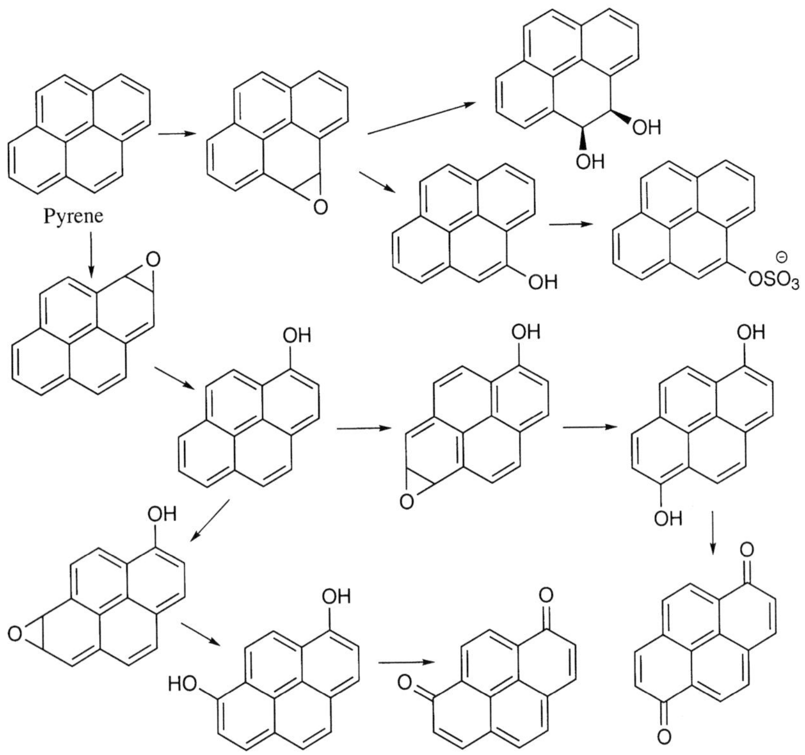

Figure 9.5. The degradation of pyrene by an *Aspergillus* species.

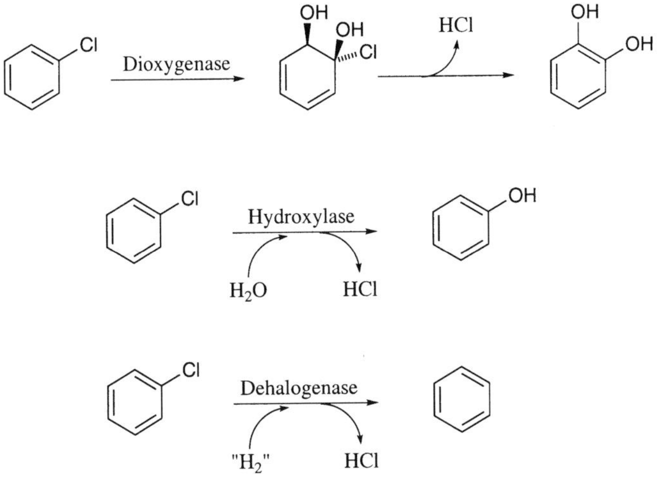

Figure 9.6. Principal dehalogenations of halogenated aromatic compounds.

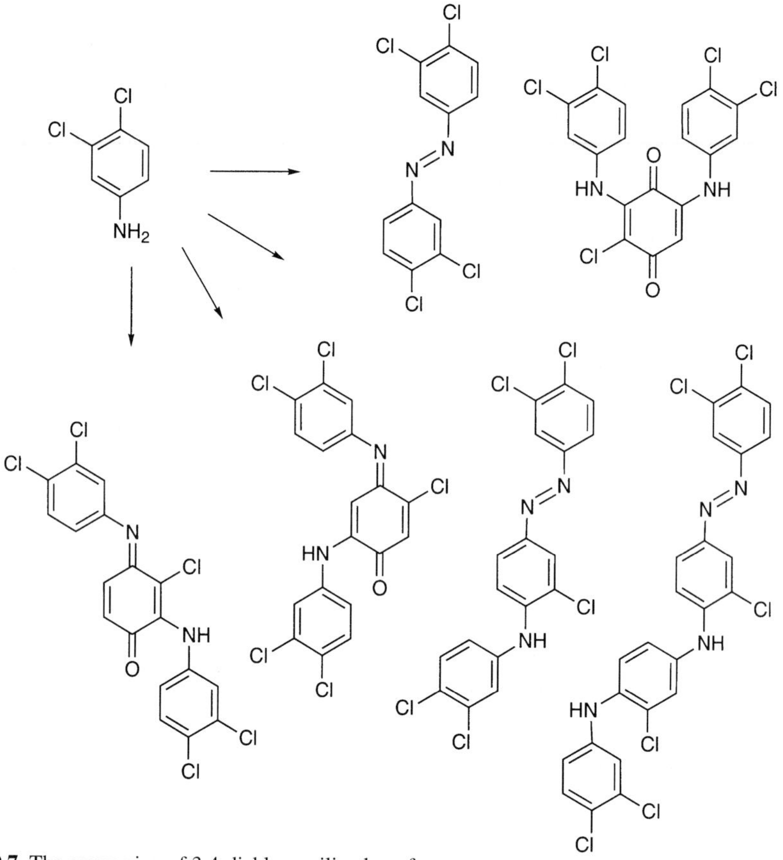

Figure 9.7. The conversion of 3,4-dichloroaniline by a fungus.

9.2 Chemical transformations

Chemical or abiotic transformations of chemicals do not involve the action of enzymes, which does not necessarily mean that they are not catalyzed. They may take place in the air, in water, or at solid surfaces, and can be divided into the major categories oxidations/reductions, hydrolyses and substitutions/additions/eliminations. In principle, all existing reactions described in the chemical literature (as well as those not yet described!) take place, although some reactions are more important. Besides catalysis, temperature is also an important factor that promotes chemical reactions. So far we have ignored it as the temperature in mammals is constant, but in the rest of the world temperature differences may be substantial. At the surface of earth heat can be generated by for example geothermic processes and sunlight. The effect of increased temperature is to increase the energy of the molecules that react, and thereby decreasing the activation energy for the reaction.

9.2.1 Oxidations

Oxidations (and also reductions) require a transfer of electrons to the oxidation agent from whatever is being oxidized, and at normal temperatures most types of oxidations have to be catalyzed. However, the presence of large amounts of oxygen in our atmosphere of course makes it oxidative in general, and easily oxidized materials will be oxidized also at ambient temperatures. We have already seen examples of such so called autoxidations, of for example hydroquinones to benzoquinones via the radical semiquinone (a process that also generates superoxide), and the oxidative reactions initiated by the reactive oxygen species (which in low concentrations occur naturally). Other examples of spontaneous oxidations with molecular oxygen are the oxidation of certain aldehydes to carboxylic acids, of thiols to dithioethers, of mercaptans to sulphoxides and sulphones, and of anilines to hydroxylamines, and nitrobenzenes. Fortunately, molecular oxygen in its ground state is not very reactive and will only oxidize certain compounds even if the energy released by the oxidation of organic compounds should promote autoxidations. However, as we shall see in section 9.4, molecular oxygen may be activated to more reactive forms by radiation.

9.2.2 Hydrolyses

We have already encountered various hydrolyses, with and without catalysts, and it is evidently an important reaction type both in biochemical conversions and chemical transformations. The most common functionality to be hydrolyzed is the carboxylic acid ester, a reaction which, as hydrolyses in general, can be catalyzed both by enzymes (esterases) as well as by acids and bases. In the absence of a catalyst, the hydrolysis of an carboxylic acid ester depends on the natures of the groups bound to the carbonyl carbon and the oxygen, respectively. As indicated by the examples given in Figure 9.8, electron-withdrawing substituents on the carbon bound to the carbonyl group will increase the rate of hydrolysis in pure water simply by making the carbonyl carbon more positively charged and more prone to be attacked by water. Electron-

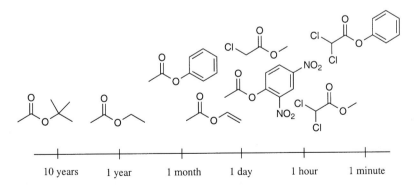

Figure 9.8. The half-lives of various carboxylic acid esters in water at 25 °C and pH 7.

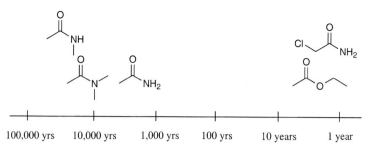

Figure 9.9. The half-lives of various carboxylic acid amides in water at 25 °C and pH 7.

withdrawing substituents on the carbon bound to the oxygen in the ester functionality will also increase the rate, especially if they can stabilize the corresponding alkoxi (phenoxi) ions (as for example in 2,4-dinitrophenyl acetate).

In addition to being hydrolyzed by water according to the classical mechanism involving an attack of the water molecule on the carbonyl carbon, the acylate ion (R-COO⁻) of a carboxylic acid esters may also be a leaving group in a nucleophilic substitution or elimination (*vide infra*). Amides are less prone to hydrolysis, as we already have noted, because amines are in general poorer leaving groups compared to alcohols and because the nitrogen in an amide has a larger tendency to donate electrons to the carbonyl group, giving the CO–N bond a partial double bond character. The half-lives of a few simple amides are compared to that of ethyl acetate in Figure 9.9, and again we see that electron-withdrawing substituents on the carbon bound to the carbonyl group increase the rate of hydrolysis.

Carbamates are widely used as insecticides and herbicides, and we will encounter an example (carbaryl) in Chapter 10. The carbamate functionality has both an oxygen and a nitrogen bound to the same carbonyl carbon making it an ester–amide of carbonic acid. Hydrolysis can take place both the ester moiety (yielding an alcohol) and the amide (yielding an amine), and as we can guess the ester is normally hydrolyzed more rapidly than the amide. In some cases both are hydrolyzed yielding an alcohol, an amine and carbonic acid. The usual rules apply (see Figure 9.10), but there is a remarkable difference in the rate of hydrolysis between carbamates that have one or two alkyl groups attached to the nitrogen, especially of the free alcohol is relatively acidic (as for example 4-nitrophenol). The reason for this is that two different mechanisms operate. Besides an attack by the water molecule on the carbonyl carbon, the hydrolysis may also be initiated by an elimination of the alcohol to form an isocyanate (see Figure 9.10). The isocyanate will then react fast with water (an addition) to form the hydrolysis product. Isocyanate formation is facilitated by the acidity of the nitrogen proton, being both attached to a heteroatom and situated α to a carbonyl group.

Another class of compounds that frequently are employed as pesticides are derivatives of phosphoric and thiophosphoric acid esters. Such esters will also be hydrolyzed, either by the attack of a water molecule on the phosphor atom followed by the departure of an alcohol moiety, or by reactions corresponding to those discussed above for carboxylic acid ester (nucleophilic substitution and β-elimination). Some examples of phosphoric and thiophosphoric acid esters that are hydrolyzed and their half-lives in water at 25 °C, are given in Figure 9.11.

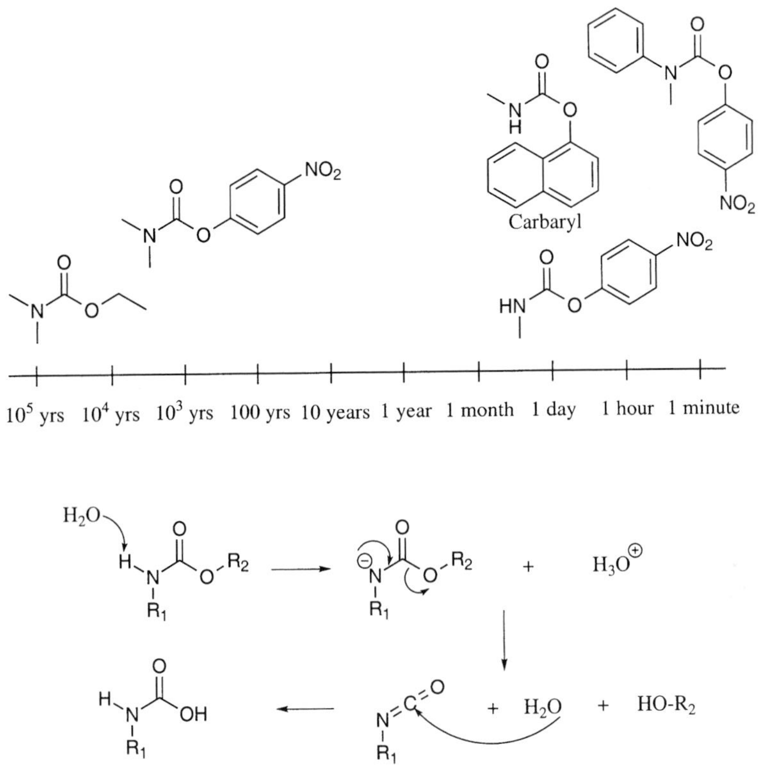

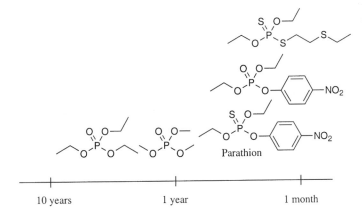

Figure 9.10. The half-lives of various carbamates in water at 25 °C and pH 7. The mechanism for hydrolysis involving an elimination and an addition is shown.

Figure 9.11. The half-lives of phosphoric and thiophosphoric acid esters in water at 25 °C and pH 7.

In addition a number of other functionalities are also subject to hydrolysis in contact with water, for example epoxides and aziridines, lactones and lactams, anhydrides and imides, carboxylic acid chlorides, carbonates and ureas, phosphonic and thiophosphonic acid esters, phosphinic and thiophosphinic acid esters, and sulphuric and sulphonic acid esters and amides.

9.2.3 Substitutions, eliminations and additions

Soil and natural waters contain, besides water itself, a number of nucleophiles that may react with electrophiles (either present as contaminations or produced by organisms) in nucleophilic substitutions. Figure 9.12 compares the nucleophilicity of the most common with that of water, and shows which approximate concentration (in M, mol/l) of a nucleophile X^- in water (the concentration of water in water is 1,000 g/l, i.e. 55.6 M) that will result in a 50:50 reaction of a electrophile like bromomethane with the two nucleophiles water and X^- (yielding 50 % CH_3–OH and 50 % CH_3–X).

Uncontaminated freshwater normally contains low concentrations of the nucleophiles shown in Figure 9.12, leaving water itself as the major nucleophile, while seawater contains for example approximately 0.5 M Cl^- and the reaction between water and various electrophiles is of minor importance. In contaminated waters the concentrations of nucleophiles (and electrophiles!) may of course be very different. As can be seen, the hydroxide ion will not react significantly in water with a pH below 10. Cyanide ions are normally present in very low concentrations, it may actually be hydrolyzed itself as discussed in Chapter 10, but hydrogen sulphide (and thereby the H_2S/HS^- couple) can be formed from sulphate ion by sulphate-reducing bacteria. The reaction between H_2S/HS^- and electrophiles will produce mercaptans (RSH/RS^-) that are even better nucleophiles. The reactivity of the electrophile will of course also influence the reaction rate, and in Figure 9.13 the half-lives of various halogenated hydrocarbons in pure water at 25 °C are shown.

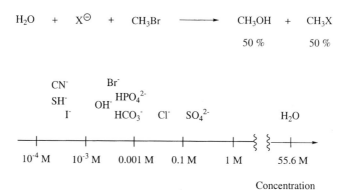

Figure 9.12. Concentrations of various inorganic nucleophiles in water that will compete with water itself for the reaction with bromomethane (see text for details).

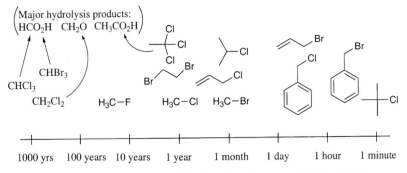

Figure 9.13. The half-lives of some halogenated hydrocarbons in water at 25 °C and pH 7.

An alternative reaction for some electrophiles, e.g. 1,2-dibromoethane and 1,1,1-trichloroethane in Figure 9.13, is the β-elimination (see Figure 9.14). This reaction will be more important if the nucleophile has basic properties and if the electrophile is sterically hindered, and some examples are shown in Figure 9.14. The alkenes produced are in general more volatile, and in some cases (e.g. the pesticide 1,2-dibromo-3-chloropropane) more reactive.

Additions are the reverse reaction to eliminations, when HX (X being for example OH, Cl, Br, SH, etc.) adds to a double bond. The combination of an elimination and an addition (in either order) will result in for example a substitution or hydrolysis.

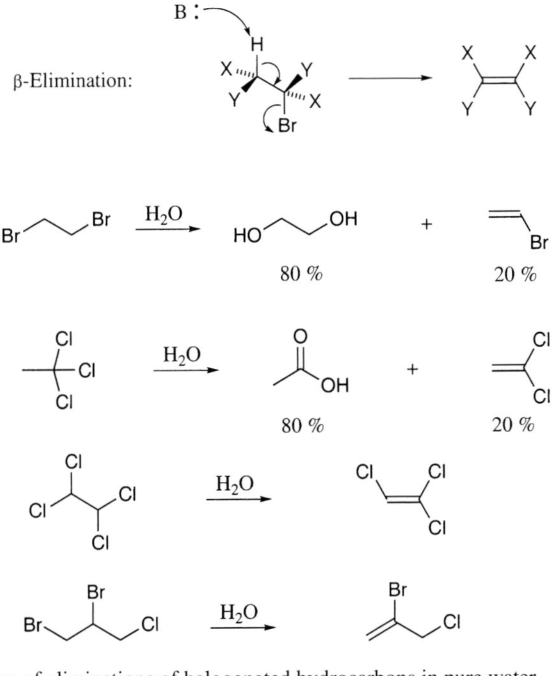

Figure 9.14. Examples of eliminations of halogenated hydrocarbons in pure water.

9.3 Transformations induced by radiation

The earth is constantly irradiated from sources in space, and the various types of radiation may affect the reactivity of the chemicals that are hit. Most important is of course the electromagnetic radiation coming from the sun, flooding the atmosphere and the surface of the hydrosphere and pedosphere with incredible amounts of energy. Actually, the yearly energy consumption by human activities is equivalent to the energy reaching the earth from the sun every 20 minutes! The electromagnetic radiation from the sun has a maximum in the visible region (wavelength 400–760 nm), but substantial parts are also in the infrared region (wavelength over 760 nm, containing less energy) and in the more harmful ultraviolet region (wavelength 100–400 nm). Fortunately, most of the ultraviolet radiation is absorbed in the atmosphere (for example by molecular oxygen which is split to two atoms of oxygen) and never reaches the ground, and as this kind of photochemistry is important for the formation and degradation of ozone this will be further discussed in Chapter 12. Radiation is the activation factor in several important photochemical reactions, and a summary is given in Figure 9.15.

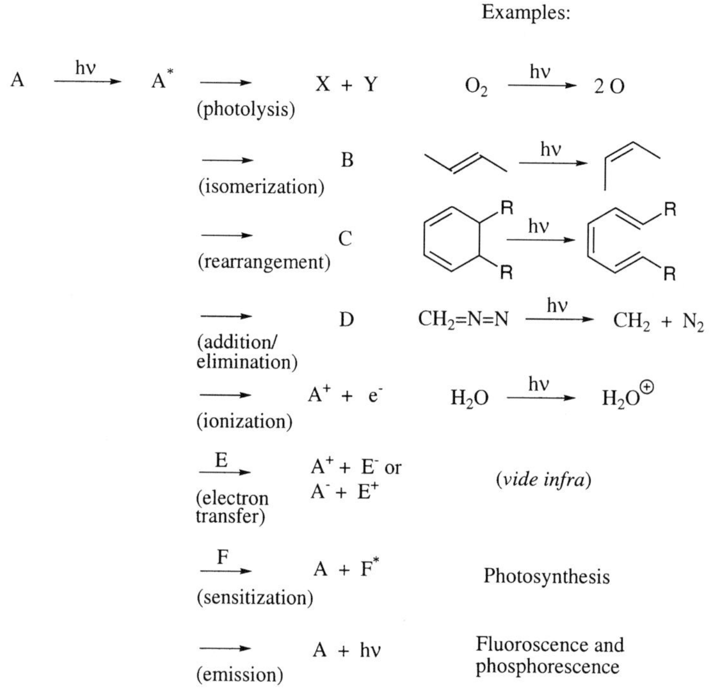

Figure 9.15. Important photochemical reactions.

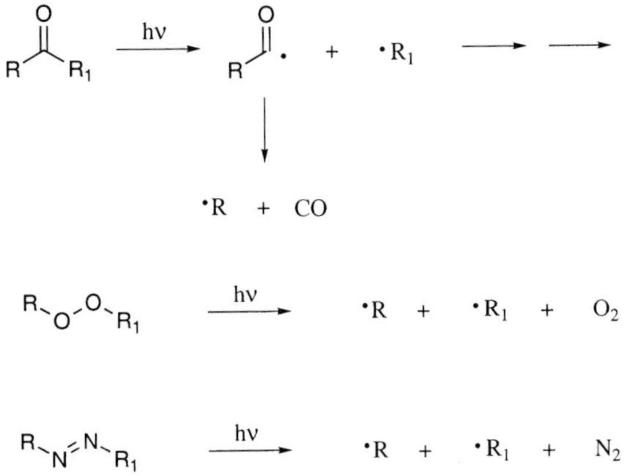

Figure 9.16. Examples of photolyses.

9.3.1 Photolysis

The energy of electromagnetic radiation is inversely proportional to the wavelength:

Energy = constant/wavelength

The energy of radiation with a certain wavelength can be calculated and compare to the energy of a bond between two atoms (the dissociation energy). If the wavelength of the radiation is so low that the energy exceeds that of the bond, the radiation will split the bond, and this will happen with molecular oxygen if the wavelength of the radiation is smaller than 243 nm. The oxygen atoms formed will react with molecular oxygen and form ozone, and we can notice this reaction (by smelling ozone) close to sources for ultraviolet radiation (e.g. ultraviolet lamps and photocopiers). While the photolysis of molecular oxygen requires radiation with relative high energy, other molecules are more easily photolyzed. Chlorine (Cl_2), for example, will be dissociated if irradiated with radiation with a wavelength of 492 nm (blue–green light) or less. Among organic molecules, ketones, peroxy ethers and aliphatic azo compounds are easily cleaved by photolysis.

9.3.2 Isomerization/rearrangement

The isomerization of double bonds is a common photochemical reaction, that has a great practical importance in that the direction of the isomerization can be decided by the frequency of the light used (see Figure 9.17). When it comes to rearrangements, there are many different reactions known and only few examples are given in Figure 9.17.

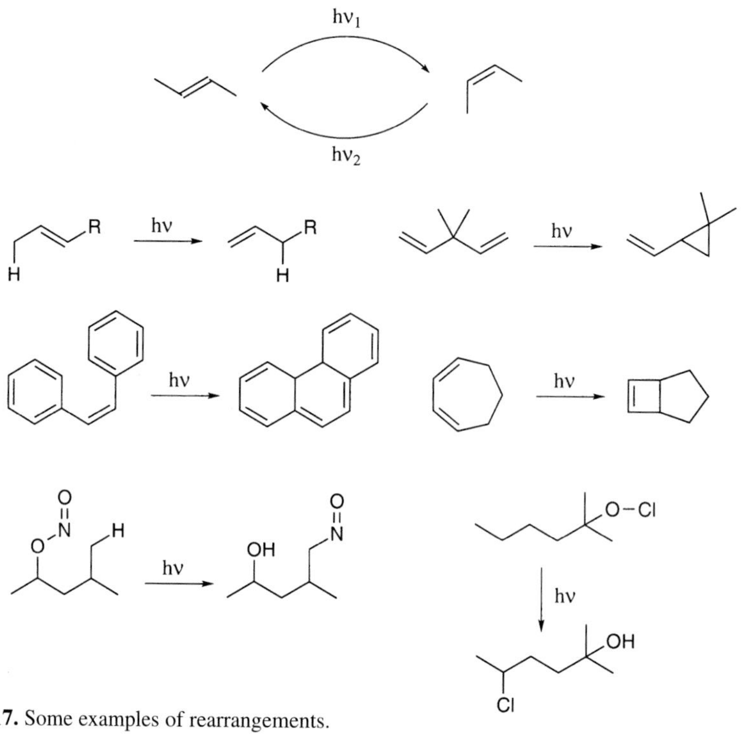

Figure 9.17. Some examples of rearrangements.

9.3.3 Additions/eliminations

The example shown in Figure 9.15 is the formation of a carbene from a diazo compound by the elimination of N_2 induced by light, and this and similar reactions are important and useful chemical reactions. The carbenes can be said to be diradicals in which the two unpaired electrons are paired, there are no unpaired electrons but also no complete octet around the carbon. Carbenes are therefore highly electrophilic and will rapidly react with compounds that can provide an electron pair, e.g. olefins, and a product from the reaction between a carbene and a olefin is the cyclopropane derivative. An example of an addition is shown in Figure 9.18, where a conjugated diene reacts with oxygen under the influence of light to form an internal peroxide.

9.3.4 Photoionization

We have already mentioned the effect of ionizing radiation, which can be electromagnetic radiation with a very short wavelength (approximately 100 nm to ionize molecular oxygen and water). However, that ionizing radiation will be hazardous is obvious, and we do not need to enter more deeply into that here.

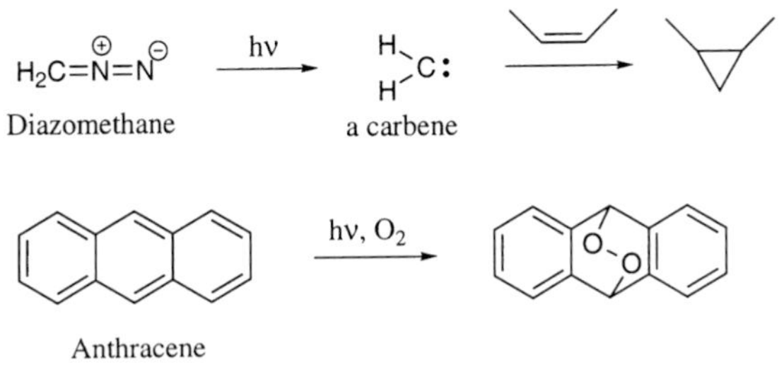

Figure 9.18. Examples of light-induced additions/eliminations.

9.3.5 Electron transfer

Simple aldehydes and ketones will (at approximately 280–320 nm) be converted to an excited state which has a diradical character (see Figure 9.19). Besides adding to carbon–carbon double bonds forming an oxetane, it can also take part in a photooxidation reaction with a hydrogen atom donor (e.g. cyclohexane in Figure 9.19) or accept an electron from another compound with a nitrogen (or sulphur) atom. The latter reaction, an electron transfer, will result in a radical anion/radical cation pair, that subsequently can react to other products.

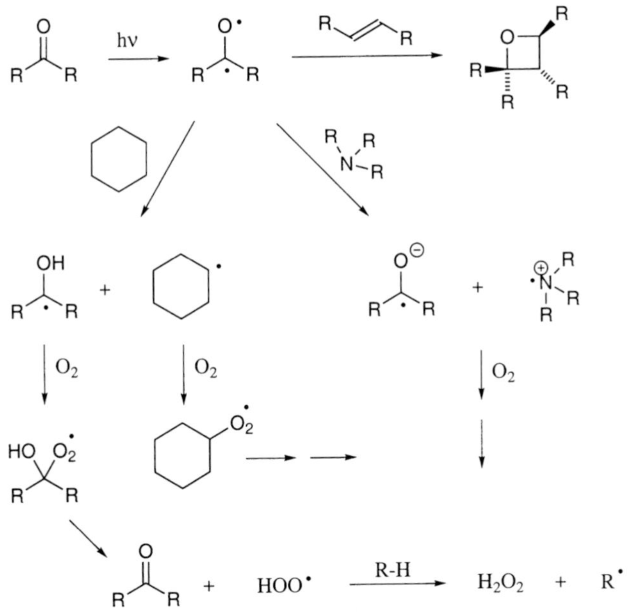

Figure 9.19. Photochemically excited aldehydes and ketones may induce an array of reactions, including electron transfers.

9.3.6 Sensitization

Photosensitization is a process when one excited molecule returns to its ground state by exciting another, which subsequently takes part in a chemical reaction. In the photosynthesis of plants the energy of sunlight is converted to chemical energy via the excited state of chlorophyll (which consequently is a photosensitizer). A common reaction is that a molecule of oxygen in its ground state is excited by a photosensitizer to singlet oxygen. The singlet state is for most molecules the ground state, but oxygen is a rare exception and singlet oxygen is an excited state (in which the electron configuration actually can be regarded as normal with a oxygen–oxygen double bond and two unshared electron pairs on each oxygen atom), and is highly reactive (the half-life of singlet oxygen in water is approximately 2 μs). Ground-state oxygen can be excited

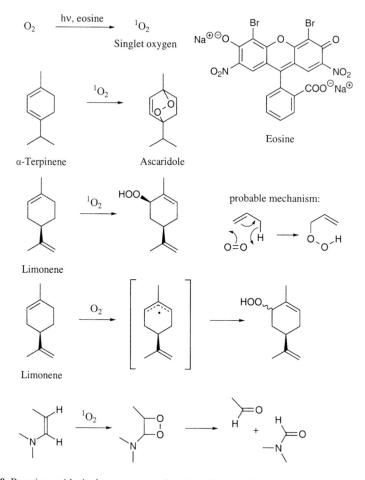

Figure 9.20. Reactions with singlet oxygen, produced by photosensitization.

by a number of photosensitizers, e.g. dyes like eosine, and in its excited form will react with for example dienes in Diels–Alder reactions or with an olefin to form hydroperoxides (see Figure 9.20).

Note that the allylic hydroperoxidation with singlet oxygen proceeds via a different mechanism compared to autoxidation with ground state oxygen, something that is evident when the stereochemical outcome of the reactions is analyzed, and a probable mechanism is shown in Figure 9.20. Also carbon–carbon double bonds may be attacked, if it has an electron-donating atom (e.g. nitrogen) attached, the product, a dioxetane, is unstable and generates two carbonyl compounds spontaneously. Singlet oxygen can be formed in the body if photosensitizers are present, and we will get back to this when we discuss the effects of the psoralenes (Section 10.2.6).

10 Toxic effects of chemicals

Chemicals exert systematic toxic effects essentially by disturbing the normal course of bio-chemical reactions, and in principle it should be possible to describe the mechanisms on a molecular level for any systemic toxicity. However, we still do not have a complete understanding for how most biochemical processes function, and it is actually only in a few cases that we fully understand what is happening. It should not be forgotten that a toxic effect observed with a chemical often is the result of several toxicities, specific and non-specific, and that it may be difficult to determine their relationships. In this chapter we will discuss the molecular mechanisms underlying some common toxic effects of chemicals, and the effects have been divided into "general toxic effects" and "organ specific toxic effects". However, it is obvious that such a division is not straight-forward. Although general toxic effects will influence all tissues in the body, one organ is always more sensitive than another (for various reasons) and will show symptoms of poisoning first. On the other hand, what we see as organ specific effects are often accompanied by effects on other organs, and general toxic effects will in some cases reappear in the organ specific. Remember also that the general intention is to link chemical properties and functionalities with toxic effects, not to give a comprehensive toxicological description of toxic effects.

10.1 General toxic effects

10.1.1 Non-specific effects of lipophilic and amphipathic compounds

The non-specific effects that relatively large amounts of lipophilic compounds have on cell membranes and the toxic effects this may cause have already been discussed briefly, and we will return to this in subsection 10.2.2. Also amphipathic compounds will interact with membranes and may disturb any biochemical activity that takes place in the membrane. Aliphatic alcohols, with one polar and one nonpolar end, will in high concentrations lyse cells because of this effect and ethanol especially is commonly used as an antiseptic, for example to kill bacteria on the skin. Phenol, which when it was introduced as an antiseptic agent in hospitals dramatically decreased the spread of infections, also kills bacteria because it is amphipathic. It is not known exactly how non-specific toxicants in very high doses cause death, because no pathologic marks are produced by this type of toxicity, but is probably due to the depression of either the CNS or the heart muscle.

10.1.2 Lipid peroxidation

This effect is caused by the oxidation, initiated by radicals and propagated by molecular oxygen, of membrane components containing unsaturated fatty acids (see Figure 10.1). The radicals are, for example, formed during the metabolism of exogenous compounds, and this was discussed in the last chapter. The initial relatively stable products formed between unsaturated fatty acid radicals and molecular oxygen are hydroperoxides, which explains the name "lipid peroxidation". The identical reactions will take place in other circumstances as well, and are for example responsible for butter and other foods going rancid when stored improperly. They are in general stimulated by temperature, oxygen and radical initiators (e.g. radicals or light). The bad smell of rancid butter is caused by the formation of shorter saturated fatty acids, e.g. hexanoic acid, as the carbon–carbon double bonds are cleaved by the oxidative processes. Another example (also mentioned in Chapter 3) of spontaneous oxidation of unsaturated fatty acids is the use of paints based on linseed oil (containing large amounts of linolenic acid) which will "dry" because they are oxidized, and a piece of cloth containing linseed oil may even autoignite because of the reaction heat developed when it oxidizes. The fact that radicals are so short-lived and difficult to study has delayed the recognition of the importance of lipid peroxidation, and it is only during the last decades that it has received proper attention. A radical may attack an unsaturated fatty acid either by abstracting a hydrogen atom adjacent to a double bond, or by adding directly to the double bond (see Figure 10.1). Remember that the reaction between a radical and a non-radical (or a diradical like molecular oxygen) always yields a new radical, and as a result radical reactions are chain reactions.

The initially formed radical may react in a number of ways, but it is not unlikely that it will react with molecular oxygen which always is available and always willing to participate in a radical reaction. That would result in peroxy radicals, which readily abstract hydrogens from, for example, other unsaturated fatty acids to become hydroperoxides. These may, as hydrogen peroxide (H_2O_2), be reduced enzymatically by glutathione peroxidase, but are also oxidation agents prone to participate in reactions that generate new radicals. The corresponding hydroxy radical may for example be formed by reduction with Fe(II), which is oxidized to Fe(III). So, in addition to being a continuous radical process constantly generating one radical for each consumed (until the chain is terminated by for instance the combination of two radicals to a product with paired electrons), lipid peroxidation also generates hydroperoxides that may produce additional radicals. The final products formed from the peroxidized unsaturated fatty acids is a complex mixture (which to some extent depends on the unsaturated fatty acid itself) of hydrocarbons, alcohols, aldehydes and carboxylic acids, formed by chain cleavage and further oxidation (see Figure 10.2). Several of the products are toxic themselves, for example the α,β-unsaturated carbonyl compounds which are electrophilic (see Figure 10.2). Especially malonaldehyde has received attention as it is known to be genotoxic, probably because it, as the enol tautomer, is a good electrophile and may react with the DNA bases (see Figure 10.2). 4-Hydroxy-2-nonenal (R=pentyl in Figure 10.2) is another product that is relatively reactive, due to the stabilization of the product as a γ-lactol (precluding the reversal of the Michael addition sometimes observed with β-substituted carbonyl compounds, see Figure 7.35 in Section 7.4.4 for an example). It can also react with primary amines to form pyrroles (see Figure 10.2).

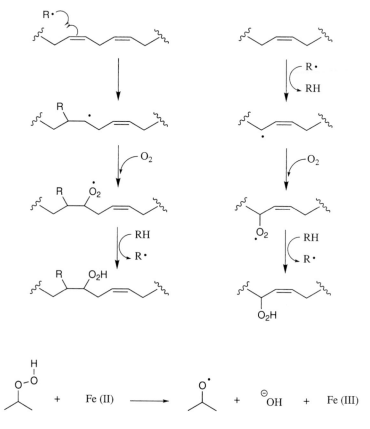

Figure 10.1. Radical induced peroxidation of unsaturated fatty acids.

Radicals that can initiate lipid peroxidation are continuously formed, for example by the additional loop of cytochrome P-450 oxidations discussed in Chapter 7. However, such formation of superoxide and hydrogen peroxide is normally kept very low by the protective enzymes, and very few radicals escape. If a person is exposed to exogenous compounds that are metabolized by cytochrome P-450, more superoxide is produced and eventually the protective systems may be overcharged resulting in increasing concentrations of reactive oxygen species. In addition, as we have seen examples of in Chapter 8, several types of exogenous compounds generate radicals directly when they are metabolized by cytochrome P-450 and other oxidative enzymes. Another source for radicals are ionizing radiation, which we constantly are subjected to from space. Ionizing radiation injures cells mainly because it will transform water molecules to the extremely reactive hydroxyl radical by knocking off an electron, and the hydroxyl radical will cause the damages observed (see Figure 10.3). (To a minor part ionizing radiation will also hit important proteins and DNA and thus directly kill a cell.)

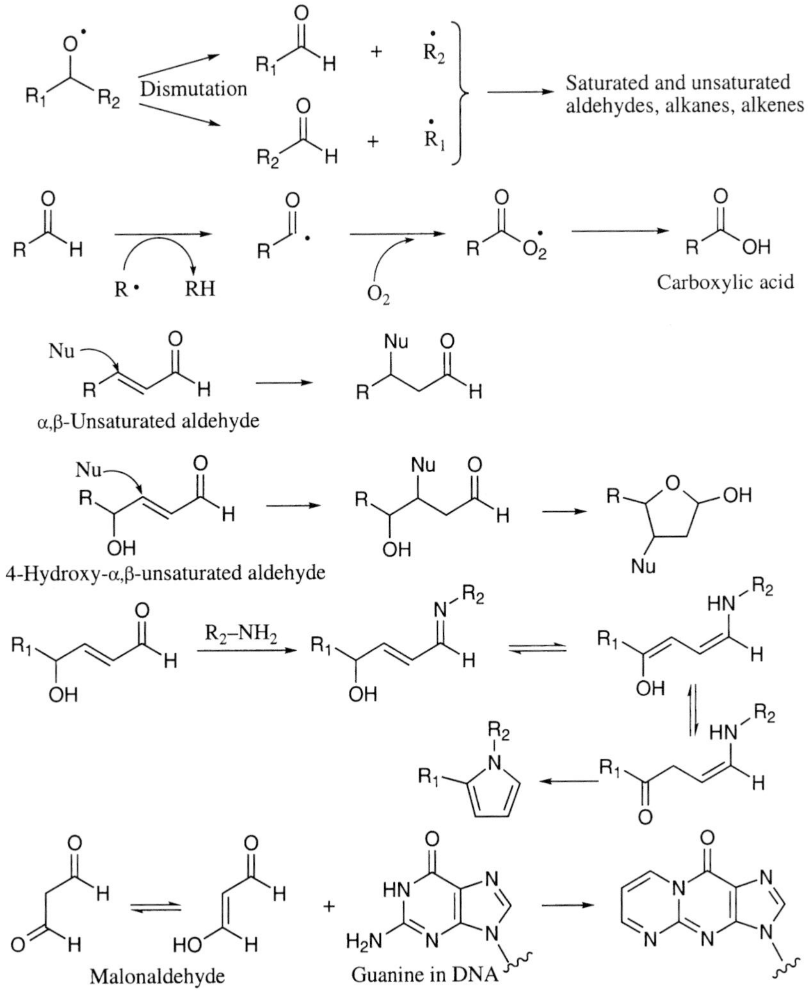

Figure 10.2. The formation of electrophilic compounds from lipid peroxidation.

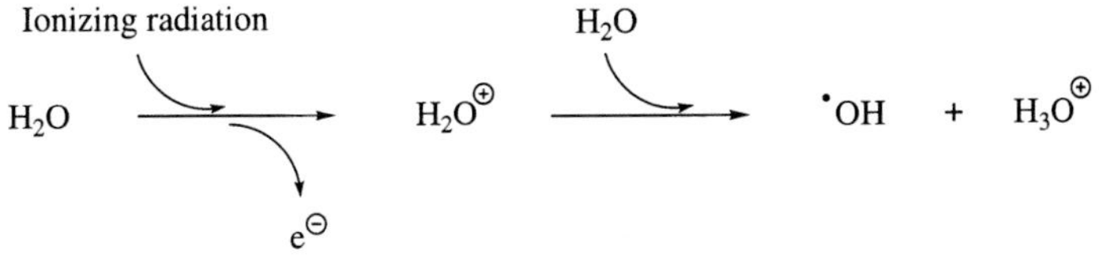

Figure 10.3. The generation of hydroxyl radicals by ionizing radiation.

Besides the enzymes superoxide dismutase, catalase, and glutathione peroxidase that will take care of superoxide and hydrogen peroxide, we are also protected against radicals by endogenous radical scavengers. Examples are α-tocopherol (vitamin E), a lipophilic but also amphipathic molecule situated in the membranes with the hydroxyl group directed towards the water solution, and the hydrophilic ascorbic acid (vitamin C). Both are essential chemicals with several important functions, and it appears as if they collaborate in their efforts of eliminating radicals in the way that α-tocopherol (and other similar tocopherols) will scavenge radicals formed in the membranes while ascorbic acid will regenerate oxidized α-tocopherol. Ascorbic acid will in its oxidized form (semidehydroascorbic acid, see Figure 10.4) either be reduced back to ascorbic acid by a reductase utilizing GSH as coenzyme, or be dismutated to ascorbic acid and dehydroascorbic acid. Glutathione obviously plays an important role in the protection against lipid peroxidation, and glutathione depletion (by for example electrophilic compounds that are conjugated with it) will promote lipid peroxidation.

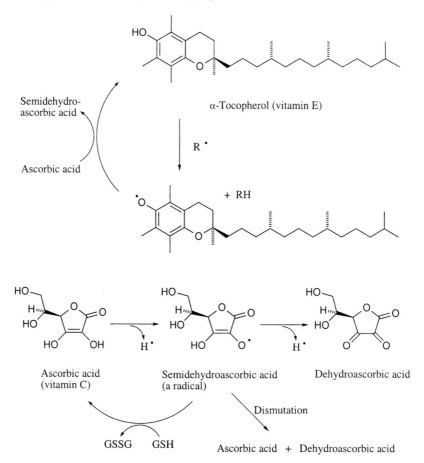

Figure 10.4. The scavenging of radicals by vitamins C and E.

Ascorbic acid is a relatively strong reductive agent and will take care of any radical including superoxide directly, as well as of singlet oxygen. However, it has been shown that it may also reduce Fe(III) to Fe(II), and as the latter is involved in the generation of hydroxyl radicals from hydrogen peroxide it is not unanimously considered that high doses of ascorbic acid are completely harmless.

In conclusion, lipid peroxidation can produce toxic effects on a cellular level in three principal ways:

1. The damage to the membranes caused by the peroxidation of its unsaturated fatty acids will, in general, be difficult for any cell. Cell membranes, internal as well as external, are designed to perform a number of important tasks that may be hampered if the structure of the membrane is changed.
2. The radicals formed during lipid peroxidation may not only react with the unsaturated fatty acids but also with the components of membrane bound enzyme systems. This will affect for example the protein synthesis, and we will later in this chapter see how this may result in a toxic effect on the liver.
3. Some of the products formed from the peroxidation of unsaturated fatty acids are non-radical but still reactive as electrophiles, and may diffuse from the membrane where they are formed to for example the nucleus of the cell and react with the DNA causing genotoxic effects.

10.1.3 Acidosis

Acidosis is a condition that is caused by a too low pH of the body fluids, while alkalosis is the opposite. The pH of any cell in the body is extremely important, no cell or organism will survive an internal pH that is significantly different from the normal value. Even a small change of the pH would have a considerable impact on the protonation of for example proteins, which in turn determines their three-dimensional form and thereby their function. In the blood of a human being, the pH is strictly regulated between pH 7.35 and 7.40 (the blood going to the lungs is more acidic due to the presence of higher concentrations of carbonic acid formed from carbon dioxide and water), and anything above or below these limits will be attended to by regulatory systems. However, the pH in the urine is more variable, and normally below 7. In the kidneys, the equilibrium between water + carbon dioxide and carbonic acid is regulated by the enzyme carbonanhydrase (see Figure 10.5), which performs an important task in resorbing sodium ions from the primary urine.

$$2\ H_2O\ +\ CO_2\ \xrightarrow{\text{Carboanhydrase}}\ H_2CO_3\ +\ H_2O\ \rightleftharpoons\ HCO_3^{\ominus}\ +\ H_3O^{\oplus}$$

Figure 10.5. Carboanhydrase catalyze the addition of water to carbon dioxide to form carbonic acid.

Sodium ions are exchanged for oxonium ions formed by the action of carboanhydrase, and the oxonium ions thus excreted give the urine an acidic pH (which may catalyze the transformations discussed in Chapter 7). The pH of the blood is regulated in several ways, for example by the buffering capacity of the blood which contains HCO_3^- ions that can neutralize acid by forming carbonic acid. The carbonic acid is excreted via the lungs, as carbon dioxide, and laboured breathing will speed up the excretion of carbon dioxide, thereby eliminating acid in the blood at the expense of HCO_3^- ions. Acid may of course also be eliminated by direct excretion with the urine, as oxonium ions are water soluble. The pH of the blood is surveyed by a biological pH meter in the brain, which will respond to a decreasing pH value by increasing the breathing frequency. Note that the different mechanisms to eliminate acid work on different time scales. While the buffering of HCO_3^- ions (and also other inorganic ions as well as proteins) acts immediately, the excretion of carbon dioxide by the lungs or oxonium ions by the kidneys will take minutes to hours. However, the regulation of pH in the body has a limited capacity, and can only neutralize relative small amounts of acid. Acidosis will result if the capacity is overloaded, and a pH of the blood below 7.0 (or above 8.0 during alkalosis) is immediately life-threatening.

Acids may in theory gain access to the body by being absorbed by the routes discussed in Chapter 6, but it is evident that acids (and bases) that are sufficiently strong to give a serious effect on the pH of the blood are ionized and thereby not readily absorbed. Instead, they give rise to local effects. Weaker acids (e.g. salicylic acid) may be absorbed, but can only cause acidosis if the amounts taken are very large (multigram doses). However, stronger organic acids may be absorbed in a masked form, that itself is not acidic but can be converted/transformed to acid inside the body (see Figure 10.6). This is in principle the case for esters of for example formic acid (pKa 3.75). A carboxylic acid functionality masked as an ester is lipophilic and esterases are relatively efficient in hydrolyzing esters to acids. However, the esters of formic acid, e.g. methyl formiate, are actually hydrolyzed so easily that they will be highly irritating locally if inhaled (the vapours), swallowed or spilled on the skin. Carboxylic halides, e.g. acetylchloride, and anhydrides, e.g. acetic anhydride, are more reactive and more easily hydrolyzed than the esters, and will give severe local effects due to their spontaneous hydrolysis upon contact with any tissue containing water. Both acetylchloride and acetic anhydride are

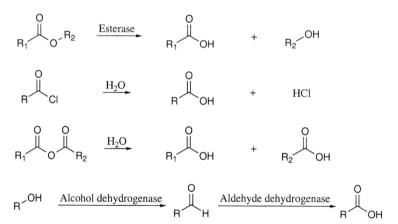

Figure 10.6. Ways that acidic carboxylic acids can be formed inside the body.

corrosive, especially in concentrated form, and besides generating acetic acid (and hydrochloric acid with acetylchloride!) the hydrolysis reaction generates heat that will aggravate the effect of the acid. As the eye is often the most sensitive part of the body that may be subjected to splashes, any handling of halides or anhydrides of carboxylic acids must be carried out with eye protection.

Another possibility for the generation of carboxylic acids inside the body is of course the oxidation of primary alcohols via the corresponding aldehyde, and this is the cause of most cases of acidosis caused by an exogenous chemicals. Methanol is the simplest of the alcohols, it is used in enormous amounts mainly as an organic solvent and a lot of people are exposed to methanol at their workplaces every day. However, when methanol causes acidosis, it is generally because it is mistaken for ethanol and consumed with the intention to get drunk. It is fairly toxic to humans, an adult may die after drinking approximately 100 ml, and at workplaces one should be aware of its high volatility and excellent ability to be absorbed through the skin. Methanol is only slowly converted in the body, and approximately 30 % of a dose is still present unchanged after 48 hours. Due to its volatility and hydrophilicity more than half of an absorbed dose will actually be excreted unchanged by the lungs and the kidneys, while the rest is oxidized to formaldehyde and formic acid. The acute effects of methanol are irritation of the mucus tissues of the stomach, an unpleasant intoxication due to the non-specific effect on the CNS, reversible and irreversible effects on the optic nerve, and acidosis. After a typical accident when a number of persons have been consuming methanol in the belief that it is ethanol, one third get away with a bad drunken fit, one third lose their eyesight temporarily or permanently, while one third dies as a result of acidosis. The effects on the optic nerve, which are unique to higher mammals as humans and monkeys, is believed to be caused by formaldehyde, possibly because the high concentrations of alcohol dehydrogenase which has an important role to play in the biochemistry of vision. In rats, methanol is oxidized mainly by the enzyme catalase and methanol will not affect the optical nerve of rats. The acidosis is caused by formic acid, the principal end-product of the oxidation of methanol in man. A lethal dose of formic acid if injected di-

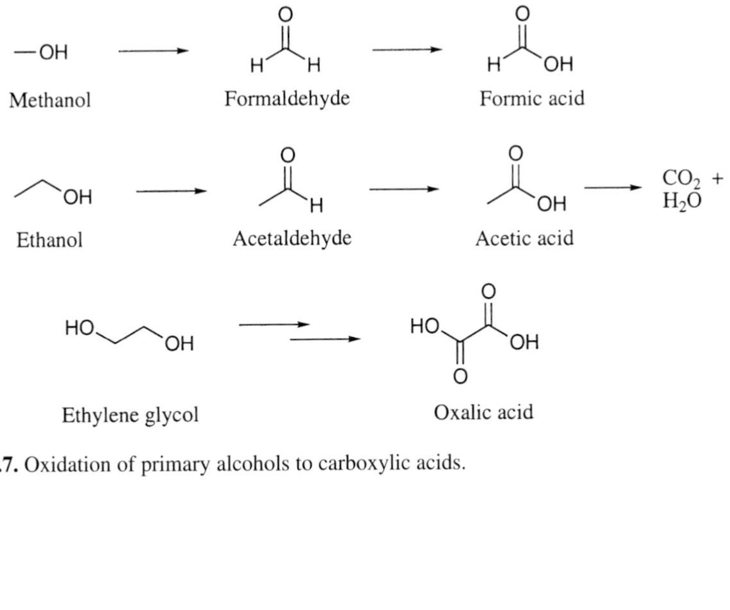

Figure 10.7. Oxidation of primary alcohols to carboxylic acids.

rectly into the blood would be approximately 3 g for an adult, corresponding to 2 g of methanol, reflecting how inefficient the conversion of methanol to formic acids actually is in the body. In other species, e.g. rats and rabbits, formic acid can be converted to carbon dioxide and water, and methanol is consequently considerably less toxic in such species.

Ethanol is oxidized by exactly the same enzymes in man, via acetaldehyde to acetic acid. Although acetic acid is a slightly weaker acid (pKa 4.75) compared to formic acid, it is still in principle sufficiently acidic to cause acidosis. However, acetic acid is the unit used by the cells for the generation of energy (*vide infra*) and our cells will have no problems to oxidize it to carbon dioxide and water. Acetaldehyde is certainly a toxic intermediate, but during normal conditions the acetaldehyde formed is immediately converted to acetic acid (several times faster than ethanol is oxidized to acetaldehyde) and is only present in low concentrations. However, individuals that for example suffer from an aldehyde dehydrogenase deficiency or are undergoing treatment with antabuse (disulfirame) or a similar drug will not be able to convert acetaldehyde, at least not as efficiently, and will suffer badly from the effects of acetaldehyde if exposed to ethanol. The conversion rates of methanol and ethanol are quite different. Ethanol, being more lipophilic and less volatile, is oxidized more rapidly (approximately 10 g per hour in an adult), and only a fraction (approximately 1 %) of the absorbed amounts are excreted unchanged (by the lungs). If both alcohols are available the alcohol dehydrogenase prefers to oxidize ethanol and leaves methanol unchanged, and this is taken advantage of when methanol poisonings are being treated. Small amounts of ethanol are continuously administered, while the excretion of methanol by the kidney is facilitated by diuretics, and/or by dialysis. In addition, the acidosis is treated by replenishing the body's supply of HCO_3^-. Besides the drug antabuse, which also is used as a rubber accelerator and vulcanizer in the rubber industry, coprine, a fungal metabolite isolated from the fruit bodies of *Coprinus atramentarius*, also inhibits aldehyde dehydrogenase. Fructose has been shown to increase the conversion of ethanol, while 4-methylpyrazole decreases the conversion by inhibiting alcohol dehydrogenase (see Figure 10.8 for structures). The latter may also be used to prevent the generation of formaldehyde during methanol intoxication.

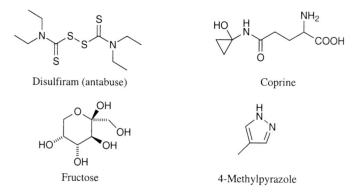

Disulfiram (antabuse)   Coprine

Fructose   4-Methylpyrazole

Figure 10.8. Compounds that affect the metabolism of alcohols to carboxylic acids.

Another commonly used alcohol that may give acidosis is ethylene glycol, perhaps most familiar as antifreeze in the cooler of cars. Ethylene glycol is approximately as toxic to man as methanol, and a large number of poisonings occur yearly. As it is often present in private garages it is available for children, who mistake antifreeze for syrup because of the colour (often red) and the sweet taste (glycol comes from glykys which is Greek for sweet). It is oxidized in several steps to the end-product oxalic acid (see Figure 10.7), which besides being a strong organic acid (pKa 1.27) is toxic to the kidney (*vide infra*). Oxalic acid is known from the kingdom of plants, it is for example present in rhubarb which consequently should not be consumed in too large quantities. Lately, propylene glycol has been launched as a substitute for ethylene glycol, and an exchange is highly recommended as propylene glycol is as efficient as an antifreeze and considerably less toxic.

10.1.4 Oxygen deficiency due to effects on the blood

A human will not survive more than a few minutes without oxygen. The energy necessary to make most biochemical reactions proceed smoothly at body temperature is produced in the mitochondria by oxidising nutrient to carbon dioxide and water, while the energy originally contained in the chemical bonds is transformed into chemical energy of ATP (adenosine triphosphate) and GTP (guanosine triphosphate). No oxygen means that energy production will stop, and when the stores of ATP are emptied the life processes will stop. Nutrients of different kinds, sugars, fats, proteins, etc., are converted to acetic acid units in the cytoplasm of the cells, and these are brought into the mitochondria as conjugates with a coenzyme called coenzyme A (CoA). In the mitochondria acetyl CoA enters the citric acid cycle, a cyclic process that combines the acetyl group with oxaloacetic acid to form citric acid, which is converted via isocitric, α-ketoglutaric, succinic acid, fumaric acid and malic acid back to oxaloacetic acid (see Figure 10.9).

Two carbons, accounting for the two added with the acetyl group, are split off as carbon dioxide during the conversions. The energy of the carbon–carbon and carbon–hydrogen bonds of the acetyl group is transferred to one molecule of GTP and several molecules of the coenzymes NADH and FADH$_2$ in their reduced forms. In the reduced coenzymes the energy is stored as a high potential for reduction, or, in other words, in electrons with high energy. This will be converted to a more useful form of energy during the electron transport in a process called the oxidative phosphorylation. In this, NADH and FADH$_2$ are oxidized while molecular oxygen is reduced, and this is coupled with the phosphorylation of ADP (adenosine diphosphate) to produce ATP (adenosine triphosphate). In the electron transport part of the oxidative phosphorylation, also called the respiratory chain, the energy stored in the electrons of the reduced coenzymes is gradually being extracted by in all four enzyme complexes via a series of coenzymes and cytochromes (see Figure 10.10) and used for the phosphorylation of ADT. In the end the electrons are combined with molecular oxygen which eventually is reduced to water, although this is not understood in detail (it takes four electrons to reduce molecular oxygen to two molecules of water, but the cytochromes only deliver one electron at the time). In ATP the energy is stored as high phosphorylation power, which can be used to activate other compounds and thereby facilitating chemical reactions by lowering the activation energy.

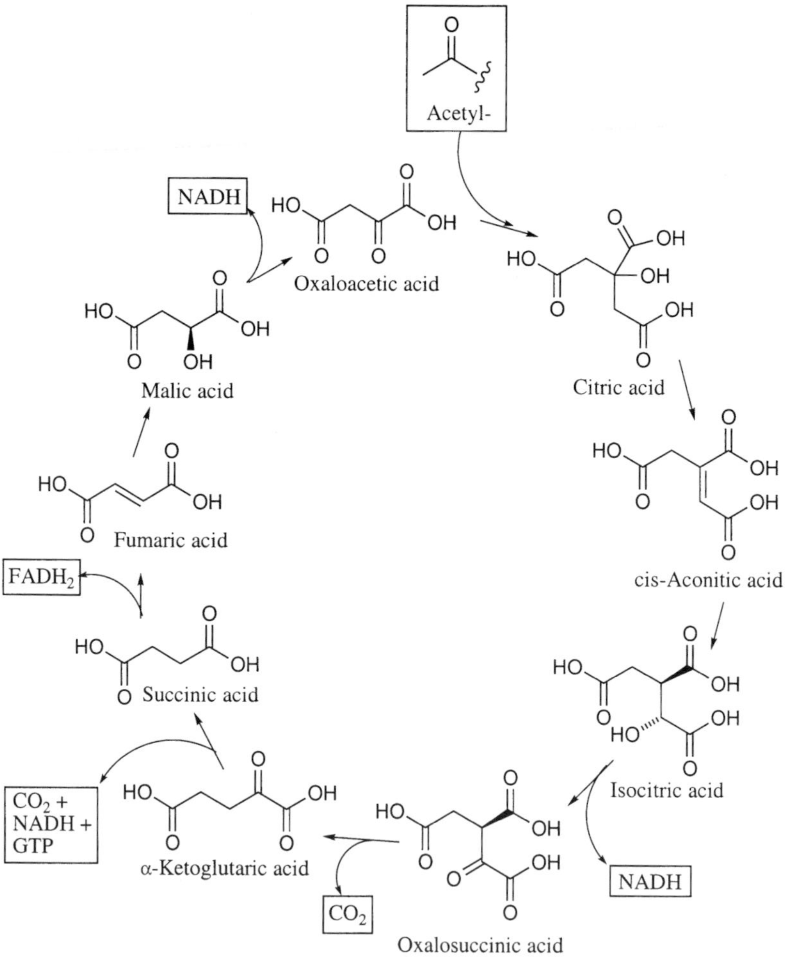

Figure 10.9. The citric acid cycle.

10.1.4.1 Carbon monoxide

The ability of the blood to transport oxygen is remarkably refined, as the haemoglobin of the red blood cells binds oxygen strongly in an environment where the oxygen concentration is high, in the lungs, and let go of it when the concentration is low, in other tissues. The red blood cells are formed in the bone marrow together with other blood cells, and exogenous compounds that affect the bone marrow may result in too few or malfunctioning red blood cells. Examples of chemicals that are toxic to the bone marrow are the solvent benzene, the insecticide lindane,

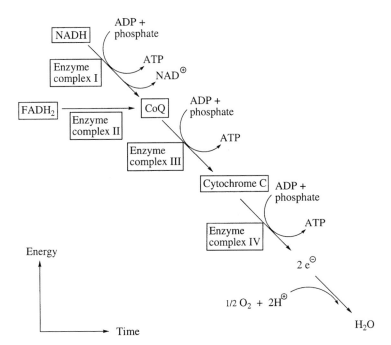

Figure 10.10. Oxidative phosphorylation.

arsenic and trinitrotoluene. A classic and very simple chemical that blocks the ability of haemo-globin to transport oxygen is carbon monoxide, CO, formed for example when organic material is incompletely burned or combusted with insufficient amounts of oxygen to oxidize the carbon to carbon dioxide. Carbon monoxide is chemically very similar to molecular oxygen and will also be bound to haemoglobin, although much (approximately 200 times) stronger, and haemo-globin having carbon monoxide bound to it will not be able to transport molecular oxygen (see Figure 10.11). If 50 % of the blood is blocked by carbon monoxide the life of a human is in danger, corresponding to approximately 0.1 % carbon monoxide in the respiration air, and higher concentrations are lethal. The toxicity of carbon monoxide is caused by the lack of oxygen that it will cause if sufficient large proportion of the blood's oxygen transporting capacity is blocked, and it will be excreted by the lungs if fresh air or oxygen gas is provided (an example of receptor

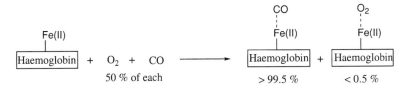

Figure 10.11. Carbon monoxide competes with molecular oxygen for haemoglobin.

antagonism as the two compounds compete for the same binding site). A common source of carbon monoxide is combustion engines. The exhausts of older cars contain several percent carbon monoxide, and to be trapped inside a closed room with an idling car was previously not an unusual lethal accident. However, the exhausts of modern cars equipped with a catalyst normally contain 0.0–0.1 % carbon monoxide. Malfunctioning heating systems based on the burning of fuels are often responsible for fatal accidents, and it regularly happens that campers bring the hot barbecue into the tent on a chilly evening and are found dead by carbon monoxide poisoning in the morning. We have also seen that carbon monoxide may be formed from the metabolism of exogenous chemicals, e.g. methylene chloride, and the poorer physical condition that smokers usually have is to some extent explained by their constant exposure to carbon monoxide. Actually, approximately 1 % of our blood is constantly blocked by carbon monoxide as small amounts of the compound are formed in the normal metabolic processes. The molecular target for carbon monoxide is the $Fe(II)$ in the haem group in the haemoglobin protein (other proteins containing a haem group with iron in oxidation state 2, e.g. cytochromes involved in the generation of ATP, are also affected by carbon monoxide, *vide supra*).

10.1.4.2 Methaemoglobinemia

Besides binding carbon monoxide strongly, the $Fe(II)$ of the haem group of haemoglobin is also, as $Fe(II)$ always is, easily oxidized to $Fe(III)$. This is not surprising, everyone is aware of that anything made of iron sooner or later will corrode, because $Fe(III)$ is the most stable form of iron in the earth's atmosphere. If the iron of haemoglobin is oxidized it is transformed to methaemoglobin, which is unable to transport molecular oxygen, and the condition is called methaemoglobinemia. In fact, the oxidation is so facile that it to some extent takes place spontaneously in our blood cells, and we all have a small portion of our haemoglobin as useless methaemoglobin. Normally this autoxidation is balanced and kept in control by an enzyme, NADH-methaemoglobin reductase, present in the blood cells that will reduce methaemoglobin back to functional haemoglobin again. Certain exogenous chemicals will be able to accelerate the oxidation of haemoglobin, and exert a toxic effect by doing so. An example is the nitrite ion (NO_2^-), used as a preservative in food for example. Nitrite is also formed when the nitrate ion (NO_3^-) is reduced by bacteria present in the mouth and the intestines, and drinking water containing high concentrations of nitrate (not unusual in drilled wells) may therefore be toxic especially to infants. (Infants are more sensitive because the bacterial flora in their intestines is different.) Nitrate is also present in for example spinach, and a soup or a stew prepared with spinach should not be left too long at room temperature because this will give bacteria a good chance to reduce the nitrate to nitrite. However, it should be noted that lethal doses of nitrite (as for example $NaNO_2$ or KNO_2) for a healthy adult are in the order of several grams. The molecular mechanism by which nitrite oxidizes haemoglobin is not understood, but aromatic amines and aromatic nitro compounds will catalyze the reaction according to Figure 10.12.

The corresponding hydroxyl amine, formed by oxidation of the amine and reduction of the nitro compound, is responsible for the reduction of molecular oxygen bound to haemoglobin, to hydrogen peroxide. Hydrogen peroxide formed in the vicinity of something as easily oxidized

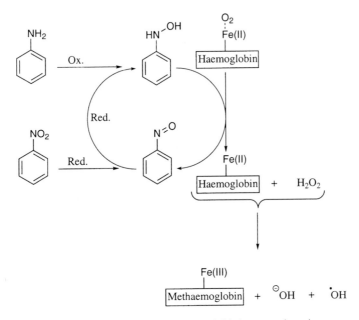

Figure 10.12. The oxidation of haemoglobin to methaemoglobin by aromatic amines.

as Fe(II) will react immediately, and oxidize Fe(II) to Fe(III). As we have noted earlier, the oxidation of Fe(II) by hydroperoxides will generate the hydroxyl radical, which immediately will start radical reactions in the membranes of the red blood cells. This leads to characteristic damage, called "Heinz' bodies", of the blood cells that can be observed in a microscope. During the process when a hydroxyl amine transforms a haemoglobin to methaemoglobin, it will be oxidized to the corresponding nitroso derivative itself. If this is reduced back to the hydroxylamine by the metabolic processes discussed in Chapter 7, it can function as a catalyst for the oxidation of haemoglobin.

It should be noted that some individuals are much more sensitive to conditions that transform haemoglobin to methaemoglobin, because they have a hereditary deficiency in the NADH-methaemoglobin reductase that normally will reduce methaemoglobin back to haemoglobin.

10.1.5 ATP deficiency due to inhibition of the primary metabolism

The biochemical processes continuously need energy in the form of ATP. ATP is an efficient phosphorylating agent, and phosphorylation is used to activate chemical functionalities so that they participate in chemical reactions more readily. Shortage of ATP will hamper all cellular processes and eventually injure all tissues, but some cell types (i.e. nerve cells) are extremely dependent on ATP and will only survive for minutes without it. Other cells, for example muscle cells, can operate under anaerobic conditions for some time, and are considerably less sensitive.

10.1.5.1 Inhibition of the citric acid cycle

An example of a chemical that efficiently inhibits the citric acid cycle is fluoroacetic acid, which resembles acetic acid so much that it can be converted to its CoA-thioester and as such accepted by the enzyme that combines acetic acid with oxaloacetate. The fluorocitric acid formed is, however, not a suitable substrate for the next enzyme in the cycle. Instead this is blocked and the whole citric acid cycle comes to a halt. The LD_{50} value for fluoracetic acid is 5 mg/kg (rat, or), comparable to that of the classic poison potassium cyanide (*vide infra*). It was originally isolated from some African plants (genus *Dichapetalum*) because it was noted that animals that grazed in areas where those plants grew often died, and has since been used as a rat poison. Interestingly, some plants have the ability to absorb fluoride ions from the ground even in fluoride-poor soils and incorporate them into its metabolites, especially in fluoroacetic acid but also in fatty acids (e.g. 16-fluoropalmitic acid, which after β-oxidation eventually will yield fluoroacetic acid and is equally toxic), and inorganic waste containing fluorides should therefore be taken care of carefully. Chloroacetic acid has the same effect although it is less potent

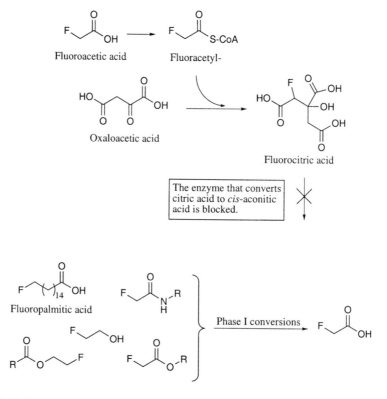

Figure 10.13. Fluoracetic acid will block the citric acid cycle.

with a LD$_{50}$ value of 75 mg/kg (rat, or). The reason for the difference between the two compounds is that the chlorine atom is bigger than fluorine and that the chlorocitric acid formed is a poorer enzyme inhibitor. Observe that chemicals that will be converted to flouroacetic acid and chloroacetic acid (e.g. 2-fluoro- and 2-chloroethanol, esters and amides, see Figure 10.13 for a few examples) by the metabolism also will affect energy production.

10.1.5.2 Inhibition of the respiratory chain

Classic poisons like hydrogen cyanide (HCN) and hydrogen sulphide (H$_2$S) bind to the Fe(III) in the haem groups of cytochromes in the respiratory chain (*vide supra*), stop the flow of electrons and thereby stop the generation of ATP. Cyanides are used for many applications, for example in the metal industry, and as the sodium or potassium salts (NaCN or KCN) cyanides are fairly easy to handle. However, should a solution of cyanides be acidified or come in contact with an acidic medium, hydrogen cyanide will immediately be formed (see Figure 10.14). Hydrogen cyanide has a boiling point of 25 °C and can be considered to be gaseous at room temperature, and will rapidly be absorbed in the lungs. This is the compound used to execute people in the gas chambers and, due to its rapid absorption and distribution to the CNS, death caused by depression of the respiration regulation will occur within minutes. If a solution (or the crystals) of cyanide salts is swallowed the transformation of the water soluble cyanides to hydrogen cyanide will take place in the acidic environment in the stomach, and hydrogen cyanide will thereafter be absorbed in the intestines. However, this will take more time and require larger amounts. To drink a glass of champagne containing potassium cyanide does not result in the immediate and dramatic death shown on TV in the mystery stories, instead it is actually possible to treat such intoxications successfully.

As indicated in Figure 10.14, the cyanide salts are not completely stable and may react slowly for example with water. If stored improperly, the crystals will absorb water and carbon dioxide from the surrounding air and be degraded to for example formic acid and ammonia. This is actually a common reason for failed suicide attempts, which when conducted with aged sodium or potassium cyanide may produce, instead of death, local effects in the mouth and throat due to the corrosive ingredients. The famous Russian religious mystic Rasputin, who strongly influenced the Tsar's family in the beginning of the 20th century, was blamed for military set-backs during the first world war and murdered by a group of Tsar court officials. Initially he was offered tea with biscuits containing large amounts of potassium cyanide, but

$$\text{NaCN} + \text{H}_3\text{O}^{\oplus} \longrightarrow \text{Na}^{\oplus} + \text{H}_2\text{O} + \text{HCN}$$

Sodium cyanide $\qquad\qquad\qquad\qquad$ Hydrogen cyanide

$$\text{KCN} + \text{H}_2\text{O} \longrightarrow \text{HCN} + \text{KOH} + \text{HCO}_2\text{H} + \text{NH}_3 + \text{K}_2\text{CO}_3$$

Potassium cyanide

Figure 10.14. Reactions of cyanide in acid and in water.

surprisingly he was not affected at all. Eventually he was first shot and then drowned in the river Neva, proving that he was not immortal, but it was still considered to be a mystery that he could withstand such large amounts of potassium cyanide. One suggestion has been that he was unable to produce sufficient amounts of hydrochloric acid in the stomach to produce lethal amounts of hydrogen cyanide from the potassium cyanide he ate. However, the pKa of hydrogen cyanide is 9.1 which means that cyanide will be protonated even at physiological pH. Instead it is likely that the potassium cyanide used to assassinate Rasputin was too old.

A completely different source for hydrogen cyanide is the glycoside amygdalin, present in bitter almonds and the stones of peach, apricot, plum and cherry. Amygdalin is most toxic if ingested, because the bacteria of the intestines possess the enzymes (glucosidases) that may hydrolyze amygdalin to its corresponding α-hydroxynitrile (see Figure 10.15). As already mentioned, α-hydroxynitriles will spontaneously liberate hydrogen cyanide, and the amounts of amygdalin present in a couple of bitter almonds will generate enough hydrogen cyanide to kill a child. Another similar glucoside is linamarin, present in several edible beans and roots consumed in large amounts (e.g. manioc or cassava). The consumers have to prepare these products in the correct way, including soaking to permit the enzymatic hydrolysis of the glucosidic bond followed by boiling to make the hydrogen cyanide evaporate, in order to avoid lethal intoxication. In Chapter 7 we have seen that nitriles may be hydroxylated by cytochrome P-450 to the α-hydroxynitrile, and, in addition, hydrogen cyanide is also formed when urethane plastics are burned.

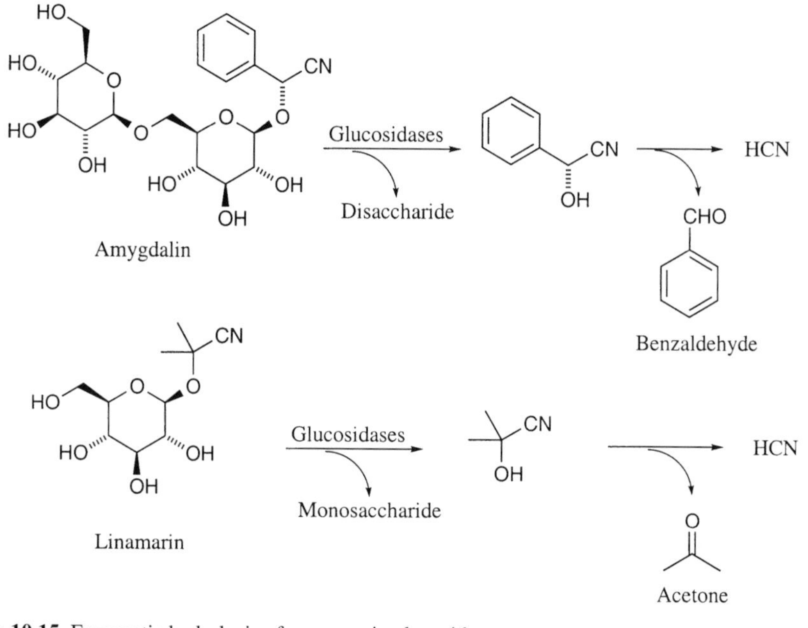

Figure 10.15. Enzymatic hydrolysis of cyanogenic glucosides.

One way of treating hydrogen cyanide poisoning is to administer an appropriate amount of sodium nitrite ($NaNO_2$, approximately 300 mg for an adult), which will cause a moderate methaemoglobinemia (see section 10.1.4.2) that by itself is not dangerous. However, even a moderate methaemoglobinemia will produce large amounts of oxidized haem groups [with Fe(III)] in the blood. This will extract hydrogen cyanide from the mitochondria of the cells to the methaemoglobin in the blood and thereby (if the amounts of hydrogen cyanide are not too large) give the body a chance to cope with the acute situation. Our bodies have a fairly good capacity for degrading HCN, with the enzyme rhodanese that converts it to a thiocyanate ion (SCN^-) which can be excreted by the kidneys, and this metabolic detoxification is facilitated by the administration of sodium thiosulphate ($Na_2S_2O_3$). Interestingly, although HCN is one of the most well-known poisons it is also an endogenous compound produced in small amounts by the phagocytes of the immune system and present in low concentrations in our exhalation air. However, its physiological function, if any, is not yet understood.

Hydrogen sulphide (H_2S) produces the same toxic effect as hydrogen cyanide. It is actually more toxic but the fact that it smells so strongly (rotten eggs) will normally warn anybody that is exposed. However, the sensitivity for the odour of hydrogen sulphide decreases rapidly during exposure. Even if the concentrations increase the odour is fading away and one should therefore not rely on this warning signal. (Also hydrogen cyanide is said to have a characteristic smell (of bitter almonds), but approximately 30 % of the population are unable to sense this and will not be alarmed by it.) Hydrogen sulphide may be encountered in high concentrations in natural and volcanic gases. Low doses will be irritating to the eyes and the respiratory system, and may cause oedema in the lung. Again, sodium nitrite may be used as an antidote, because the hydrosulphide anion (SH^-) forms a complex called sulphmethaemoglobin with methaemoglobin. However, hydrogen sulphide will be oxidized to sulphite (SO_3^{2-}) and sulphate (SO_4^{2-}) in the body and will not accumulate.

10.1.5.3 Uncoupling of the oxidative phosphorylation

Chemicals that has this effect are called uncouplers, because their effect is not to bring the respiratory chain to a halt but instead to let the energy produce heat instead of being used for the production of ATP. This will result in a lack of ATP as well as in a rise in temperature, and as the lack of ATP results in a stimulation of the energy production in the cells more and more heat will be produced. (One can compare this with the clutch of a motor engine as it uncouples the engine and replaces the flow of useful energy to the wheels with the production of heat.) Lethal poisoning by uncouplers will cause death by fever. The exact molecular mechanism for the inhibition of phosphorylation is not understood, although it is known to be a specific effect on the oxidative phosphorylation as other phosphorylations are not affected. A number of phenolics having electron withdrawing groups in the benzene ring are efficient uncouplers, e.g. dinitrophenol and pentachlorophenol, and have been used as both slimming agents and pesticides. They work as slimming agents because the metabolism increases and more nutrients (fat) than necessary are oxidized to carbon dioxide and water. The side effect at therapeutic doses is a slight fever, but frequent overdosing resulted in a large number of accidental deaths.

10.2 Specific toxic effects to organs (or equivalents)

10.2.1 Effects on the blood

The blood is one of the most important targets for toxic chemicals, but as the major acute effect is inhibition of the ability of the blood to transport oxygen we shall confine the discussion to the effects already described in Section 10.1.4.

10.2.2 Effects on the nervous system

The nervous system, especially the CNS, is frequently the target organ for toxic, exogenous compounds. We have already noted that non-specific effects of lipophilic and amphipathic compounds in high concentrations will affect membranes in general and nerve cell membranes in particular, causing for example dizziness or unconsciousness. Although it is difficult to prove it is nevertheless reasonable to assume that such compounds will be partitioned to the membranes and that they will change the chemical properties of the membranes slightly. As the three-dimensional form of a membrane protein, e.g. a receptor for a neurotransmitter in the nerve cell membranes (*vide infra*), is very dependent on the constitution of the membrane, such a change could affect their function. This is believed to be the way ethanol affects us, and also anaesthetics like chloroform, diethyl ether and halothane. In addition, many compounds that are toxic to the nervous system interact specifically with the receptors and affect communication between nerve cells. Toxic chemicals that in one way or another inhibit the production of ATP (*vide supra*), in principle a general toxic effect, can also be said to have a specific effect on CNS because this is the organ first damaged.

The role of the nervous system is to relay signals, from one end of a nerve cell to the other (i.e. within a cell) as well as between cells (mainly between different nerve cells but in general between cells with excitable membranes). This is done by two principally different mechanisms, an electric and a chemical (see Figs. 10.16 and 10.17). A resting nerve cell will create a chemical concentration gradient between the inside and the outside of the cell, by pumping out sodium ions in exchange for potassium ions (the sodium/potassium-pump). As the potassium ions to some extent may diffuse out of the cell through special ion channels, a small electrical potential (approximately 80 mV) is created over the cell membrane. The initiation of a signal at the membrane of a nerve cell is accompanied by the opening of ion channels for both sodium ions and potassium ions, which quickly diffuse through the membrane in order to equalize the concentrations on the two sides. As the gradient for sodium ions is slightly steeper sodium diffuses more rapidly and this causes a local electrical depolarization of the membrane. This in turn results in an opening of further ion channels in the vicinity and the depolarization propagates over the entire cell membrane as depicted in Figure 10.16.

After the depolarization the ion channels close and the sodium/potassium-pump re-establishes the original membrane potential, and the whole process takes only a few milliseconds. The depolarization is normally initiated by a signal from another cell, and normally results in the discharge of a signal to other cells. The chemical signal between cells is conveyed by a

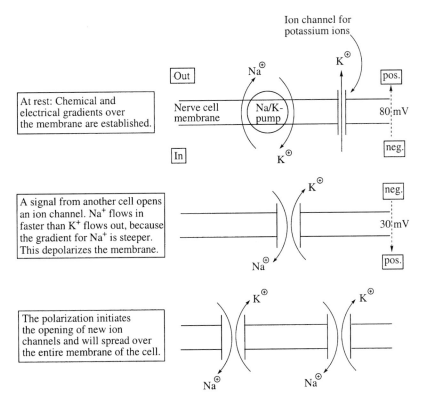

Figure 10.16. The depolarization of a nerve cell membrane.

special compound called a neurotransmitter, which is released by the signalling cell and binds to specific receptors in the cell membrane of the receiving cell (see Figure 10.17). The neuro-transmitter/receptor complex may then initiate a response in the receiving cell (for example the opening of ion channels leading to the depolarization of its membrane). However, it should be noted that neurotransmitters not always are activating, stimulating a depolarization of the mem-brane of the receiving cell, but also can be inactivating, making the receiving cell less prone to depolarize its membrane. The junction between two nerve cells where the chemical signal is transmitted is called a synapse, and the distance between the two cells is short in order to keep diffusion times for the neurotransmitter at a minimum. The number of synapses that a nerve cell has is considerable, it is normally more than 1000. A nerve cell receives a number of both activating and inactivating signals, and whether it fires or not depends on the balance between the incoming signals. This is of course very complex, and not understood in detail.

In addition, the neurotransmitter must be removed from the synapse after the signal has been transmitted, otherwise it will continue to stimulate the receiving cell as soon as it has been restored. This is done either by resorbing and reusing the neurotransmitter by the signalling cell, or by converting it to an inactive form. One of the most important neurotransmittors, ace-

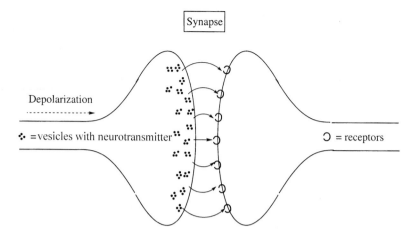

The neurotransmitter is resorbed by the signalling cell or degraded in the synaptic cleft.

Figure 10.17. The transmission of a chemical signal between two nerve cells.

tylcholine, responsible for the signals in the cholinergic system, is inactivated by the hydrolysis of the ester bond by the esterase acetylcholine esterase (see Figure 10.17). The conversion is very rapid and the choline formed is completely inactive.

Exogenous chemicals may specifically affect the nervous system in a number of ways:
a) The release or reabsorption of the neurotransmitter by the signalling cell may be affected. The botulin toxins are toxins produced by the bacterium *Clostridium botulinum* in food poisoning, and they block the release of acetylcholine. The effect is paralysis, which is lethal if it reaches the muscles involved in respiration. The incredible toxicity of the botulin toxins (LD_{50} approximately 0.01 µg/kg in man) indicate how sensitive the nerve system can be to toxic effects. Cocaine (see Figure 10.18 for structure) is active on the CNS because it blocks the resorption of noradrenalin, another neurotransmitter. The effect of cocaine is stimulating, as it makes noradrenalin stays longer than needed in the synaptic cleft.
b) Chemicals may have the same effect on the neuroreceptor as the natural neurotransmitter, and be able to stimulate cells to fire (if it is an activating neurotransmitter). Such compounds are called agonists, and a typical toxic effect that agonists may give is convulsions. Mus-

carine, one of the toxic principles of the mushroom fly agaric (*Amanita muscaria*) binds to the receptors that normally bind acetylcholine (which is reasonable when the two structures are compared). Not only does it bind, it also triggers the receptors to respond (i.e. an agonist) and results in an additional stimulation of the cholinergic part of the nervous system.

> The fly agaric has in primitive civilizations in Sibiria been used as a intoxicant, and as muscarine largely is excreted unchanged the urine of the intoxicated was collected and reuse by others.

c) The receptors may also be blocked by compounds that resemble the neurotransmitter enough to bind to the receptor but not enough to provoke the normal neurotransmitter/receptor complex response. Such compounds are called antagonists, and are toxic because they prevent the normal signals going through. This typically causes paralysis. Examples of compounds that are antagonists is hyoscyamine, isolated from

> The plant got its name belladonna because women in former times used it to make themselves beautiful. A drop of the juice of *Atropa belladonna* in the eye opens the iris completely and gives anybody a dreaming expression which was considered to make women beautiful. They would not be able to see much, but beauty has its price!

the plant *Atropa belladonna*, and similar alkaloids, which bind to the acetylcholine receptor. Hyoscyamine and atropin (atropin is a mixture of hyoscyamine and its benzylic epimer) are still used to relax the muscle controlling the iris in the eye, to make the iris open maximally and facilitate inspection of the retina.

The activity of hyoscyamine and atropin is quite different compared to that of cocaine, although their structures show several similarities.

d) If the neurotransmitter is converted to an inactive form in the synapse, any compound that inhibits the enzyme responsible for the conversion may of course have dramatic effects. If the neurotransmitter is activating, it will not disappear from the synaptic cleft but continue to stimulate the receiving cell to fire, and a normal toxic effect of such compounds is again convulsions. Very efficient inhibitors of the enzyme acetylcholine esterase has been discovered, and among the most active are the organophosphates (e.g. derivatives of phosphoric acid and phosphonic acid esters and corresponding phosphorothioic acids) of which many have been and are still used as insecticides (e.g. parathion and phosphamidon). Sarin is a chemical warfare agent that fortunately not has been used in any larger conflict (it was produced already during the second world war), because it is extremely toxic (lethal dose for man may be as low as 0.01 mg/kg). The structure of sarin is shown in Figure 10.18, while the mechanism by which organophosphates in general inhibit acetylcholine esterase is presented in Figure 10.19. Another class of compounds that have the same effect is the carbamate esters, although they are less potent, and they are also mainly used as insecticides. The perhaps most common of the carbamate insecticides is carbaryl (the insecticide produced from naphthol and methyl isocyanate in Bhopal).

e) The ion channels through which the sodium and potassium ions are flowing in the depolarized membrane may be blocked by compounds, that may fit both in size and charge (as a

The puffer fish (Fugu in Japanese) is considered to be a delicacy in Japan, but because of the extreme toxicity of tetrodotoxin it may only be prepared by especially trained cooks that know exactly which organs containing the poison should be discarded. Still, there is a number of fatal accidents yearly. Tetrodotoxin is not produced by the fish itself, but by a bacterium that is part of the same food chain as the puffer fish.

cork in a bottle) but also may affect the ion channels by interacting with the membrane. Several local anaesthetics and barbiturates are active in this way, and another well-known example is tetrodotoxin, the poison of the puffer fishes (also present in a other fishes and octopuses). Tetrodotoxin efficiently blocks the flow of sodium ions into the cells, which inhibits the depolarization of the membrane.

The insecticide DDT acts on the same ion channel, but affects its closing and thereby the time it takes for a repolarization of the membrane.

As indicated in Figure 10.19, the acetylcholine esterase inhibitors react covalently with the enzyme as it attempts to hydrolyze them as well. They add to the hydroxyl group of a serine moiety, and even though the acetyl group that normally ends up in this position when acetylcholine is hydrolyzed is easily removed by the enzyme, other groups will be more long-lived. The alkylphosphates especially react almost irreversibly, and are consequently very toxic. However,

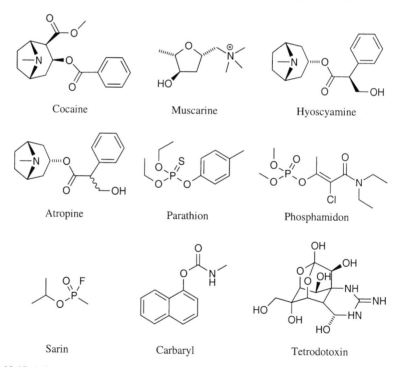

Figure 10.18. A few compounds that are toxic to the nervous system.

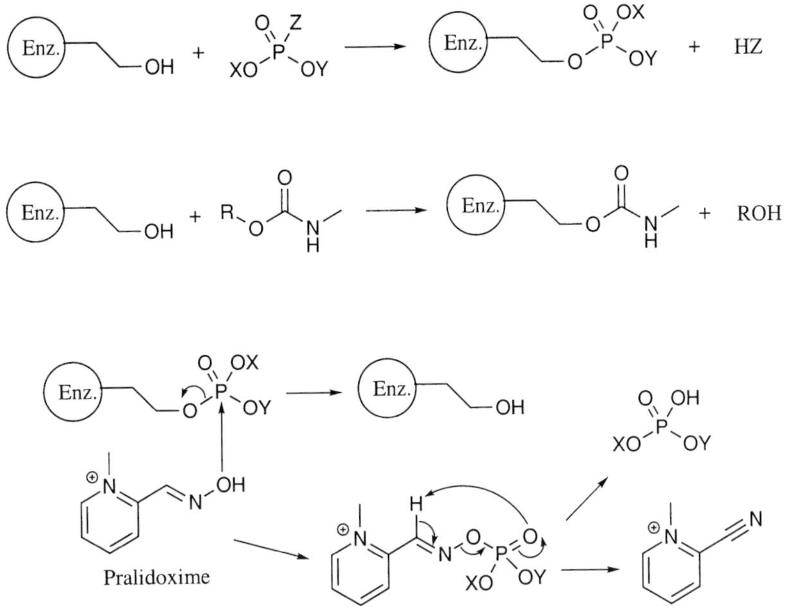

Figure 10.19. The inactivation of acetylcholine esterase by an organophosphate and by a carbamate ester, and the reactivation of the inhibited enzyme by pralidoxime.

it has been discovered that some compounds may regenerate the functional enzyme again, and such compounds are useful as antidotes. An example is the oxime pralidoxime, used by many military forces as personal protection against organophosphate warfare agents. It fits to the same binding site as acetylcholine and is able to react with the organophosphate residue so that the serine becomes free again (see Figure 10.19). In addition, poisoning with organophosphates is often also treated with atropine (an acetylcholine esterase antagonist, *vide supra*), which reduces the effect of the surplus acetylcholine present.

10.2.3 Effects on the liver

The liver plays a central role in both the excretion of chemicals from the body and the metabolism of both endo- and exogenous compounds. The metabolic capacity is also responsible for two typical toxic effects to the liver, fatty liver and liver cancer, because chemicals that are metabolized to reactive compounds are normally formed in the largest amounts in liver cells. Chemicals that are reactive as they are, not requiring metabolic activation, have no preference for the liver as a target organ. Synergistic effects are commonly observed with liver toxicity, as the mechanism for synergism often depends on the induction of cytochrome P-450.

Certain compounds excreted by the liver via the bile damage the bile ducts as they pass on the way to the bile bladder. In some cases, e.g. allylic alcohol, *para*-dimethylaminoazobenzene (butter yellow), coumarin, ethionine and thioacetamide (see Figure 10.20 for structures), toxicity to both liver and bile ducts are observed, while other compounds, e.g. α-naphthylisothiocyanate, and sporidesmin A, are only toxic to the bile ducts. In a few cases (e.g. sporidesmin A) the effect on the bile ducts is considered to be caused by the parent compound, while in most cases metabolites formed in the liver are the true toxic principal. The bile flow may also be reduced, by bile stones (normally composed of cholesterol and calcium salts), by inflammatory processes in the bile ducts, and by interference with the production of bile in the liver cells. A number of chemicals, e.g. anabolic steroids such as testosterone, the immunosuppressive drug cyclosporin A (a cyclic undecapeptide, structure not shown), and manganese (Mn), are known to give this toxic effect, but little is known about the mechanisms.

The accumulation of fat in the liver (steatosis or fatty liver) is produced in an individual whose liver is producing more fat (triglycerides) than it can export. To produce fat from surplus nutrients is a normal task for the liver, the fat is transported to the adipose tissue for storage and may be reused when for example the need for energy exceeds production. The fat is released from the liver cells into the systemic circulation combined with a lipoprotein, and a toxic effect that lowers the concentration of this lipoproteins may result in a fatty liver. As the proteins are synthesized in the endoplasmic reticulum where exogenous compounds also are metabolized,

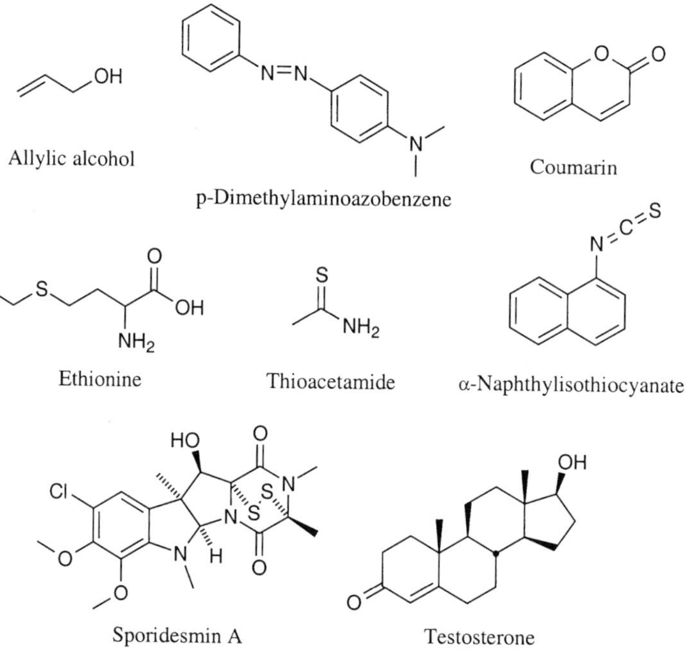

Figure 10.20. Compounds toxic to the liver.

reactive metabolites may of course affect the protein synthesis. Especially efficient are compounds generating radicals, as these give rise to lipid peroxidation that may destroy the membrane and the proteines associated with it. This is the mechanism by which for example carbon tetrachloride, which is reduced by the metabolism to the trichloromethyl radical, gives fatty liver. A fatty liver is generally considered to be a reversible effect, and not serious, but chronic exposure to a compound such as carbon tetrachloride will eventually result in liver cirrhosis (an irreversible liver damage) and liver necrosis (death of liver tissue). The reactive metabolites formed enzymatically or chemically as a result of the metabolism of exogenous chemicals in the liver may also give rise to liver cancers, and this will be discussed in Chapter 12. Besides cancer, reactive compounds may of course also kill liver cells and induce serious liver damage (e.g. paracetamol in large amounts, Section 8.2).

10.2.4 Effects on the kidneys

The kidneys are the most important excretion organ, and their function has been briefly discussed in Chapter 6. Due to the intense activity of the kidneys, they require delivery of substantial amounts of oxygen and nutrients and are sensitive to chemicals that produce anoxia. A decrease in the blood pressure or blood volume (as a result of haemorrhage for example) will also be more harmful to the kidneys compared to other organs. The fact that the kidneys resorb approximately 99 % of the primary urine permits the body to control the volume and composition of extracellular fluids, but small effects on this resorption by toxicants will have large consequences (1 % depression results in the additional daily loss of approximately 2 litres of water from an adult). During the concentration of the urine there is also a possibility that chemicals eventually are present in toxic concentrations that may injure the tubular cells, or are no longer soluble and form crystals. Ethylene glycol, frequently used as antifreeze in cars, may, in addition to give acidosis, also be toxic to the kidneys. Oxalic acid, the end product of ethylene glycol metabolism, reacts with calcium ions to form calcium oxalate which is excreted in the kidneys as small crystals together with low concentrations of oxalic acid. As the urine is concentrated the crystals may adhere and form stones that can obstruct the tubuli. Several metals, in particular mercury and cadmium, give severe toxic effects in the kidneys, and will be further discussed in Chapter 12.

10.2.5 Effects on the respiratory system

The lungs are responsible for the exchange of oxygen and carbon dioxide between the blood and the air, and some important toxic effects associated with chemicals in the respiration air have already been discussed in Chapter 6. Especially important is lung cancer, which is one of the major causes of mortality among cancer deaths. This is largely due to smoking, although in approximately 20 % of all lung cancers other factors are believed to be responsible. In addition, other factors (e.g. exposure to asbestos fibres in the workplace) act synergistically with tobacco smoke. There are many carcinogenic principles in tobacco smoke, of which most require metabolic activation, and the lung cells are consequently able to metabolize and activate exogenous compounds. An interesting example of a specific toxic effect on the lung cells is the nonselective contact herbicide paraquat. Regardless of by which route paraquat is administered it always damages the lungs, and this specificity is caused by two effects: First, the alveolar cells are able to accumulate paraquat by a special transport system (the diamine/polyamine transport system, a system that was not discussed in Chapter 6), second, the concentration of oxygen is relatively high in the lung cells. Paraquat is readily reduced by the coenzyme NADPH to form a radical, which rapidly is oxidized back to paraquat by molecular oxygen which in turn is reduced to the superoxide anion (see Figure 10.21). The toxic effect depends on a combination of both the depletion of NADPH in the affected cells, and on the generation of radicals with subsequent lipid peroxidation.

The pyrrolizidine alkaloids (e.g. monocrotaline in Figure 10.21) are plant metabolites that injure the liver and the lungs, and a large number of intoxications have been reported. They require metabolic activation via the pyrrole derivative shown in Figure 10.21 and further by cytochrome P-450 oxidation of the ring to a bifunctional electrophile of yet unknown structure. This activation is mainly carried out by the liver cells, and not in the lung cells. The reason that the lung is the target organ for some of the pyrrolizidine alkaloids is instead that the electrophilic metabolite is fairly unreactive and survives transport from the liver to the next organ in the

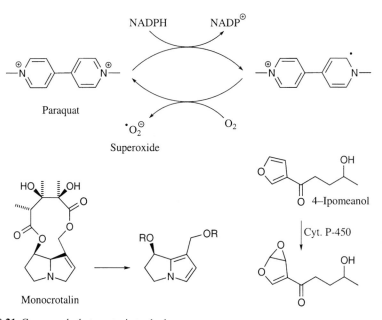

Figure 10.21. Compounds that are toxic to the lung.

circulatory system, the lung (although high doses will also damage the liver). Another example of a compound that is toxic to the lung is 4-ipomeanol, a furan that is responsible for the toxic effect of mouldy sweet potatoes given to cattle. 4-Ipomeanol is oxidized (probably to the epoxide shown in Figure 10.21) by cytochrome P-450 isoenzymes present in high concentrations in lung cells.

10.2.6 Effects on the skin

The skin is our barrier to the rest of the world, it protects us from external threats such as microorganisms, chemicals and radiation and it helps to preserve the body fluids (quantitatively as well as qualitatively). It is an organ system that frequently comes in contact with exogenous chemicals, which in some cases can penetrate the skin and be absorbed by the body, and it is most frequently the target for chemicals that give local effects. Strongly reactive chemicals, electrophiles, acids, bases, etc., will corrode the skin (chemical burns), which in severe cases can be life-threatening. In some cases chemicals react with the moisture of the skin, to produce reactive products (e.g. acetylchloride and acetic anhydride as well as $TiCl_4$ and $SnCl_4$ which both liberate HCl in contact with water), or generate heat that will burn the skin in contact with water (e.g. quicklime, CaO). In such cases one has to take care if rinsing with water is used to get rid of the chemical from the skin.

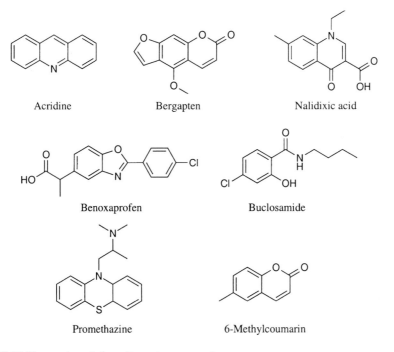

Acridine   Bergapten   Nalidixic acid

Benoxaprofen   Buclosamide

Promethazine   6-Methylcoumarin

Figure 10.22. Phototoxic and photoallergenic compounds.

An important toxic effect to the skin is allergy in the form of contact dermatitis, and this topic is covered in section 10.3. In addition, chemicals that are absorbed through the skin or by any other route may be activated by exposure of the skin to sunlight, to excited states that may affect the skin. The process is called photosensitization. When the result is a general toxic effect on the skin (e.g. blistering) it is specifically called phototoxicity while the term photoallergy indicates that a delayed-type hypersensitivity reaction (*vide infra*) has taken place. Phototoxic chemicals in their excited forms can either react directly with macromolecules and produce a toxic effect, or react with molecular oxygen which in turn generates oxygen radicals. Examples of phototoxic chemicals are the psoralenes (plant metabolites, e.g. bergapten present in bergamot oil), acridine (used for the manufacture of dyes), benoxaprofen (an analgesic), and nalidixic acid (an antibacterial agent), while halogenated salicylanilides (e.g. buclosamide, an antifungal agent), promethazine (an antihistamine agent), and coumarin derivatives (e.g. 6-methylcoumarin) are photoallergenic (see Figure 10.22 for structures).

10.3 Chemical allergens

Chemical allergens are toxic to the immune system. The central functions of the immune system are to protect the organism from infectious agents (microorganisms and viruses) and to survey the cells of the organism and neutralize cells that have changed significantly (e.g. tumour cells). Besides infectious agents the immune system will also sense the presence of and attack certain structures including large molecules that do not belong in our bodies. An allergy is at hand when the immune system reacts too strongly, so that the response not only takes care of whatever should be taken care of but in addition is experienced as troublesome. The mechanisms that are involved are too complex to be described in any detail here, but consist in principle of a nonspecific and a specific part. The nonspecific reacts immediately on chemical signals, for example emitted by the invading microorganism or by other cells of the immune system, and attacks and destroys the agent by phagocytosis. This causes what we recognize as an inflammation. The specific part of the immune system has aquired the ability to determine if chemical components are "self" or "foreign", and will attack anything considered foreign (containing an antigen which is the chemical component that is recognized as foreign). There are many mechanisms for the specific part of the immune system to react and deal with antigens, but they all depend on either the production (by so called B cells) of specific antibodies, or immunoglobins, which are proteins that bind to antigens, or the development of specifically sensitized lymphocytes (T cells, T for "thymus dependent") that contains receptors in their cell membrane that recognize antigens. Antibodies will bind to the antigens of for example a bacterium and attract effector cells that attack it, and this response comes quickly. Examples of allergic responses that depend on the production of antibodies against an antigen are the immediate hypersensitivity, or anaphylaxis, resulting for example in asthma or rhinitis (in the respiratory system), food allergies (in the gastrointestinal tract) and anaphylactic shock (in the vasculatory system). The allergic response of the T lymphocytes is called delayed-type hypersensitivity and results mainly in contact dermatitis (a rash or irritation, possibly including blistering of the skin that comes in contact with the allergen). This is the most common form of allergy to chemicals. (The immune system is described schematically in Figure 10.23.)

For obvious reasons, the proteins that identify chemical structures as "self" or "foreign" are designed to recognize large structures such as proteins and polysaccharides, and cannot identify small molecules. A certain size and a certain chemical complexity is required, and as a general rule it is usually considered that only molecules with a molecular weight of more than 5000 will be able to become antigens. Nevertheless, a number of small molecules are known to be allergenic, and they must then combine with bigger, endogenous molecules (e.g. proteins) before they are able to stimulate the immune system. Small molecules that do this are called haptens, and by reacting with and/or binding to for example a protein this will be transformed from a "self"-protein to a "foreign"-protein. Therefore, the general characteristic of en allergenic chemical is that it, either directly or after spontaneous chemical transformation or enzymatic conversion, is able to change the structure of a macromolecule. Electrophilic or pre-electrophilic compounds are potentially allergenic, as are radical-generating compounds and metal ions that are able to form stable complexes with for example proteins.

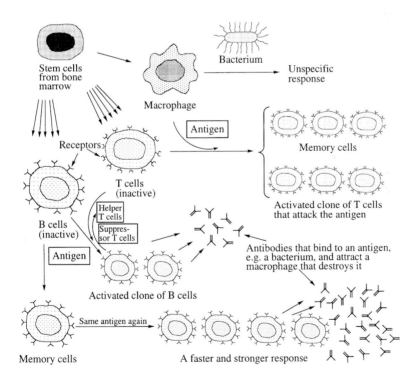

Figure 10.23. The principles of the immune system.

The mechanisms by which the immune system learns which chemical structures are "self" are not understood, but the process probably takes place already during the functional maturation of the immune system in the foetus. In general, an antigen (e.g. macromolecules from a microorganism, or an endogenous protein modified by a reactive chemical) will be picked up by special cells of the immune system that process and present them to lymphocytes in for example the lymph nodes or the thymus. If the antigen is recognized as foreign, the lymphocytes will develop to effector cells (either B or T cells) that are specific to that very antigen. In addition, long-lived memory cells are formed that in principle enable the body to remember that this antigen is foreign for the lifetime of the organism. This process is called sensitization and may take anything from days to years, but any contact with the antigen after sensitization will result in a quick response because the immune system has a "memory" (the memory cells) of this antigen. The immediate hypersensitivity (e.g. hay fever) responds very fast (minutes), while the delayed-type hypersensitivity takes between 12 and 48 hours to respond. An individual that is sensitized to a chemical is often extremely sensitive to its presence even in trace amounts, and for chemicals like formaldehyde and nickel that are difficult to avoid the allergy may become a life-long problem. In addition to the B and T lymphocytes responsible for antibody production and recognition of antigens, a number of other subpopulations of lymphocytes that play important roles in the immune response exist. Examples are the killer cells (K cells) and natural killer cells (NK cells) that are important for defence against cancer cells and helper (T_H) as well as supressor (T_S) cells that modulate the response of the immune system so that it is appropriate and balanced. In addition, there is a number of chemical factors (e.g. interleukins and interferons) that activate, amplify and regulate various functions of the immune system.

10.3.1 Immediate hypersensitivity

Immediate hypersensitivity is a quick response of the immune system to an agent that previously has been recognized as "foreign", and may in some instances be life-threatening (e.g. the anaphylactic shock resulting from a bee-sting). Some allergens encountered in work-places are sufficiently big to function as antigens as they are (for example dust of flour that may give bakers rhinitis and asthma but most act as haptens and conjugate in some way to an endogenous protein. Trimellitic anhydride (see Figure 10.24) is used for the manufacture of certain plastics and paints, and is allergenic. The most common response is antibody-mediated asthma, but it also gives rise to delayed-type hypersensitivity. Phthalic anhydride will give a similar response. The isocyanates are notorious reactive chemicals, used extensively for the preparation of polymers and paints. Especially the diisocyanates, containing two isocyanate groups (e.g. diphenylmethane-, hexamethylene- and toluene-diisocyanate) are potent allergens causing both asthma and contact dermatitis. Interestingly, individuals that are allergenic to toluenediisocyanate may develop cross-reactivity to other diisocyanates to which they never have been exposed. Ethylene oxide and formaldehyde, both highly reactive, also induce antibody-mediated immediate hypersensitivity. Treatment of infections with penicillin and other β-lactam antibiotics (see Figure 10.25 for the structure of benzylpenicillin) has shown that these may be allergenic in rare cases. Especially the immediate type of allergy resulting in anaphylaxis is serious and

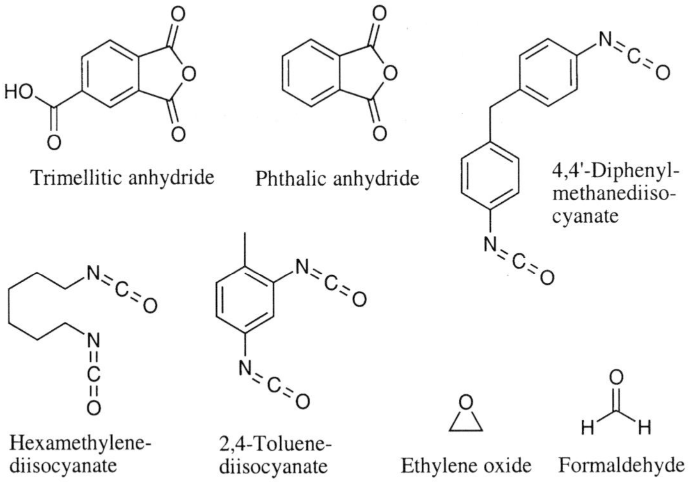

Figure 10.24. Examples of chemicals that give immediate hypersensitivity.

may be life-threatening, and hypersensitivity to penicillin is actually the most common cause of anaphylaxis in man. It is probably metabolites of the β-lactams that are responsible for their allergenic effect, although this has not been clarified in detail. Several metals are also important allergens that give both types of responses, and this will been discussed further in Chapter 12.

10.3.2 Delayed-type hypersensitivity

That a compound is allergenic by this mechanism and not by immediate hypersensitivity depends on a multitude of factors, and not only on the chemical itself. As has been noted above, many allergenic chemicals will induce both types of hypersensitivity, which is reasonable. Several of the most allergenic compounds that give contact dermatitis are present in preparations that are applied on the skin, e.g. plasters, ointments, perfumes, etc., and it is a good guess that other reactive (or proreactive) chemicals would also give the same effect if used in a similar way so that they come in close and prolonged contact with the skin. Important chemicals that may induce a delayed-type hypersensitivity are metals like nickel, chromium, cobalt and organomercury compounds (see Chapter 12), catechols (for example 3-(8',11',12'-pentadecatrienyl)-1,2-benzenediol from the plant poison ivy), primine from the plant primrose, aromatic amines, epoxides, derivatives of acrylic acid, and formaldehyde, but also penicillin and other β-lactam antibiotics (*vide supra*). As indicated in Figure 10.25, hydroquinones are relatively easily oxidized to reactive benzoquinones (also discussed in Chapter 8), and hydroquinones (as well as compounds that yield hydroquinones after metabolic activation) are frequently allergenic.

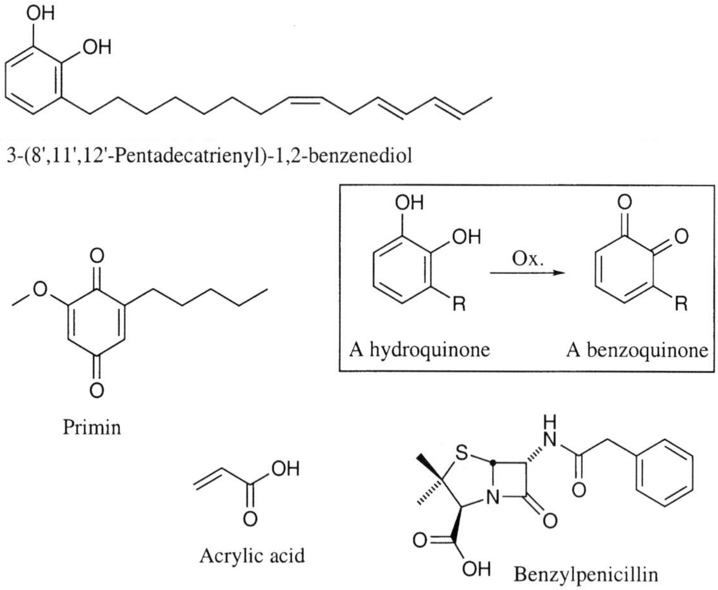

3-(8',11',12'-Pentadecatrienyl)-1,2-benzenediol

Primin

A hydroquinone   A benzoquinone

Ox.

Acrylic acid

Benzylpenicillin

Figure 10.25. Examples of chemicals that give delayed-type hypersensitivity.

In addition to allergy, chemicals can also be toxic to the immune system in other ways. Immunosuppression, which eventually may lead to immunodeficiency, may be caused by the exposure to solvents like benzene, by environmental hazards like PCBs and PBBs, dibenzodioxins, other halogenated aromatic hydrocarbons, ozone, PAHs, several pesticides, etc. It is possible that the immunosuppressive activity of such agents is important for their toxicity, although it is not known to what extent it modulates other toxic effects. Another toxic effect to the immune system is autoimmunity, resulting in the attack of the effector cells of the immune system on endogenous components. Little is known about the mechanisms for autoimmunity, but in several cases it has been linked to the exposure to chemicals (e.g. vinyl chloride, perchloroethylene and epoxy resins). In such cases it is believed that the chemical modifies an endogenous protein that is recognized as foreign, and that the response to this is incompletely specific and also attacks non-modified protein.

10.4 Chemical teratogenicity

Teratogenic agents are toxic to the embryo (what the human offspring is called during the first 8 weeks after implantment in uterus) or the foetus (the rest of the time up to the birth) and cause structural or functional abnormalities in offsprings after exposure to either parent before the conception or to the female during the pregnancy. The abnormalities, or birth defects, can be anything from extensive malformations to minor structural defects and metabolic disorders as well as growth or mental retardations, and can result in anything from the death of the embryo to insignificant defects. Obviously, it is difficult to define exactly what is a birth defect, as well as to quantify small and apparently insignificant structural or functional abnormalities. It is also hard to decide whether an abnormality was the result of an event taking place before or after birth. An example is the cancers that affect children, which are probably initiated already during the gestation although they are normally impossible to observe at the time of birth. It is known that a large proportion of all fertilized ova will not develop normally, which in most cases results in the reabsorption or a spontaneous abortion so early that the pregnancy has not been recognized. About 3 % of liveborn infants have birth defects that are recognizable during the first year of life. Obviously, as is the case for mutations, spontaneous factors are responsible for a significant portion of these effects, but the proportion between spontaneous and external factors is not known. Also potentially responsible for the interruption of gestation are the inheritable genetic defects that many of us, perhaps all, have. They may be the result of mutations that took place several generations ago. They are normally not expressed and therefore not noticed but are incompatible with certain genetic combinations that may arise in conception.

The gestation takes approximately 38 weeks (humans) and can be divided into the following major processes; histogenesis (formation of body tissues, during the first weeks), organogenesis (formation of organs, during the first months), functional maturation and growth of organs and tissues (last two thirds of the gestation). It starts with the fertilization of the ovum by a sperm cell, and after a number of cell divisions the blastocyst (the term used before differentiation has started, and before the term embryo) formed is implanted in the uterus approximately one week later. The organogenesis, a period when the basic structure of the organs and tissues are established, takes place from week 3 to week 9, and it is during this period that teratogenic agents give rise to malformations. The whole gestation is of course characterized by intense cellular activities in the embryo/foetus, besides the rapid cell division necessary, and these are directed by signals (e.g. hormonal) from both the mother and generated internally. The histogenesis and the organogenesis require that the correct parts of the genome are turned on or off at a precise moment (day), the maturation of the various functions in the right time require that things happen in the predestinated order, and there are really no margins for the disturbance or delay of any process.

The nature of a teratogenic agent can be physical and biological, as well as chemical, just as is the case with mutagens and carcinogens (see Chapter 11). The physical and biological are ionizing radiation (e.g. X rays) and viruses, and an example of the latter is the rubella virus which is a fairly common source for birth defects. 15–20 % of the women that are infected with the rubella virus in the beginning of their pregnancy will have a miscarriage or a stillbirth, and the risk for severe malformations in the surviving foetus is considerable. Also the infection as late as during the third or fourth month frequently results in for example hearing defects. A large number of chemicals that are teratogenic in animals are known (>1,000), but only a few (approximately 30) are proven to be human chemical teratogens. However, it is important to remember that chemicals not only may have negative effects on the development of an embryo, but some are absolutely necessary for the biochemical processes to take place. Consequently, the lack of sufficient amounts of essential chemicals due to maternal nutritional deficiencies (e.g. fasting) may also be teratogenic. For example, the lack of iodine may result in cretinism, a birth defect that is unusual today because iodine is added to table salt.

Besides interfering with the specific processes that take place in the embryo/foetus, chemical teratogens may also give their effects as a result of normal toxicity that also will affect adults. The embryo/foetus is often more sensitive to toxic effects, as for example the energy consumption and the cell division rate are high. Mutagenic and cytotoxic chemicals are therefore generally teratogenic, as are chemicals affecting the production of ATP and the oxygen transportation as well as nucleic acid analogues. Several commonly used solvents, e.g. toluene and ethers of glycol, have been shown to give teratogenic effects in animals, as has even moderate alcohol consumption and the use of cocaine. Several phenoxy acids [e.g. (2,4,5-trichlorophenoxy)acetic acid] used as herbicides are teratogenic to several animals, and TCDD is a potent teratogen in animal experiments. So far, probably due to the limited exposure that has taken place there are no evidence that TCDD also is a human teratogen, but it is strongly suspected to be so. Inorganic forms of mercury, lead, cadmium and arsenic are also teratogenic in animal experiments, and the exposure of pregnant women in Minamata (see Section 12.7) to methyl mercury resulted in severe birth defects (CNS malformations). Many of the human chemical teratogens known are drugs, and the most well-known is thalidomide (marketed under a number of different trade names) which was a sedative and hypnotic that appeared to be non-toxic to man. It was introduced 1956 and withdrawn in the beginning of the 60s. It was prescribed to calm the nausea associated with pregnancy and caused abnormalities in more than 10,000 foetuses. Most conspicuous are the limb deformations, but malformations of the cardiovascular, intestinal and urinary systems were also observed.

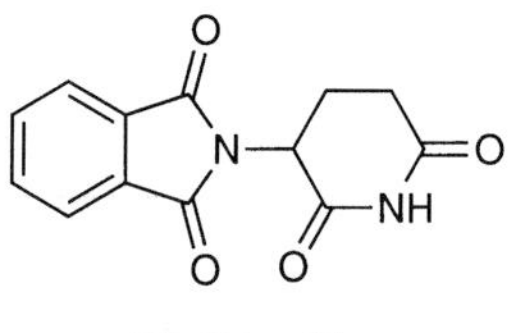

Thalidomide

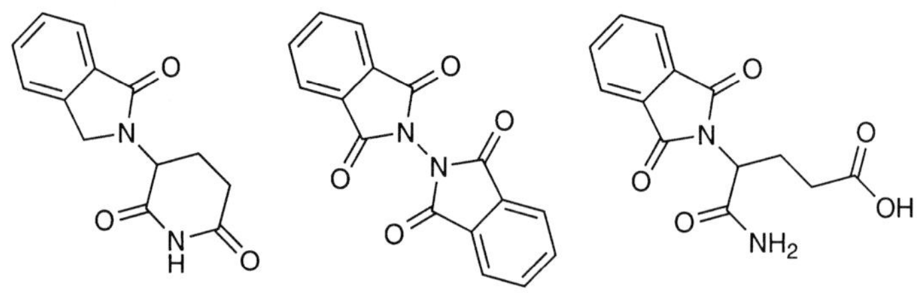

Teratogenic analogues of thalidomide

Figure 10.26. Thalidomide and some analogues that retain the teratogenic activity.

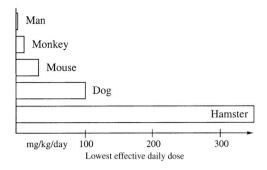

Figure 10.27. The daily doses it takes for thalidomide to provoke a teratogenic effect in various mammals.

Surprisingly, some mammals (e.g. rat) are resistant to the teratogenic effect of thalidomide, and for some animals (e.g. rabbit) different strains show highly variable sensitivity. Figure 10.27 gives an impression of how the sensitivity towards the toxicity of thalidomide varies among different species.

Despite extensive research to find out the molecular mechanism for the teratogenic activity of thalidomide, it is still not understood how it acts. Because of the variations observed in different strains of the same species it is unlikely that it for example is activated by the metabolism, although this possibility cannot be ruled out. Among the many analogues and derivatives of thalidomide prepared and tested, only a few have the same effect as the parent compound (see Figure 10.26). Lately, thalidomide has received attention again because of its ability to inhibit the replication of HIV (the virus causing AIDS), and efforts are being made to develop the molecule into something useful.

The effect of teratogenic chemicals depends not only on how the teratogenic agent interferes with the biochemistry of the embryo/foetus, but also on the period of exposure, and this complicates the study of teratogenic agents. For example, the limb malformations in humans caused by thalidomide are caused by exposure only during a fortnight, between day 23 and 38 after conception. While malformations obviously are caused during the period of organogenesis, cancer cells are predominantly formed later during the growth period, and the different effects could very well be caused by the same chemical. Early childhood tumours are generally considered to be formed already in the foetus, which actually is quite sensitive to carcinogenic agents. This is because the cell division rate is much higher in foetal cells compared to normal somatic cells (a factor that facilitate the transformation of a normal cell to a cancer cell) and because the immune system that can detect and kill abnormal cells (e.g. cancer cells) is poorly developed in the foetus.

The maternal and foetal blood are not mixed, but are separated by the placenta through which nutrients and waste products are transported by the mechanisms discussed in Chapter 6. Lipophilic exogenous chemicals rapidly diffuse through placenta, but also water soluble molecules (with molecular weights less than 800) have been shown to pass the human placenta (although much less efficiently). The latter transport takes place by diffusion through water-filled channels with fixed diameters. Lipophilic compounds that are transformed/converted to

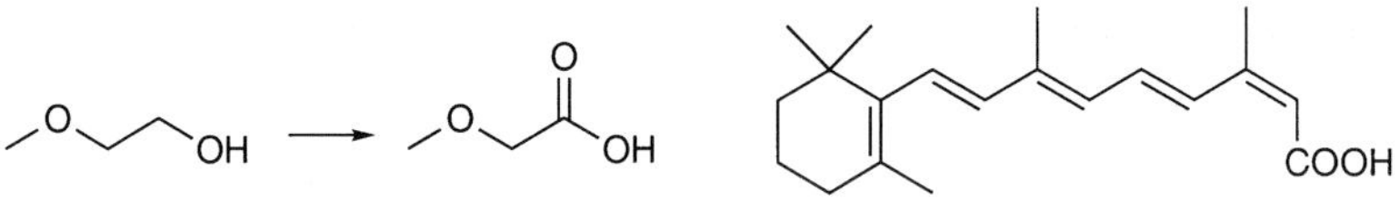

| Ethylene glycol monomethyl ether | Methoxyacetic acid | Isotretinoin |

Figure 10.28. Additional teratogenic compounds.

more polar forms may be concentrated in the embryo/foetus, and this has been discussed as a possibility for thalidomide (which relatively easily is hydrolyzed to more polar compounds in contact with water). It has been shown that the pH of the blood of rodent embryos is higher than that of the maternal blood, and weak acids would consequently be trapped and concentrated in the embryos. Several chemical teratogens are in fact weak acids, e.g. methoxyacetic acid (the teratogenic metabolite of the solvent ethylene glycol monomethyl ether, see Figure 10.28), isotretinoin (a synthetic retinoid used to treat certain forms of acne, see Figure 10.28), thalido-mide (the hydrolysis products are weak acids) and valproic acid (a drug against epilepsy, see subsection 7.3.1.4 for structure and metabolic activation), and this pH gradient may play an important role for these teratogens. The placenta has a limited capacity to metabolize exog-enous compounds and is normally considered to be of minor importance in this respect, al-though it has been shown that smoking for example will induce the metabolic capacity (e.g. cytochrome P-450) of the placenta considerably. The foetus also has limited metabolic capacity, although it increases constantly as it grows. Normal concentrations of for example cytochrome P-450 will be attained approximately one year after birth. Elimination from the foetus takes place via the placenta or via the kidneys to the amniotic fluid, the latter only being a temporary solution as this fluid constantly is swallowed.

Although few human teratogens are known, the same principle as with toxicity in general also applies to teratogenic effects: Any chemical is able to harm the development of the embryo/foetus if the dose is sufficiently high. The thalidomide tragedy has made us aware about the potential potency of chemical teratogens, and has had an enormous impact of the science of teratology. Thalidomide escaped the warning systems of the 50s, but although the test protocol for new drugs has been changed today and constitutes a considerably more fine-meshed net, there are no guarantees that the story will not repeat itself. It should be remembered that the reason for the rapid disclosure of thalidomide as a teratogen was due to the fact that it causes unusual malformations (the limb deformations). If it had caused more common malformations it would have been much more difficult and have taken a much longer time to establish the connection, and even more children would have been afflicted.

11 The molecular basis for genotoxicity and carcinogenicity

Genotoxic chemicals will affect the genetic material in particular. The effects may be that a gene is injured so badly that it no longer can fulfil its tasks, or that the information encoded in the sequence of the nucleic acids in DNA is altered so that the protein product is modified. The first effect, if the gene struck is essential for the cell, will lead to severe malfunction or death of the cell, while the second is called a mutation and generates permanently changed cells with slightly different properties. Note that there is a significant difference between "damage to DNA" and "mutation". Mutations are the basis of a number of very serious effects of chemicals, e.g. hereditary defects, tumours and teratogenic effects, and the fact that they affect not only individuals but also the whole of mankind draws special attention to genotoxic chemicals. Although mutations have played a key role in the evolution of life on earth, and may continue to do so even if other factors today are believed to be more important, we should consider mutations to be harmful in general and particularly to individuals. Mutations happen spontaneously, although at a very low rate, due to built-in "shortcomings" in the chemistry of nucleic acids and of the enzymatic systems synthesizing and handling DNA. Such mutations are inevitable, but the additional ones, caused by external agents, should be avoided as far as possible.

11.1 Chromosomes, genes and mutations

The chemistry of DNA has been discussed in Chapter 4, and Figure 11.1 shows how the DNA is organized in larger structures called chromosomes. (How the genetic information is extracted and used for the synthesis of proteins will not be discussed here. Those that are not familiar with this are recommended to have a look in a textbook in biochemistry or molecular biology.) The chromosomes are visible in a light microscope when they are condensed, which is in connection with cell division when it is absolutely necessary that they are compact, transportable structures that can be physically moved from the mother cell to the daughter cells. In preparing for a cell division the genetic material of a cell is duplicated, and in the "chromosome man" (see Figure 11.1) with two arms and legs that we can see in a microscope, one arm and one leg is a copy. Our cells contain 23 pairs of chromosomes, one chromosome in each pair originating from our mother and the other from our father.

A defect, caused by a genotoxic event, affecting a specific function in a cell is naturally considered to be associated with the gene(s) that encode for the protein(s) involved directly in the function in question. However, it should be remembered that, for example, the loss of the ability to carry out a certain enzymatic conversion because the cell has been exposed to a genotoxic

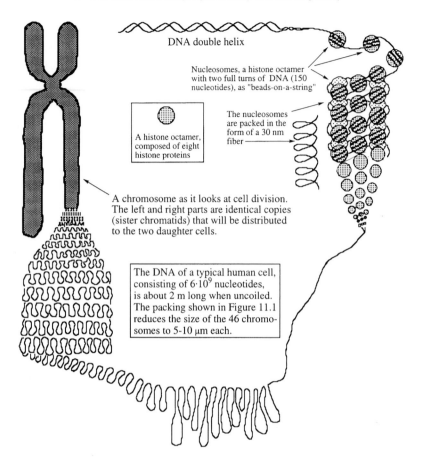

Figure 11.1. The condensation of DNA into a chromosome.

chemical does not necessarily mean that the gene for that enzyme has been destroyed or mutated. Another possibility is that genes that are involved in the regulation of when and how much of a protein that should be formed have been affected. Such genes may be located in completely different chromosomes, and we shall see examples of this later. In Figure 11.2, an outline of a simple regulation of the encoding of the gene for an enzyme is presented.

Say that the enzyme is called X and that its task is to convert compound Y to compound Z. As long as the cell contains no Y, there is no need for the enzyme and it is not formed. In this situation the encoding of the gene is blocked by a repressor, that binds to the stretch of DNA just in front of the gene (this stretch is called the operator of the gene) and hinders the encoding enzymes to do their job. However, if the concentration of Y increases, the repressor leaves (for example by forming a complex with Y that is chemically different and no longer can bind to the

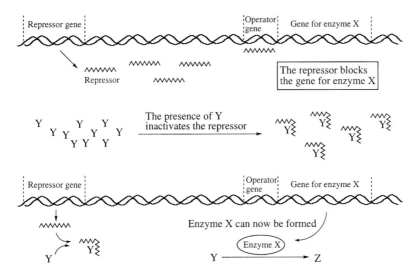

Figure 11.2. How the encoding of a gene for an enzyme can be regulated.

operator) and the information in the gene is translated to produce enzyme X (see Figure 11.2). If the conversion of Y to Z in a cell does not work because of a mutation, it could of course be the gene for enzyme X that has been mutated. But it could also be the operator that has been changed by the mutation, and if the new operator does not bind the repressor enzyme X will always be formed which can be a big problem for a cell. Alternatively, the repressor may bind too well, not letting go as the concentrations of Y increase, resulting in a constant lack of enzyme X. The repressor, a protein that is encoded by a gene that is situated somewhere else in the genome, may also be changed if it is the repressor gene that has mutated. The mutated repressor may bind too loosely or too strongly, creating similar problems as already mentioned. The fact that we in principle have two copies of all genes (maternal and paternal copy) is valuable; if one copy is damaged, the other can often cope alone.

Another technique that the cells use to mark out which genes should be encoded and which should not, is to methylate certain positions (e.g. the carbon opposite to the carbonyl group in cytosine, something that will not affect the base-pairing of cytocine) making use of SAM (Section 7.4.5) and specialized enzymes. Other enzymes can demethylate DNA, so it is a reversible process. The methyl groups function as "flags" that direct the encoding machinery, which of course may be confused if an exogenous methylating agent inserts methyl groups at random.

11.2 DNA as a molecular target

DNA is an extraordinary molecule, cleverly designed for its function as the carrier and mediator of information in cells and organisms and thereby also the master. One can imagine that evolution has tested many different molecular constructions for the job, before settling for DNA! Not only chemicals can damage DNA, but also various forms of electromagnetic radiation (e.g. UV-light) or ionizing radiation emitted from radioactive materials. At the molecular level DNA is characterized by a rich content of heteroatoms, oxygen in the backbone and nitrogen in the DNA bases. The bases are reasonably efficient nucleophiles and react with electrophiles forming chemically modified bases. In addition, DNA is an extremely large molecule and a portion of DNA can form a complex with another molecule or with metal ions, and this may be significant for the mechanisms by which some mutagenic and carcinogenic materials act (e.g. metals like chromium, nickel and cadmium).

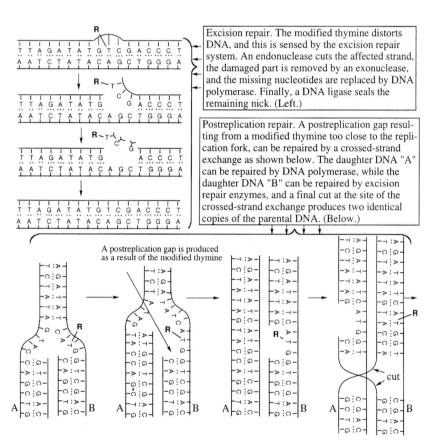

Excision repair. The modified thymine distorts DNA, and this is sensed by the excision repair system. An endonuclease cuts the affected strand, the damaged part is removed by an exonuclease, and the missing nucleotides are replaced by DNA polymerase. Finally, a DNA ligase seals the remaining nick. (Left.)

Postreplication repair. A postreplication gap resulting from a modified thymine too close to the replication fork, can be repaired by a crossed-strand exchange as shown below. The daughter DNA "A" can be repaired by DNA polymerase, while the daughter DNA "B" can be repaired by excision repair enzymes, and a final cut at the site of the crossed-strand exchange produces two identical copies of the parental DNA. (Below.)

A postreplication gap is produced as a result of the modified thymine

Figure 11.3. The repair of chemically modified DNA.

While our exposure to genotoxic chemicals certainly has increased during the last decades, radiation has always been around and constantly posed a serious threat to life. As a response, our cells have developed several enzymatic systems that detect and repair damage to DNA. As the double helix is such a condensed structure, an additional group attached to a base will give rise to a distortion of the helix and the repair systems are very skilful in detecting this. DNA repair is an important aspect of genotoxicity, and Figure 11.3 gives a schematic presentation of two such repair systems. The so called excision repair will probe along the double helix and sense structural irregularities caused by a chemically modified base, and the affected nucleotide (possibly together with a few neighbours) will be enzymatically removed. Because it leaves the opposite unmodified strand as it is, other enzymes can use this as a templet, add the missing nucleotides and restore the DNA completely. This repair is almost faultless, and will take care of the majority of all alterations of DNA. However, some damages are not so easily repaired, for example those that take place immediately before the DNA of a cell is duplicated in preparation of cell division, or that are caused by bifunctional electrophiles crosslink the two DNA strands. There are repair systems capable of handling also such events (see Figure 11.3), but they are more error-prone and the risk for an unsuccessful repair leading to a mutation increases.

11.3 Chemical effects on DNA and DNA-handling systems

11.3.1 Modification of the DNA bases

DNA bases that have been modified chemically have problems fitting into the double helix in the correct way and forming the hydrogen bonds to the complementary base in the opposite strand. There is a certain probability that such chemical changes take place spontaneously, for instance by hydrolysis (deamination) of cytosine via the addition of water to the double bond followed by elimination of ammonia to form an enol that rapidly is converted to a ketone (demethylthymine). Another possibility is the spontaneous hydrolysis of a nucleoside followed by the β-elimination of the open form of the deoxyribose (the two forms are in equilibrium with each other). This will result in a break in one strand as shown in Figure 11.4 (see Figure 11.8 for a discussion about the mechanism).

These and other spontaneous chemical conversions of DNA are not rapid reactions, but in view of the high number of DNA bases present in our cells a few per day will nevertheless be affected. Such spontaneous changes may lead to mutations according to the mechanisms discussed in the following section, and this is part of the evolution of life on earth. More important are exogenous chemicals that are able to react with and modify DNA, and as the DNA bases are nucleophilic any electrophile will in principle be genotoxic. A wide range of electrophilic chemicals and compounds that are converted to electrophiles by the metabolism are known to react with the DNA bases, both *in vivo* and *in vitro*, and the properties of electrophiles and their formation during the metabolic conversions have been discussed in Chapters 3 and 7. The positions in the DNA bases in double-stranded DNA that most frequently react with nucleophiles are indicated in Figure 11.5.

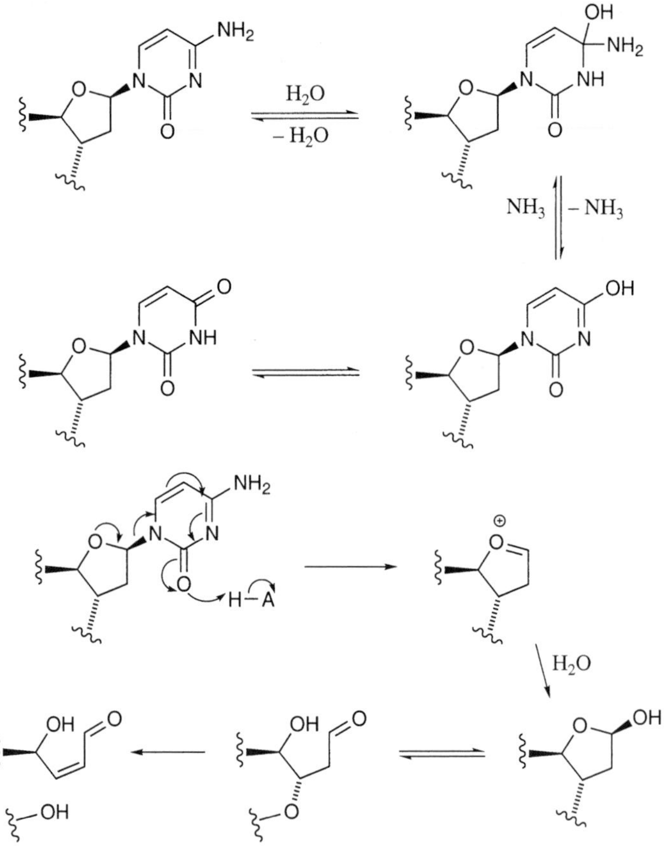

Figure 11.4. Spontaneous chemical reactions that may take place in DNA.

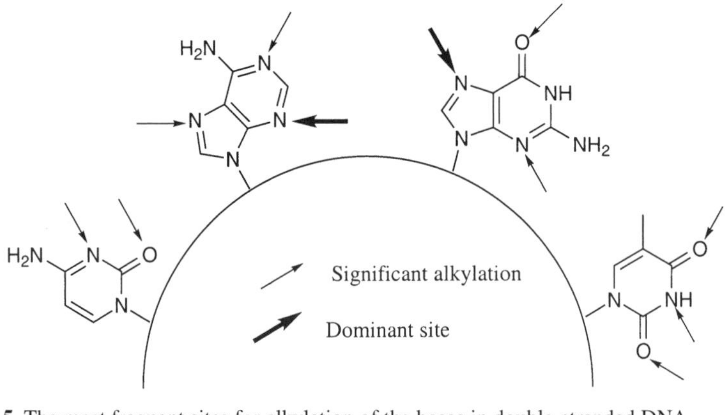

Significant alkylation

Dominant site

Figure 11.5. The most frequent sites for alkylation of the bases in double-stranded DNA.

Due to the closed structure of double-stranded DNA, the bases will only react when the helix is dissociated during for example decoding. The monomeric nucleosides, which are building blocks for the enzymatic systems synthesizing and repairing DNA, are more prone to reaction with electrophiles, and although the enzymes normally reject modified nucleosides they could in principle be used, resulting in a chemically modified DNA. However, the likelihood for this is so small that the reaction between a free nucleoside and an electrophile instead can be regarded as a protection for more sensitive biomolecules. The fate of unnatural or modified bases in DNA is in most cases that they are detected by the repair systems (*vide supra*) and exchanged. Some modified bases will be chemically unstable and eliminated from DNA (*vide infra*), and will be replaced by the repair systems. Chemical lesions to DNA that for some reason are difficult to detect, difficult to repair or formed just before the cell replicates its DNA in preparation for a cell division, may cause a mutation by the mechanisms discussed in the next section. Examples of DNA lesions that are difficult to handle are those caused by bifunctional electrophiles, as they besides modified bases also can form intra- and interstrand cross links (see Figure 11.6) which in general are more difficult to handle for the repair systems.

In addition, although little is in fact known about this, it is reasonable to believe that chemicals that are reactive as for example radicals (formed for example during lipid peroxidation) or oxidizing agents (e.g. metal ions) also may modify DNA in various ways.

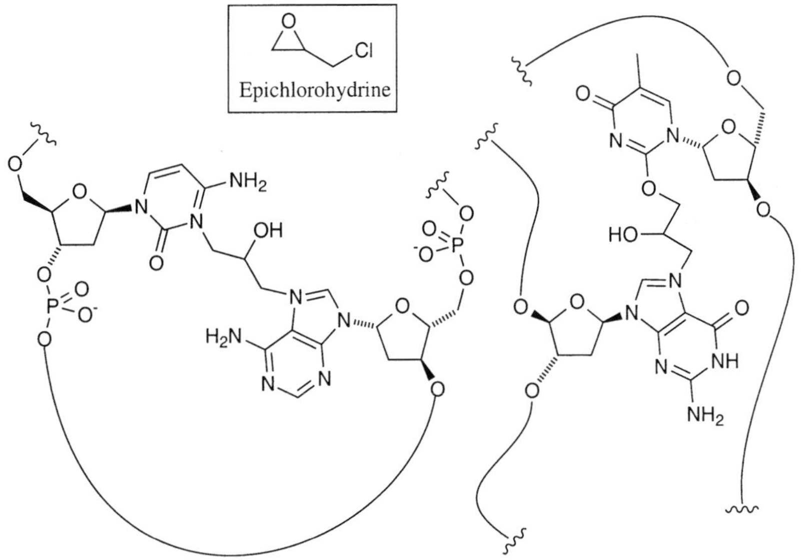

Figure 11.6. Intra- and interstrand cross links by the bifunctional epichlorohydrine, which to the left has reacted with adenine and cytosine in the same strand, and to the right with guanine and thymine in opposite strands.

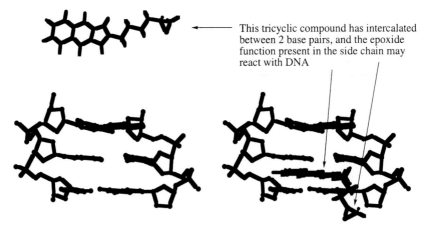

This tricyclic compound has intercalated between 2 base pairs, and the epoxide function present in the side chain may react with DNA

Figure 11.7. Intercalation.

11.3.2 Intercalation

Intercalators are flat molecules often consisting of several aromatic rings that can fit between two base pairs in the double helix, as the meat in a hamburger is positioned between two slices of bread. Intercalation as a phenomenon has been known for quite some time, as it easily can be measured that the length of DNA increases in the presence of intercalators. As many polyaromatic hydrocarbons as for example aromatic amines are mutagenic and carcinogenic, it was initially believed that their genotoxicity is caused by their ability to intercalate with DNA. However, the expanding knowledge about the metabolic activation of such compounds to reactive electrophiles has changed our view of interrcalation. We know that it takes place although it *per se* no longer is considered to be important. It can of course direct a compound containing an additional electrophilic functionality to a suitable position and facilitate its reaction with DNA, something that is believed to be the case with the carcinogenic mycotoxin aflatoxin B_1 (see Figure 11.17 for structure). An example of intercalation is shown in Figure 11.7.

11.3.3 Loss of modified bases

In DNA, the bases are linked to deoxyribose with a N–C bond, and the carbon atom has also an oxygen bound to it. In Chapter 3 we learned that molecules containing a carbon atom that has two heteroatoms bound to it by single bonds will be hydrolyzed in contact with water, generating the free carbonyl group. Consequently, DNA in water should be hydrolyzed, and in Figure 11.4 a mechanism for this is shown. However, the rate of hydrolysis is very slow. At pH 7 and 37 °C the half-life for adenine and guanine in DNA is approximately 15,000,000 hours. (This appears reassuring until one calculates how many adenines and guanines our DNA contains ...) Chemical modification of a DNA base may decrease the stability of a nucleoside dramatically,

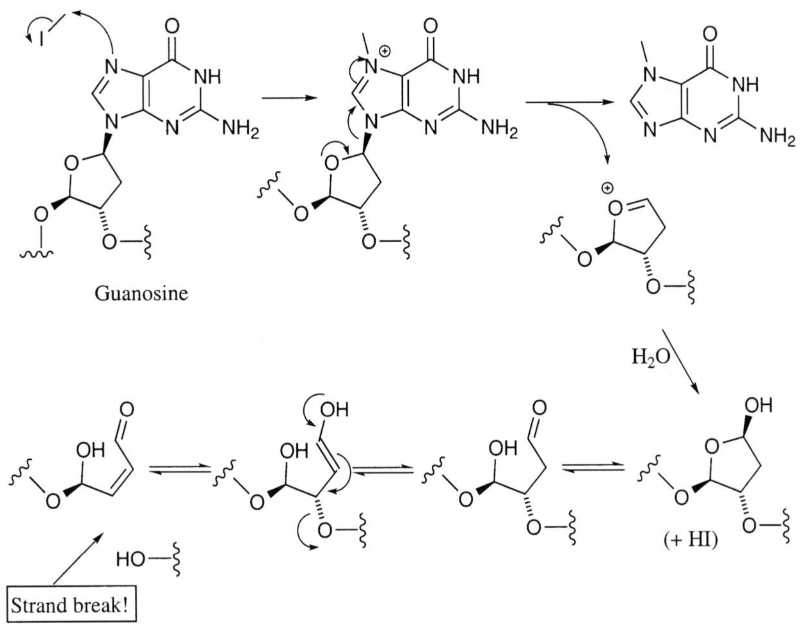

Figure 11.8. The methylation of guanosine with iodomethane promotes the hydrolysis.

and the corresponding half-life of adenine and guanine (in DNA) that has been alkylated in position 7 (the major position for alkylations, see Figure 11.5) is less than 200 hours. As free nucleosides, 7-alkylated adenine and guanine are even less stable, with half-lives of a few hours. The reason for this is evident from the deficit of electrons that such an alkylation will cause, and the mechanism for the hydrolysis is shown in Figure 11.8. Note that the hemiacetal produced by this hydrolysis is in equilibrium with it's open form, the γ-hydroxyaldehyde, which may undergo β-elimination according to the mechanism shown in Figure 11.8 (via the corresponding enol). This will result in a strand break, a serious DNA damage for any cell to handle.

11.3.4 Modification of other macromolecules

A large number of enzymes and other proteins are involved in the handling of DNA, the synthesis and repair of DNA and the transport of the chromosomes to the two daughter cells during cell division. If these enzymes are not fully functional, for example because they have been affected by chemicals, they may make mistakes leading to mutations. Again, little is known about this, but it is reasonable to assume that the fidelity of enzymes will decrease if they are slightly modified by chemicals. Electrophilic compound will of course be able to affect proteins, by reacting with the nucleophilic functionalities present, and radicals will as well. In addition, these enzymes are considered to be important targets for mutagenic and carcinogenic

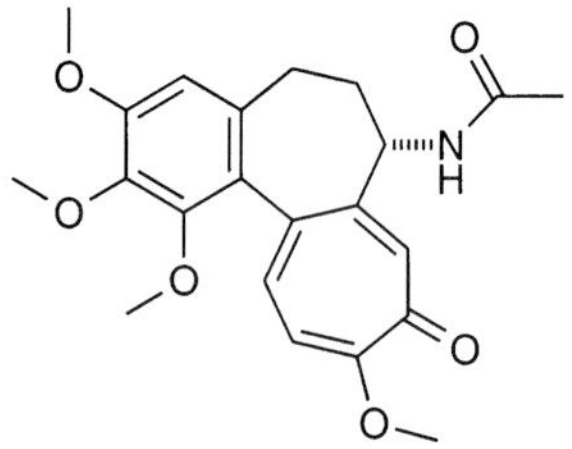

Figure 11.9. Colchicine.

metals (e.g. cadmium, chromium and nickel). Metal ions may bind specifically to proteins by complexing, leading to a change in the three-dimensional protein structure and consequently to a modification of its function.

The transport systems for the chromosomes are still poorly understood, they are complex and can be disturbed by many chemicals. One of the best studied is the toxic plant alkaloid colchicine (see Figure 11.9). Colchicine disturbs the microtubules which are protein structures that function like threads during the cell division and pull the chromosomes to the two cell nuclei being formed. Any exogenous chemical affecting these mechanisms may besides being very toxic also be genotoxic.

11.4 Different types of mutations

11.4.1 Point mutations

Point mutations are small, they only affect one or a few base pairs. They can be divided into two types; base-pair substitutions and frame shift mutations.

11.4.1.1 Base-pair substitution mutations

As the name implies, base-pair substitutions arise if one base-pair in the base sequence, e.g. G–C, is exchanged for another, C–G, A–T or T–A. A base-pair substitution will only change the sequence in one triplet of bases, and if the triplet is part of a gene encoding for a protein, one of the amino acids in this protein (consisting of perhaps several hundreds of amino acids) will at the worst be exchanged in the mutated cell. In the case that this amino acid plays a crucial role for the function of the protein, and the protein plays a crucial role for the function of the cell, the mutation will be serious and possibly lethal for the cell. However, many individual amino acids in a protein may be exchanged without affecting the function of the protein significantly. Base-pair substitutions are therefore, as indicated in Figure 11.10, in most cases not very serious for a cell.

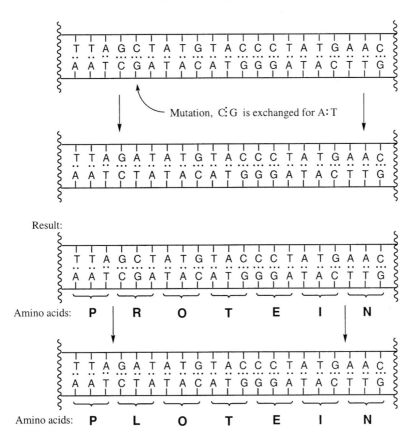

Figure 11.10. The effect of a base-pair substitution.

11.4.1.2 Frame shift mutations

A shift in the frame of a double helix means that one or several base-pairs are added or removed, resulting in a mutated cell with one or more extra or fewer bases. A frame shift mutation in a gene will in most cases be much more serious for the cell compared to a base-pair substitution. If the number of base-pairs added or removed is not divisible by three, the reading frame for the encoding enzymes will be distorted from the position of the mutation and in principle all triplets and thereby all amino acids will be changed. This will result in an useless protein, and the effect for the cell is more obvious as indicated in Figure 11.11.

If the number of base-pairs added or removed is a multiple of three, the immediate effect of a frame shift mutation will be more comparable with that of a base-pair substitution. Besides affecting genes for proteins, mutations can also change base sequences used as for example a

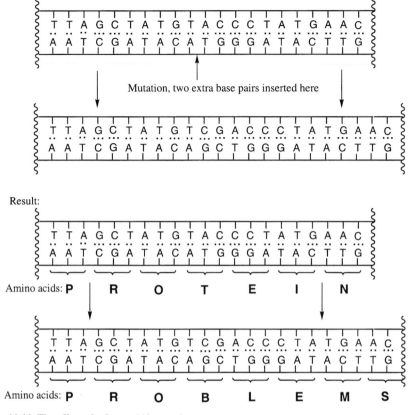

Figure 11.11. The effect of a frame shift mutation.

"stop" signal, or to which enzymes or other proteins bind in order to initiate or block the encoding of a certain gene (*vide supra*). However, there are no general rules about the effects of various point mutations in such cases.

11.4.2 Chromosome aberrations

Genetic alterations that are so extensive that they can be observed when the condensed chromosomes are observed in a microscope are called chromosome aberrations. They include the loss of a chromosome or a part of a chromosome from the genome of a cell, the transfer of a part from one chromosome to another, or the duplication of a part or a whole chromosome (see Figure 11.12).

Cromosome aberrationsnormally result in the death of the afflicted cell, and this is certainly to be expected if genetic information has been lost. However, although no information has been

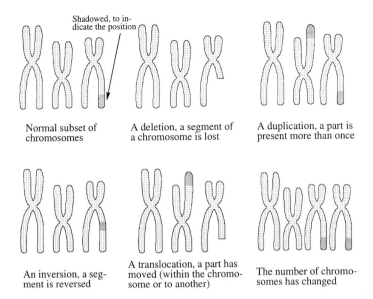

Shadowed, to indicate the position

Normal subset of chromosomes

A deletion, a segment of a chromosome is lost

A duplication, a part is present more than once

An inversion, a segment is reversed

A translocation, a part has moved (within the chromosome or to another)

The number of chromosomes has changed

Figure 11.12. Examples of chromosome aberrations.

lost after a transfer of a piece of a chromosome from one position to another, this is nevertheless a serious mutation because the cell's possibilities to control how the information in the transferred part should be used has changed. Even if one whole chromosome has been duplicated and exists as a free-standing perfect copy this will give rise to an slight imbalance in the genetic information that does not leave the cell untouched. However, chromosome aberrations are not necessarily lethal, and for example in the case of C-21 trisomy (Down's syndrome) in which all cells of a human have an additional copy of the 21st chromosome, the differences with a normal individual are not that conspicuous.

In addition to point mutations and chromosome aberrations it is reasonable to assume that intermediate mutations, affecting one or a few hundreds of base-pairs, also take place, but we have at the moment very little knowledge about such mutations.

11.5 Chemical mutagenesis

As has been discussed above, the most likely event that will result from a chemical modification of DNA is that it is repaired, and very few chemical lesions actually give rise to mutations. This happens when the repair systems fail because the lesion is too complicated, when there are too many injuries so that the repair systems are saturated, or when the cell is dividing rapidly. The process when chemically modified DNA is transformed to mutated DNA may proceed via

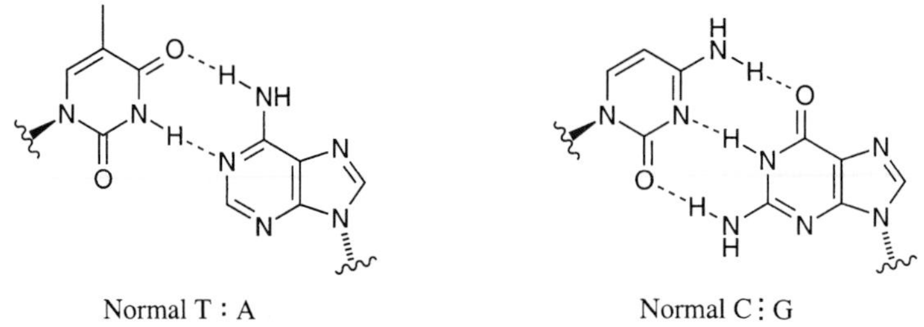

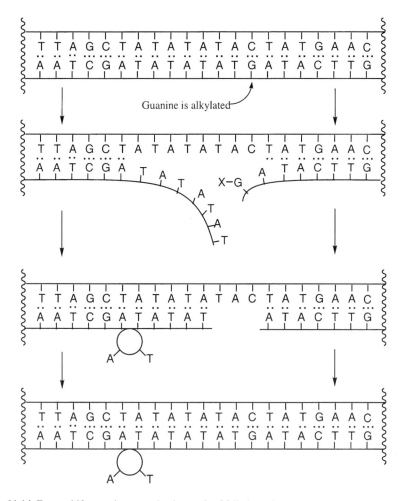

Figure 11.13. The hydrogen bonds between the normal DNA base pairs.

Normal T : A

Normal C : G

Guanine is alkylated

X–G

Figure 11.14. Frame shift mutations may be the result of failed repairs.

different mechanisms, and this is another area where our knowledge is not complete. Base-pair substitutions can obviously be the result of a mismatch, when a chemically modified base pairs up with an incorrect base. Modifications that affect the hydrogen bonding ability of the normal base pairs (see Figure 11.13) could in principle lead to a mismatch that eventually results in a base-pair substitution.

Chromosome aberrations could be the result of damage that cut both strands in DNA, and an example of a compound that is able to do this is the antitumour antibiotic calicheamicin γ_1^{Br} (Section 5.4.2).

Any type of mutation can be caused by unsuccessful attempts to repair a chemical injury to DNA, and this is how frame shift mutations are probably induced. Frame shift mutations are most common in repetitive stretches in the DNA, containing the same or the same combination of bases over and over again as exemplified in Figure 11.14. If one base in such a stretch is modified and subject to the efforts of for example the excision repair system (*vide supra*), the deoxyribose-phosphate polymer is hydrolyzed close to the modified base. This induces some motility of the ends close to the cut, and if one end dissociates for a second and then binds back again, a loop containing one or several bases can be formed. Meanwhile, if the modified base has been hydrolyzed all is set for adding the missing bases (with the complementary strand as template) and sealing the deoxyribose-phosphate chain. As shown in Figure 11.14, the result is that one strand in the double helix now contains additional bases, and if the cell containing this DNA enters the cell division cycle before this has been recognized one of the daughter cells will have a frame shift mutation.

11.6 Effects of mutations

We have already noted that chromosome aberrations are more serious to a cell compared to point mutations, and that frame shift mutations are more serious than base-pair substitutions. However, we are composed of so many cells of so many different functional types (see Chapter 3), and the risks for multicellular organisms are obviously not the same as for individual cells. For a discussion about the effects of mutations, it is convenient to distinguish between the three principally different cell types germ cells (in mammals sperm and ova and their precursors), embryo cells (of which an embryo consists) and somatic cells (the cells that make up a complete individual) (see Figure 11.15).

11.6.1 Effects of mutations in germ cells

Germ cells, sperm and ova in mammals, participate in conceptions and may result in new individuals. In an conceived ovum that will develop into an embryo and further into a foetus (discussed in Chapter 10), mutations in either the ovum or the sperm that participated are very likely to create difficulties. The major reason for this is that in principle all genetic information present in these cells is important and will be used very soon. Mutations in germ cells therefore

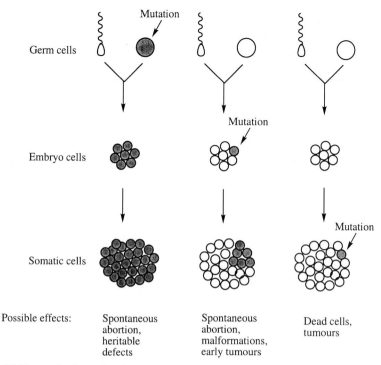

Figure 11.15. The result of mutations in various cell types.

have a relative high probability to create a problem, and problems in an embryo frequently lead to a miscarriage within a short time after conception. In some cases, when the mutation is insignificant for the development of the embryo and the foetus, it may survive and is then of course present in every cell of the whole individual including its germ cells (see Figure 11.15). If the mutation does not prevent the individual reaching sexual maturity and does not interfere with reproductivity, the mutation will be transferable to future generations and is in principle eternal.

11.6.2 Effects of mutations in embryo cells

The cells in a growing embryo need more of the genetic information available compared to somatic cells, and are therefore more sensitive to mutations. In addition, the fact that they divide more frequently enhances the risk that a chemical lesion will be transformed into a mutation. The earlier during a pregnancy a mutation takes place, the more likely is it that the result is a miscarriage. Various malformations can also be the result of mutations in embryo cells, and this was discussed in Chapter 10. In addition, it is believed that many cancers afflicting children are caused by mutations disturbing the control of the cell division that occurs in embryo cells.

Cancers, as will be discussed below, normally appear in older people because it takes time for most cancer cells to develop, but if the transformation starts already in the womb the process may be finished at an early age.

11.6.3 Effects of mutations in somatic cells

The cells of a complete individual are generally the least sensitive, simply because they needs little of the available genetic information to fulfil their tasks and because most somatic cells are relatively unimportant for the organism. If a mutation occurs in a somatic cell, the most likely result is that it takes place in a part of the genome that is not used by this cell and therefore will pass unnoticed. If the mutation takes place in an active part of the genome and affects information that is vital for the cell, the cell will not work as well as it should and may die. In that case it will most likely be replaced by a neighbouring cell that divides. Some cells, e.g. nerve cells, are not replaced, and lethal mutations in such cells will slightly deteriorate the function of the individual. The worst thing that can happen is that mutations in somatic cells change them in a way that they evade the control systems necessary for the cooperation of various cells in a multicellular organism. For example, the rate of cell division has to be kept under strict control, this control is based on genetic information and mutations may transform a cell to a tumour cell (*vide infra*). However, even if the diseases that result from mutations in somatic cells may be lethal, it is important to remember that the effect is restricted to that individual.

11.6.4 Which mutation is most serious?

Chromosome aberrations are more serious than point mutations, for any cell. However, if one instead considers organisms, like human beings, the question is not so easily answered. Is a chromosome aberration in an epithelial cell in the lung that results in the transformation of this

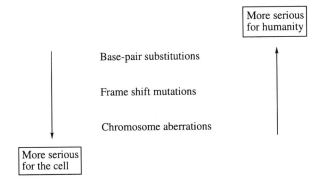

Figure 11.16. The effect of different kinds of mutation.

cell to a tumour cell worse than a frame shift mutation in a germ cell that will induce a miscarriage? If the perspective is further expanded to the human species, or life on earth in general, the question is simplified again. For humanity the base-pair substitutions in germ cells are the most serious, because they have the greatest chance to pass from one generation to another and become part of the human genome. Those are the mutations that really matter in the long run, and that we should be most concerned about!

11.7 Chemicals and cancer

While mutations have been absolutely necessary for the development of life on earth, and a prerequisite for the existence of man, there are no positive effects in a short run. In organisms that reproduce sexually there are also other mechanisms for creating new sets of the genome. Man would therefore continue to evolve also in the absence of mutations and it is reasonable to assume that mutations are harmful. One of the most dreaded effects for man is that an individual cell loses its ability to control its rate of division and is transformed to a cancer cell. The loss of control over the rate of division is typical for a tumour cell, although it should be remembered that tumour cells can be benign or malignant. While benign tumours, made up by benign tumour cells that divide frequently, are seldom dangerous because they only grow to a certain size and do not spread, the malignant tumour cells grow faster and infiltrate other tissues (metastasis) and will generally kill its victim if not treated. Cancers are malignant tumours. Anything that is carcinogenic is then by definition able to speed up the process of transformation of a normal cell to a cancer cell.

It is well known that many chemicals are carcinogenic, and this is at least partly responsible for the "chemophobia" that is widespread in modern society. Our knowledge about the carcinogenicity of chemicals is based on either epidemiologic investigations, when large groups of exposed individuals (cigarette smokers, workers in certain chemical industries, etc) are compared to unexposed individuals, and on the results of animal experiments, during which laboratory animals are exposed to fairly large doses of the tested chemical. Epidemiologic investigation, which is the only way to establish that a chemical is carcinogenic to man, suffers from very low sensitivity, as only the normal exposure levels can be studied. In order to ascertain a statistically significant excess of tumours in a group of exposed people, the group has to be large (e.g. smokers), the chemical has to be potent (e.g. bis-chloromethyl ether, previously used for the manufacture of resins), or it has to give rise to a very unusual tumour (e.g. vinyl chloride, still used for PVC production). Largely due to this insensitivity, only 30–40 chemicals are proven carcinogens in man. We can assume that chemicals that are carcinogenic to animals also are carcinogenic to man, although as discussed in Chapter 5 this is not necessarily true, and data from reliable animal experiments increase the number of chemical carcinogens to a few hundreds. However, even non-smokers and people not working with chemicals are afflicted by tumours, and one may ask what is causing the tumours of apparently unexposed people. Besides chemicals, various forms of radiation (e.g. ultraviolet, x-ray and ionizing radiation) and some viruses may also be carcinogenic, and for some types of tumours there is a strong hereditary disposition, so the significance of chemical exposure for the total incidence of tumours is not evident.

One way to study the causes of tumours is to compare how common various types of tumours are in different parts of the world. Surprisingly, some tumours are more than 100 times more common in one place (the high-risk area) compared to another (low-risk area). In general, every country has its own tumour profile, and while, for example, England is a low-risk area for liver cancer it is a high-risk area for lung cancer. Such differences may be explained by the different environment, habits and background exposure in different countries, the English have for example always been relatively heavy smokers and the high incidence of skin cancer in Australia may be caused by the large proportion of fair-complexioned individuals not adapted to the strong sunlight of this continent. However, another explanation is that different people have slightly different genomes, and that, for example, the English are predisposed for lung cancer. The importance of genetic factors for the incidence of tumours has been estimated by comparing how tumour trends change for populations that emigrate from one part of the world to another. Several investigations have shown that such groups in a few generations adopt a similar tumour pattern to the genuine inhabitants of the new home country, even if they during this time keep a substantial part of their original genome intact. All this information suggest that the majority of all tumours in industrialized countries are caused by external carcinogenic agents, while approximately 20 % depend on hereditary factors. However, only few tumours are solely due to hereditary factors, instead these will modify the impact of an external carcinogen by for example not being able to detoxify a carcinogenic chemical efficiently or to repair damage to the DNA. Radiation in various forms is responsible for a few percent, comprising most cases of skin cancers. The impact of viruses is not known, but a reasonable guess is a few percent. In addition, the deficiency of essential chemicals (e.g. vitamins and selenium) due to unbalanced diet is associated with an increased risk for tumours, although it is difficult to estimate its importance. Also, tumours may of course arise from spontaneous mutations that take place in all cells. The rest of the tumours of mankind are in principle caused by chemicals, and the two major sources for the carcinogenic chemicals are tobacco smoke and food. While most people are aware of the hazards of smoking, fewer are aware about the impact of food for the incidence of tumours. However, the amounts of food ingested during a lifetime are large and even minute concentrations of carcinogenic chemicals in food will with time build up to significant amounts. Chemical carcinogens in food originate from, for example, the food itself (several natural products present for example in edible plants and mushrooms are carcinogens), from the ways we prepare food (frying or grilling meat will transform carbohydrates and amino acids to potentially carcinogenic heterocyclic compounds), or are formed in badly stored food due to the presence of microorganisms (e.g. moulds) that produce potent carcinogens (see Figure 11.17 for examples).

Another possibility is the production of carcinogenic chemicals by the intestinal flora as a response to an imbalanced diet, giving some bacteria the possibility to proliferate at the expense of others. The fecapentaenes are potent mutagens and suspected carcinogens formed by certain bacteria, and were discussed in Chapter 6. Besides the chemicals that we are exposed to from smoking and eating, carcinogenic chemicals are also encountered at workplaces and in the environmental in general, occurring naturally or as pollutants, and although the local variations

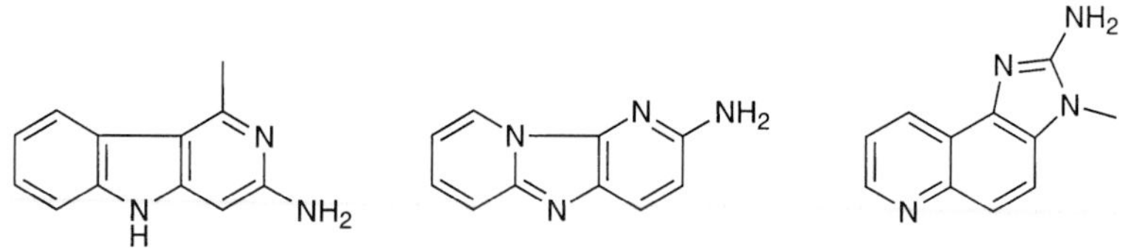

Genotoxic compounds formed in grilled food

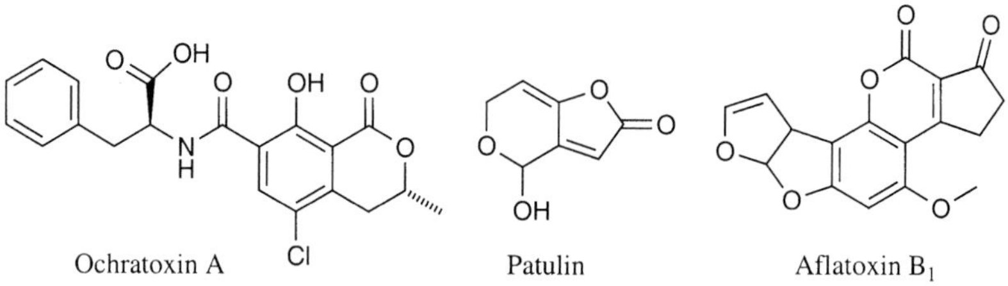

Ochratoxin A   Patulin   Aflatoxin B$_1$

Genotoxic metabolites of fungi that may infect foods

Figure 11.17. Carcinogens that may be present in grilled or mouldy food.

are substantial due to for example the presence of a ground rich or poor in minerals that are carcinogenic (e.g. cadmium and radon (Rn) which is radioactive) or essential (e.g. selenium), and polluting industries, such factors are considered to cause a few percent to the tumours of the industrialized countries.

11.8 Chemical carcinogenesis

A cancer cell is completely different from a normal cell, even compared with the cell that it originates from. Apparently, several major changes must have taken place, either as a result of multiple mutations or from the activation/deactivation of silent/active parts of the DNA. Remember that somatic cells still contain the entire genome although they only make use of a small part, and the parts that are not used are "shut off" in various ways. However, if reactivated, by for example a chromosome aberration, proteins that the cell should not produce are suddenly formed and this may have dramatic consequences. The fact that it normally takes a long time (decades) between the exposure to carcinogenic chemicals to the appearance of a diagnosticable tumour may reflect the necessity of several discreet changes in the developing cancer cell, although the time scale is also influenced by the ability of the immune system to detect and kill abnormal cells and thereby probably eliminate most tumour cells formed, or at least constantly reducing the number of cells in a growing tumour. A central feature of a cancer cell is that it is dividing continuously. Normal cells can also divide, but do so only when there is a need for example during the healing of a wound after an injury. The information regulating cell division is stored in the genome, and although it is not understood in detail how cell division is regulated external signals must through some kind of signal transduction be translated into an effect on the cells DNA that makes it respond by initiating the process of cell division. This can be imagined as either the activation of genes that produce proteins that are important for the process or the deactivation of genes producing proteins that repress it (by repressors, *vide supra*), or both, and will be further discussed in section 11.8.2. As the changes that transform a cell to a cancer cell are hereditary, they must have a genetic base and in fact be mutations. Chemicals may cause such changes either by being reactive (directly or after metabolic conversion/chemical transformation) and interact directly with DNA to produce mutations (as described earlier in this chapter), or affect DNA in indirect ways (discussed below). However, it should be remembered that many chemicals are carcinogenic by more than one of the mechanisms discussed in this section.

11.8.1 Reactive or pre-reactive chemical carcinogens that act on DNA

There is obviously a strong link between chemical mutagenesis and chemical carcinogenesis, and compounds that are mutagenic according to the mechanisms discussed earlier in this chapter should at least be strongly suspected to be carcinogenic as well. Differences may occur when mutagenicity has been assayed in other organisms, for example bacteria or yeasts which are more convenient to use as test organisms than animals or mammalian cells, because of differences in the secondary metabolism, cell membrane, etc. To summarise, reactive or pre-reactive chemicals that act on DNA may be divided into the following categories:

11.8.1.1 Electrophiles

Electrophilic compounds, either as they are or after metabolic activation. Such compounds will react with DNA to transform it to chemically modified DNA, which either may result in base-pair substitutions, frame shift mutations or chromosome aberrations. Most chemical carcinogens known today belong to this class of compounds.

11.8.1.2 Radicals or reactive products formed after radical reactions

In general, radicals are too reactive compounds to be responsible for systemic effects, if not formed inside the cells by for example the metabolism of xenobiotics or by lipid peroxidation. The radical reactions in the cells will not only produce new radicals that by themselves may react with for example enzymes and DNA, but also electrophilic products of the oxidative degradation of the unsaturated fatty acids that may be mutagenic.

11.8.1.3 Metal ions

Several metal ions are known to possess mutagenic and carcinogenic activities, e.g. Ni^{2+}, Cd^{2+} and Cr^{6+}. The chemical mechanism for their activity is, however, not yet understood. One can imagine several possibilities, for example that the metal ions make a strong three-dimensional complex with the heteroatoms of DNA and that this complex is recognized as chemically damaged DNA by the repair systems that initiate a repair process that results in a mutation. Other possibilities are that the metal ion while bound to DNA by electrostatic forces will catalyze redox reactions that result in a chemically modified DNA, possibly via the formation of reactive oxygen species, and that the metal ions are complexed by the enzymes that synthesize/repair DNA (*vide infra*).

11.8.2 Chemical carcinogens affecting DNA by indirect mechanisms

While chemical mutagens are likely to be carcinogens as well, the reverse is not necessarily true. Chemicals may also be carcinogenic by for example generating mutations by indirect mechanisms, or to influence the way the genetic information is apprehended and interpreted. There are several ways for our cells to regulate which parts of the genome that should be active or inactive, for example by the selective methylation of certain bases or by the influence by external factors that may be hormones. Anything that interferes with such processes, for example by demethylating methylated DNA-bases, may activate "foetal" parts of the genome (used during the early developments) which in turn may accelerate cell division. Besides the obvious effect that such events may lead to the loss of the control of how the cell should divide, and as discussed in section 11.7.2.3, cell division *per se* may generate mutations and further develop the emerging cancer cell.

11.8.2.1 Chemicals acting on the components of DNA metabolism

The biochemical components involved in the synthesis, transport, protection and repair of DNA and the nucleic acids may also be the targets. Reactive compounds in general may besides affecting DNA also react with other biomolecules, and if they react with the proteins and enzymes that take part in the synthesis, transport or repair of DNA this may result in the loss of their fidelity and consequently the risk for mutations increase. In addition, one can imagine that the free DNA-nucleotides that are used for synthesis or repair of DNA are modified by reactive compounds, and if these pass the proof-reading systems they may induce mutations as well. Chemicals that reduce the level of gluthathione, for example by simply reacting with it, will also be harmful and indirectly enhance the possibility of another chemical carcinogen to interact with DNA.

11.8.2.2 Alteration of gene expression

Mg^{2+} and Zn^{2+} play important roles in the transcription of genetic information, and several metals are able to induce and/or inhibit the formation of proteins by interacting with the regulation of genes. Any chemical that has such an effect may be carcinogenic, by for example inducing oncogenes/repressing anti-oncogenes (see section 11.9), or inducing enzymes that activate procarcinogens to carcinogens.

11.8.2.3 Mitogenic chemicals

Several carcinogenic chemicals have been shown to possess virtually no chemical reactivity, but instead stimulate the cells to divide. This may be caused by a specific mechanism, by which compounds bind to receptors on the outside of the cell and convey a signal to the nucleus that "it is time for a division". Such signals are absolutely critical when cells should divide, but only then. Examples of compounds that may fit into such receptors are phorbol esters and TCDD, both of which are carcinogenic without binding significantly to DNA.

The phorbol esters are present in the oil of croton seeds (*Croton tiglium,* Fam. Euphoribaceae) and the latex of other plants of the same family. They are extremely irritating and will cause blisters and burns if they contact the skin. Ingested in very small amounts, croton oil is strongly catharactic and has been used for this purpose, in larger doses (approximately 1 ml) it will cause severe and potentially lethal effects on the oesophagus.

Although in principle all chemical mutagens should be mitogenic, chemical mitogens does not have to be mutagenic and mitogenesis will stimulate cell division without necessarily resulting in critical mutations. In that case, if the exposure to the chemical mitogen stops, things will simply go back to normal (= as they were before the exposure to the mitogen). However, the increased cell division rate will increase the risk for spontaneous mutations. It has been calculated that approximately 10,000 chemical modifications of DNA takes place in every cell

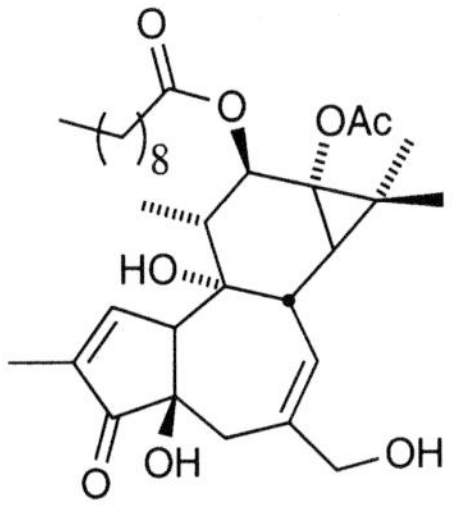

One of several naturally
occurring esters of phorbol

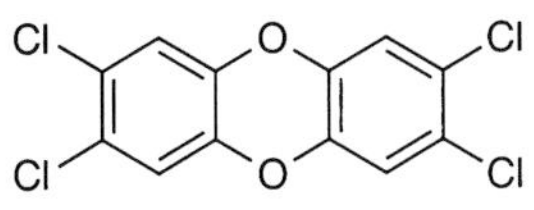

TCDD

Figure 11.18. Specific mitogenic compounds.

every day (which is not much if the size of DNA is considered). These are normally efficiently repaired, but if the cell constantly divides it will not have time to carry out all reparations with the required fidelity. In this way, one can say that cell division is genotoxic! Besides being specific, chemical mitogens can also be non-specific and irritate cells so that they divide. This is nothing but a general toxic effect, and such compounds are mitogenic while the concentration is cytotoxic. Many compounds, if not most, will have this effect in high doses, and it has been questioned if it is reasonable to assay carcinogenic effects at cytotoxic doses.

If a mammal is exposed to mitogenic compounds together with or after a genotoxic compound, the effect of the genotoxic compound will be enhanced because the cell's possibility to repair DNA damages is restricted by the mitogen. In such cases, the mitogen is called a cocarcinogen or a promotor (*vide infra*).

11.8.2.4 Cocarcinogens and promotors

A cocarcinogen is something that enhances the effect of a genotoxic compound if the two are given together, while a promotor enhances the effect of a genotoxic compound if it is given later. There are several mechanisms for cocarcinogens and promotors. Compounds that induce cytochrome P-450 or other enzymes that may be involved in the metabolic activation of a genotoxic compound are cocarcinogens, and ethanol together with other simple alcohols is an example of this. In addition, compounds that react chemically with another compound to form a genotoxic adduct are also cocarcinogenic, and we have previously seen that sodium nitrite can react with secondary amines to form carcinogenic *N*-nitrosoamines. All genotoxic compounds will be cocarcinogens and promotors, because they are toxic, and for example liver infections will promote liver cancer while asbestos and tobacco smoke will promote the genotoxic activity of certain components in tobacco smoke leading to lung cancer. The classic picture of chemical carcinogenesis, established by extensive animal experiments decades ago, is shown in Figure 11.19. The process starts with the initiation, which is an irreversible mutation brought about by a genotoxic compound. The mutation strikes at a critical disposition, that is central in the transformation of a normal cell to a cancer cell, but this mutation alone is not enough. Other changes are also required, and they can be caused by the action of genotoxic compounds as well, or by "spontaneous" mutations happening if the mutated cell is stimulated to excessive cell division. This is where promotion comes in, and although it was never understood exactly what cellular changes were required and in which order they should take place, promotion was considered relatively harmless if it did not go beyond a critical point. The discontinuation of promotion will then only lead to the loss of the redundant cells and in principle the reversal of the process to the situation after the initiation (mutations will of course not be corrected). However, if promotion (or exposure to genotoxic agents) continues, perhaps for years or decades at a slow rate (e.g. smoking), it may eventually induce all the changes that are necessary to convert the cell to a cancer cell (a process called transformation). After transformation it is all down-hill, from the point of view of the cancer cell, and it will rapidly divide and spread via the blood vessels and the lymphatics to other tissues and organs where metastases (daughter tumours) are formed. The metastasis is essentially the reason why cancers are so difficult to treat.

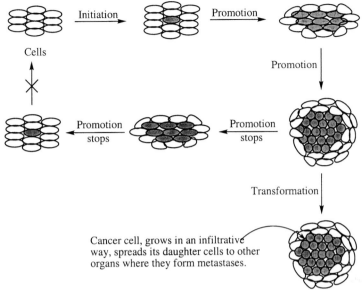

Figure 11.19. The classic view of chemical carcinogenesis.

Today the events that are important for the transformation are better understood, and a brief outline of this is given in section 11.9.

11.8.2.5 Immunosuppressive agents

Immunosuppressive agents are compounds that decrease the ability of the immune system to respond to stimulation, and are used to prolong allograft survival after organ transplantations and to treat autoimmune disorders. Although it has limited importance, it has nevertheless been observed that for example renal transplant patients treated with immunosuppressive drugs are more likely to develop tumours. As one of the duties of the immune system is to control and limit the growth of new tumour cells, it is not surprising that the imbalancing of the surveillance functions of the immune system increases the possibilities of cancer cells to proliferate. Several of the compounds discussed above that are believed to be carcinogenic by indirect mechanisms, e.g. TCDD (see Figure 11.18), also possess immunosuppressive activity which in fact may increase their carcinogenicity.

11.8.2.6 Hormones

Several oestrogens have been shown to promote the liver carcinogenicity of mutagenic compounds, either by increasing the mitotic activity of the liver cells or by inducing certain cyto-

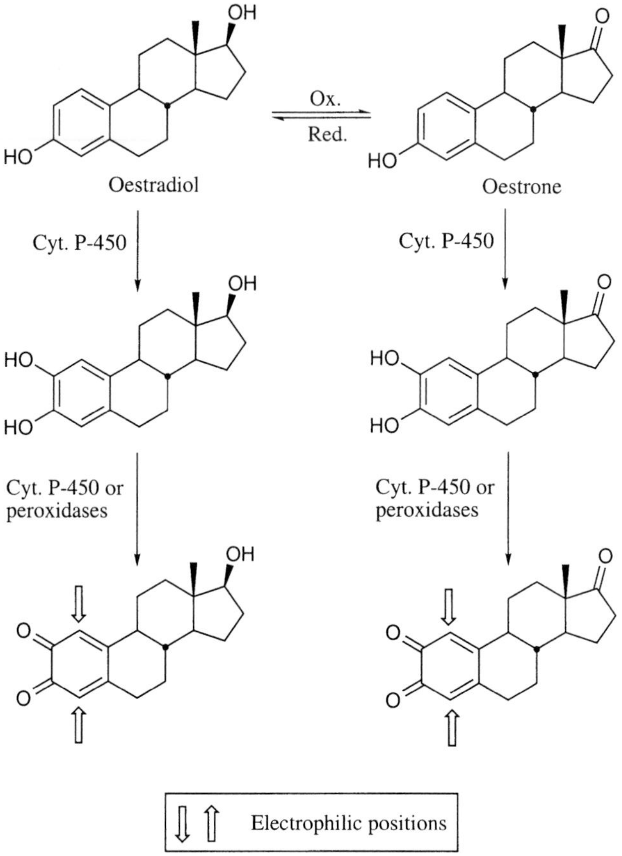

Figure 11.20. The metabolic activation of oestrogenic hormones to quinone derivatives.

chrome P-450 enzymes. However, oestradiol, which is the most potent of the mammalian oestrogenic hormones produced by the ovary, and oestrone, which is its oxidation product, may also be metabolized themselves to reactive products.

11.9 Oncogenes, proto-oncogenes and anti-oncogenes

Oncogenes are genes that will make a cell lose control over its division rate (i.e. turn a normal cell into a tumour cell) when activated (i.e. being transcribed from DNA to RNA and translated from RNA to protein) in the cell. They were first discovered when carcinogenic viruses were investigated. Because a virus only contains a few genes it was possible to separate

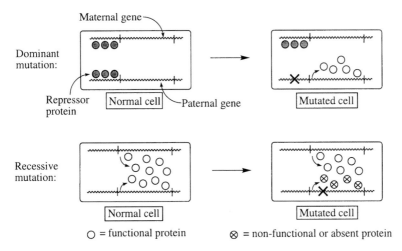

Figure 11.21. The difference between dominant and recessive mutations.

the viral genes and determine which one (i.e. the oncogene) was associated with the loss of cell division control when the virus infected a mammalian cell. Interestingly, oncogenes were also found to be present in the genome of all human cells, as identical but inactive copies of the oncogenes found in viruses or as almost identical copies in which one or a few base pairs are different, and such genes are called proto-oncogenes. The conversion of a normal cell to a tumour cell may then simply be a question of turning on proto-oncogenes, by activating their transcription/translation or, in the case the proto-oncogene is not an identical copy of an oncogene, by a mutation in the proto-oncogene in the correct position so that it becomes an oncogene. In the case the expression of an proto-oncogene is supressed by a repressor, a mutation in the DNA sequence that the repressor binds to could affect its binding and initiate the transcription. This would be a dominant mutation, and the effects of dominant mutations are observable directly (see Figure 11.21). Another possibility is that the regulation of the proto-oncogene is affected by a chromosome mutation, which moves it from an inactive region in a chromosome to an active region, perhaps in another chromosome, and this has been observed in a human cancer called Burkitt's lymphoma.

The proteins encoded by the oncogenes are in most cases enzymes with protein kinase activity, which phosphorylate certain amino acids (tyrosine and serine/threonine) in other proteins. Phosphorylation of proteins is an important tool used by the cells to initiate and control cellular processes, as a protein (enzyme) may behave very differently if it is phosphorylated or not. One phosphorylating enzyme may phosphorylate many other proteins and thereby bring about several changes in a cell, including expression of genes. Protein kinases are certainly involved in the signals that make a cell divide, and the activation of oncogenes coding for protein kinases is believed to be a key step in chemical carcinogenesis. However, for multicellular organisms the regulation of cell division is crucial, and the evolution has provided mammalian cells with additional control systems.

Consequently, cell division can also be supressed by proteins produced by the cell, and the genes coding for such proteins are called anti-oncogenes. They were initially discovered when a hereditary tumour of the eye (retinoblastoma), that almost exclusively afflicts infants, was investigated. The cells of this cancer are characterized by the lack of a protein, that normal cells produce and that acts as a suppresser of cell division. (This protein is a repressor, that blocks the transcription of a gene important for cell division.) This anti-oncogene is continuously expressed in cells that do not divide, in the cells of both hereditary afflicted individuals and normal individuals. However, what runs in the families where the risk for retinoblastoma is high is a defect gene for the repressor protein, but this defect is not noticed as long as the complementary gene, originating from the other parent, is functional. The mutation that has caused this defect is called recessive, it can be transferred from generation to generation and will not be noticed until the complementary gene is damaged (see Figure 11.21). In individuals that are hereditary tainted for retinoblastoma, one of the two anti-oncogenes discussed above is defect, and the risk that a mutation strikes at the other and thereby stops the production of the repressor protein in a cell is much higher compared to the risk that two independent mutations inactivate both anti-oncogenes in a cell of a normal individual (see Figure 11.22).

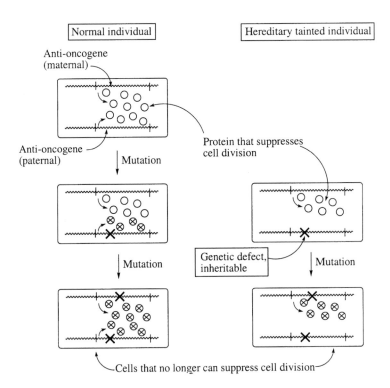

Figure 11.22. Mutations that will stop the production of a protein that suppresses cell division, in normal individuals and in individuals hereditary tainted for retinoblastoma.

Instead of the concepts of initiation, promotion and transformation, the process when a normal cell develops into a cancer cell can now be regarded as a series of genetic changes that involves the activation of proto-oncogenes and inactivation of anti-oncogenes. It is still not understood what other changes, if any, are needed, or how many and which proto-oncogenes/anti-oncogenes have to be changed, or in what order, but at least some of the gene targets of chemical carcinogens have been identified.

12 Environmental effects of chemicals

The environmental effects of pollutants are often very complicated and difficult to foresee, as sensitive balances in ecosystems are affected. An example is, again, the use of DDT to kill mosquitoes that spead malaria, which besides the long-term effects also resulted in dramatic local effects in the areas where it was used. In Borneo for example, the insects killed by DDT were eaten by lizards which were intoxicated and became easy preys for cats. The cats in turn became ill and died, the absence of cats resulted in an explosion of the rat population, and the rats consumed the local crops and spread dangerous diseases. The government of Borneo eventually had to reintroduce cats to cope with the problems! Chemicals that we consider to pollute the environment may be anthropogenous, resulting from human activities, or natural. This chapter will mainly deal with the anthropogenous, which are those we can do something about, but it is important to relate these to the natural emissions which for several classes of chemicals are substantial.

12.1 Natural emission of chemicals

12.1.1 Inorganic chemicals

Volcanoes are the major polluters of our environment. Since the beginning of time volcanic activity has more or less continuously belched out huge amounts of for example sulphur compounds and ash into the atmosphere where it is distributed worldwide by the weather systems. A massive volcanic eruption is extremely dramatic and dangerous. Several of those that happened during the last centuries immediately killed more than 10,000 persons, and the emission of chemicals may under such circumstances be considered to be a secondary problem. The eruption of "El Chichon" in Mexico in 1982, which did not cause that much damage in terms of lives lost, is nevertheless estimated to have brought 500,000,000 m^3 of ash, pumice and dust into the atmosphere, and 40,000 tons of HCl into the stratosphere (accounting for an immediate 40 % increase of the HCl concentration). The local effects of major eruptions are of course massive, but global effects can also be observed. During the Mount Pinatubo eruption in the Philippines in 1991, approximately 10 million tons of sulphur in the form of sulphur dioxide were blasted into the stratosphere. The sulphur dioxide is oxidized by molecular oxygen (via reactive oxygen species such as the hydroxyl radical and hydrogen peroxide) to sulphur trioxide, which reacts with water to form sulphuric acid. The sulphuric acid condenses to small droplets with a diameter around 1 µm, which are spread around the globe and can stay in the stratosphere for a long time (more than a year). Such aerosols not only reflect part of the irradiation from the sun, but they will in addition act as cloud condensation nuclei. Clouds will also scatter the sun's rays, but will in addition absorb and to some extent reflect the heat radiation from the earth, and the overall effect of a major vulcanic eruption on the temperature will be a global cooling.

Lightning releases large amounts of energy in the atmosphere that transforms molecular oxygen and nitrogen to nitrogen oxides, especially nitric oxide (NO) and nitrogen dioxide (NO_2), and ozone (O_3). Nitric oxide, which also is formed during the combustion process in for example car engines, is easily oxidized in air to nitrogen dioxide (*vide infra*).

Winds will transport dust, and especially in desert areas huge quantities will be moved. Finer particles, with a diameter of less than 50 µm, will form an aerosol that can be transported even between continents. A few hundred million tons of mineral dust are transported to the atmosphere as an aerosol yearly.

The contiuous movement of the water in the oceans produces small water droplets that are carried away by the wind and dried. The remaining salt particles (mainly chlorides but also sulphates) that result from this process form an aerosol that transports approximately 1 billion tons of sea salts per year to the atmosphere. Most of it comes down again close to the spot from which it emerged, with the rain, but the wind may of course transport it for substantial distances and part of it ends up on land.

Forest fires release both inorganic and organic chemicals, and among the inorganic gases carbon monoxide, carbon dioxide, nitric oxide and nitrogen dioxide are the most apparent.

12.1.2 Organic chemicals

More complex and highly hazardous organic compounds are emitted in forest fires, for example mixtures of polyaromatic hydrocarbons, polychlorinated dioxins and polychlorinated dibenzofurans. Yes, the famous "doomsday chemical" TCDD is also formed naturally and has been present on earth long before man arrived. The amounts formed in forest fires compared to those formed anthropogenously are difficult to estimate, but several investigations have suggested that they are in fact comparable. It is likely that polychlorinated dioxins and polychlorinated dibenzofurans also are formed *in vivo* by water and soil microorganisms, as it has been shown that chlorophenols (e.g. 1,4,5-trichlorophenol) can be converted to polychlorinated dioxins and polychlorinated dibenzofurans by peroxidase enzymes, and as halogenated phenols are formed by organisms, although this remains to be demonstrated. Also polychlorinated biphenyls (PCBs) are formed naturally, as demonstrated by a study of the ash from the 1980 eruption of mount St. Helens. The dioxins and the PCBs will be further discussed in section 12.6.2.

Living organisms produce large amounts of organic as well as inorganic chemicals that are released into the environment. Some of these compounds, natural products of which we have encountered several examples throughout the text, may be highly toxic. Others possess less biological activity but are more interesting because of the amounts emitted, but in general one can say that natural products are relatively easily biodegradable.

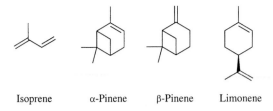

| Isoprene | α-Pinene | β-Pinene | Limonene |

Figure 12.1. A few of the many volatile terpenes that are produced by plants.

12.1.2.1 Hydrocarbons

Simple hydrocarbons (also emitted from petroleum sources) are important because they participate in several of the effects discussed in this chapter. The simplest, methane, is mainly produced by anaerobic microbial degradation of organic matter in for example lake bottoms, landfills and in the stomachs of ruminant animals. Other hydrocarbons, for example the terpenoids, are emitted in large amounts (estimated amounts 1 billion tons yearly) from plants, especially in coniferous forests. A few examples of some common hemiterpenes (with 5 carbons), monoterpenes (with 10 carbons) and sesquiterpenes (with 15 carbons) are shown in Figure 12.1.

12.1.2.2 Sulphur compounds

Sulphur is an important element for organisms, and especially important for some microorganisms (e.g. the chemoautotrophic sulphur bacteria which produce energy by oxidating sulphur to sulphate). Volatile sulphur compounds, mainly hydrogen sulphide, carbon disulphide (formed by anaerobic fermentation), carbonyl sulphide (from volcanic emissions), mercaptanes, sulphides and disulphides, are released by microorganisms as well as by plants, and it is estimated that in total 65 million tons are emitted yearly. The major contributors are the plankton of the oceans, producing approximately 43 million tons of dimethylsulphide. All volatile sulphur compounds will be oxidized to sulphur dioxide in the atmosphere, which in turn will be transformed to sulphuric acid (as described above) and contribute to acid rain (see Section 12.4).

12.1.2.3 Halogenated compounds

Although generally considered as typical synthetic products, halogenated compounds are produced in enormous quantities in nature. We have already encountered a few, some halogenated fungal metabolites that resemble synthetic pesticides (in Chapter 4), and two toxic fluorinated plant metabolites (in Chapter 10). Figure 12.2 shows some examples of compounds produced by marine organisms which are believed to be the major producers of halogenated or-

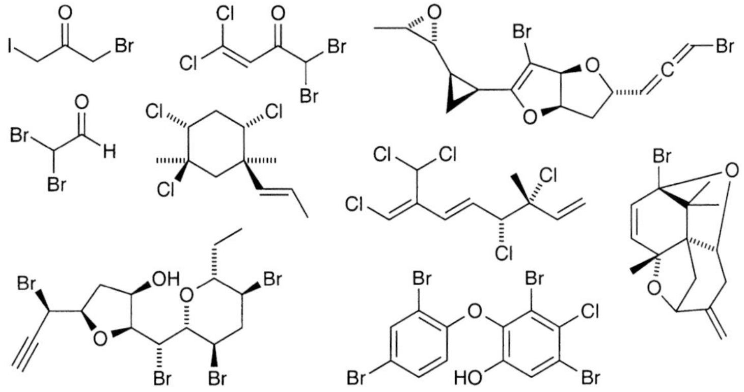

Figure 12.2. A few halogenated organic compounds produced by marine organisms.

ganic compounds in nature. The quantitatively most important halogenated compounds are of course the most simple, and it is estimated that the natural production of chloromethane and iodomethane is in the order of several million tons per year which by far exceeds the emission from anthropogenic sources. As these compounds are volatile they are important for the global transport of halogens, especially from the oceans to land.

Besides fungi and marine organisms, all other organisms also produce and excrete halogenated compounds, and large amounts are also emitted by volcanoes.

12.2 Smog and other air pollution

Smog (smoke and fog) is a typical local problem, affecting an area within 100 km of the place of emission. It is a mixed aerosol, formed as a result of the heavy pollution of sulphur dioxide and soot mainly due to the burning of coal together with an inversion of the temperature layers (a weather condition characterized by cold air close to the ground and warmer air surrounding it that prevents the air pollutants and water vapour escaping). It was previously a big problem in densely populated areas and industrial regions. The smog that reigned for a week in London in 1952, with sulphur dioxide concentrations of approximately 0.7 ppm and 1.5 mg soot per m^3, killed approximately 4000 persons. Happily, as the use of coal has decreased, the incidence of smog has as well.

However, inversions regularly occur over cities, and even if modern heating systems no longer emit soot and particles there is still a lot of air pollution in the form of for example nitrogen oxides and hydrocarbons from traffic that may give rise to what is called photochemical smog. This is a process for which sunlight is essential, and that generates photochemical oxidants like ozone (*vide infra*).

12.2.1 Degradation of trace gases in the troposphere

Most of the UV radiation is filtered by the ozone in the stratosphere, but some amounts pass though and come down to the troposphere, where it may transform some of the few molecules of ozone present to molecular oxygen and an oxygen atom. The oxygen atom thus generated is in an electronically excited state (if the wavelength is below 310 nm), it may react with a molecule of water to form two hydroxyl radicals, and this reaction is the major generator of the very important hydroxyl radicals in the atmosphere. Other examples are the reaction of superoxide with ozone or nitric oxide, or the homolytical splitting of hydrogen peroxide by light. In the troposphere, although present in very low amounts (100–1000 times less) compared to in the stratosphere, the hydroxyl radical will act as a cleaner and react with trace gases that are present and transform them into more hydrophilic forms that are washed away by the rain. Examples are methane and carbon monoxide, which together are responsible for the degradation of 90 % of the hydroxyl radicals in the troposphere in reactions that generate ozone (see Figure 12.3).

Both methane and carbon monoxide is oxidized to carbon dioxide, methane via formaldehyde and carbon monoxide (see Figure 12.3), while the hydroxyl radical transforms other (minor) air pollutants to water (molecular hydrogen), sulphuric acid (hydrogen sulphide and sulphur dioxide), hydrogen chloride (chlorinated organic compounds), and nitric acid (nitrogen dioxide and ammonia). The products formed after the reaction of these gases with the hydroxyl

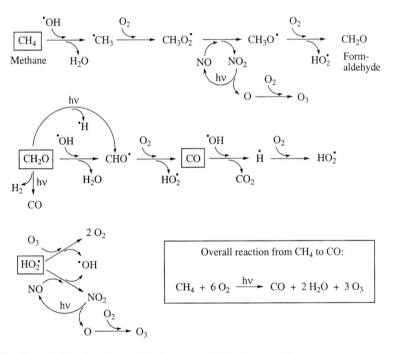

Figure 12.3. Degradation of methane and carbon monoxide in the troposphere.

radical are all water soluble and washed away by the rain. More complex organic compounds present in the atmosphere, for example terpenes emitted by trees, will also be attacked by the hydroxyl radical and transformed by similar reactions into more water soluble products.

12.2.2 Photochemical oxidants

The mechanism for the oxidation of methane and carbon monoxide with the hydroxyl radical in the troposphere is shown in Figure 12.3. The oxidation is coupled with the oxidation of nitric oxide to nitrogen dioxide, which is split by light back to nitric oxide, and an oxygen atom that forms ozone with molecular oxygen. Normally, the ozone formed by this photochemical reaction will be consumed by the oxidation of nitric oxide to nitrogen dioxide. But in the presence of oxygen radicals, e.g. superoxide or alkylperoxyradicals nitrogen dioxide will be formed anyway and there is a net formation of ozone. Ozone is the most important photochemical oxidant, but also for example peroxyacetylnitrate (PAN), peroxypropionylnitrate (PPN) and peroxybenzoylnitrate (PBN) (see Figure 12.4), formed by oxidation of the corresponding aldehydes as shown in Figure 12.4, are important (although their concentrations are much lower).

Ozone is a strong oxidation agent, of which most (90 %) of the total amount in the atmosphere is present in the stratosphere where it plays an important role in the absorption of ultraviolet radiation from the sun (the subject of the discussion in Section 12.4). The remaining 10 % is found in the troposphere, where it is more of a nuisance, but most unfortunately the good ozone (in the stratosphere) is decreasing while the bad (in the troposphere) is increasing (with approximately 0.7 % per year in the lower layer of air). The effects of photochemical smog are essentially due to ozone, and photochemical smog can be highly irritating, especially for individuals suffering from bronchitis or asthma, although not as acutely lethal as the London smog. Plants will be affected by ozone, and many polymers as well as natural materials will be attacked. In addition, ozone is a greenhouse gas (discussed in Section 12.5).

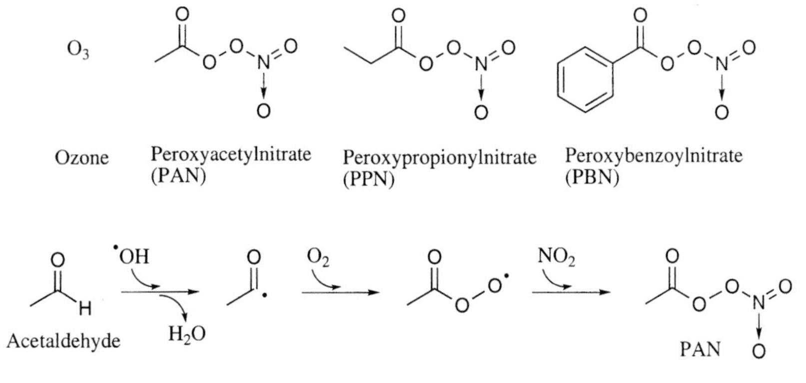

| Ozone | Peroxyacetylnitrate (PAN) | Peroxypropionylnitrate (PPN) | Peroxybenzoylnitrate (PBN) |

Figure 12.4. Major photochemical oxidants, and the formation of PAN.

12.3 Acid pollution

Besides the biotic production of volatile sulphur compounds (see Section 12.1), even bigger amounts (approximately 100 million tons per year) are produced by human activities. The main contribution comes in the form of sulphur dioxide from the burning of fossil fuels, especially coal, but for example the smelting of sulphide minerals is also an important source. The sulphuric acid eventually formed from the anthropogenous emissions add to those formed from the natural emission of volatile sulphur compounds (all forms of sulphur, organic and inorganic, reduced and oxidized, will be transformed to sulphuric acid in the atmosphere). Although the formation of sulphuric acid from various sulphur species can take a number of routes, it can be seen as an initial formation of sulphur dioxide that is oxidized by reactive oxygen species (hydroxyl radical, ozone, superoxide ...) in either the gas phase or in water droplets, to sulphur trioxide that reacts with water to sulphuric acid. The sulphuric acid will be dissolved by the water in rain and clouds and increase the concentration of cloud condensation nuclei, and will eventually come down with the rain as acid rain. Also rain in completely unpolluted areas will be weakly acidic (pH 5.0–5.5), because carbon dioxide will be present and it forms carbonic acid in contact with water; but rain in polluted areas can have pH values down to 4. Because the sulphuric acid is formed quickly in the atmosphere and washed out by rain, the acid rain will essentially come down close (within 1000 km) to the emission source and this is consequently a typical regional problem. Besides sulphuric acid which is responsible for 80–85 % of the acidity in acid rain, nitric acid (from nitrogen oxides formed for example during combustion) and hydrochloric acid (from hydrogen chloride emitted for example by volcanoes) also contribute.

Nitric acid is formed from the nitrogen oxides NO and NO_2 (often called NO_x together). NO is oxidized in the atmosphere by reactive forms of oxygen to NO_2, which in the presence of water (and oxygen) generates nitric acid by the reactions shown in Figure 12.5.

$$3\ NO_2\ +\ H_2O\ \longrightarrow\ 2\ HNO_3\ +\ NO$$

$$2\ NO_2\ +\ 0.5\ O_2\ +\ H_2O\ \longrightarrow\ 2\ HNO_3$$

Figure 12.5. The formation of nitric acid from nitrogen dioxide.

The nitrogen oxides NO and NO_2, both of which are toxic gases that generate nitric acid in the tissues they contact, are formed naturally in the stratosphere (by the oxidation of nitrous oxide (N_2O), see Section 12.4), by lightning (*vide supra*), by oxidation of ammonia (NH_3) present in the atmosphere, and by microorganisms in the soil. However, the natural production is exceeded by the anthropogenic. The main anthropogenic sources are the burning of fossil fuels (traffic, heating and generation of electricity) during which molecular nitrogen of the air is oxidized, and the burning of biological material which contains organic nitrogen. Minor contributions come from the conversion of fertilizers and urea by microorganisms. Although nitric oxide is quickly oxidized to nitrogen dioxide, nitrogen dioxide is rapidly degraded by photolysis (wavelength below 420 nm) to nitric oxide and oxygen (which may combine with molecular oxygen to ozone), and the concentrations of NO and NO_2 in the atmosphere are therefore comparable. On average, the concentrations are 30–50 ppb, in remote maritime areas far from land very low amounts (1 ppt) are present but in large cities with heavy traffic values around 1 ppm have been reported.

The effects of the increased acidity due to the acidic gases (sulphur dioxide, the nitrogen oxides and hydrogen chloride) are in some cases evident and in others a matter of discussion. Some ecosystems are more sensitive than others, for example lakes in areas where the pH buffer capacity is limited due to low amounts of dissolvable carbonates in the ground. Such lakes will rapidly have their flora and fauna changed, and especially fishes will be among the first to disappear. The direct effects of acid rain on forests are not that clear. The "Waldsterben" that was believed to be wiping out forests and received much attention during the 1980s turned out to be an exaggeration. It is clear that a low pH will affect plants directly and even kill them if it is low enough, as shown by the devastating effects on the forests in the vicinity of industrial complexes in former eastern European countries, but the effects observed in for example western Germany are not so clearly correlated with emission of acid gases. It is a complex situation where several effects (e.g. eutrophication, acid rain, other pollution, soil quality, parasites, etc) have to be considered together. Perhaps most important are the effects on the ground in which plants grow, from which the for plants crucial calcium and potassium ions are leached by acid rain.

An economically important effect of acid rain is the damage it causes on materials used in buildings, for example metals, polymers, protective paint, etc. This costs society enormous sums yearly, money that could have been used on more constructive projects.

12.4 Depletion of stratospheric ozone

Life on earth is sensitive to UV radiation, and depends on the protection that the stratospheric ozone provides. Ozone is a reactive and toxic gas whose formation and effects in the troposhere have been discussed in Section 12.2.1, but it is at the same time the only compound in the stratosphere that absorbs UV radiation with wavelengths between 200 and 300 nm. In the stratosphere it is formed from molecular oxygen, which continuously is degraded by photolysis (wavelength <242 nm) to two oxygen atoms that either combine back to molecular oxygen or react

$$O_2 \xrightarrow{\text{hv} < 242 \text{ nm}} 2\,O$$

$$O + O_2 \longrightarrow O_3$$

$$O_3 \xrightarrow{\text{hv} < 1180 \text{ nm}} O + O_2$$

$$O + O_3 \longrightarrow 2\,O_2$$

Figure 12.6. The natural formation and degradation of ozone in the stratosphere.

$$O_3 \longrightarrow O + O_2 \qquad\qquad O_3 \longrightarrow O + O_2$$

$$Cl + O_3 \longrightarrow ClO + O_2 \qquad\qquad NO + O_3 \longrightarrow NO_2 + O_2$$

$$ClO + O \longrightarrow Cl + O_2 \qquad\qquad NO_2 + O \longrightarrow NO + O_2$$

$$\text{Overall: } 2\,O_3 \longrightarrow 3\,O_2 \qquad\qquad \text{Overall: } 2\,O_3 \longrightarrow 3\,O_2$$

$$\text{Byreaction: } \quad ClO + NO \longrightarrow Cl + NO_2$$

Figure 12.7. The catalytic degradation of ozone by chlorine atoms and nitric oxide.

with molecular oxygen to form ozone (see Figure 12.6). Ozone is also photolyzed, to molecular oxygen and an oxygen atom, and can in addition react with an oxygen atom to form two molecular oxygens. There is consequently an equilibrium between molecular oxygen and ozone in the stratosphere, which under normal conditions maintains a steady concentration of protective ozone.

The concentration of ozone would not be affected by chemicals in the stratosphere that react with ozone and are destroyed. Instead the species that seriously disturb the equilibrium between molecular oxygen and ozone act as catalysts and establish new equilibria. Another criterion is that such chemicals, or their immediate precursors, must be fairly long-lived in order to be able to reach the stratosphere. The chlorine atom and nitric oxide are examples of species that will degrade ozone catalytically, as shown in Figure 12.7.

In addition, other species are also involved, and an example is the hydroxyl radical. These three, the chlorine atom, nitric oxide and the hydroxyl radical are active at different heights. The concentration of ozone is highest at approximately 30 km above sea level, and this is also where nitric oxide is responsible for around 70 % of the degradation of ozone. The NO/NO_2 cycle is consequently the most important for the degradation of ozone. The chlorine atom contributes most to the degradation at 45 km, while the hydroxyl radical is most important at the extremes (10 and 70 km above sea level) of the stratosphere.

The CFCs are named "R xyz", where R stands for refrigerant, x the number of carbons -1, y the number of hydrogens $+1$, and z the number of fluorine atoms. R 11 is $CFCl_3$ (it should really be R 011), R 12 is CF_2Cl_2, R 13 is CF_3Cl, and R 14 is CF_4. The halons have a similar nomenclature, Halon xyzst, where x gives the number of carbons, y the number of fluorines, z chlorines, s bromines and t iodines. Halon 1211 is consequently CF_2ClBr while Halon 2402 is $C_2F_4Br_2$.

NO/NO_2 formed at sea level (see Figure 12.3) is too unstable to be transported to the stratosphere, but two major sources for stratospheric nitric oxide are the exhausts of airplanes and nitrous oxide (N_2O). Nitrous oxide is a non-toxic and in the troposphere very stable (half-life 150 years) gas that is used as an anaesthetic (laughing-gas). It is produced naturally by organisms in both the soil and the oceans, but also formed in substantial amounts by the degradation of fertilizers and by burning fossil fuels. It can be transported to the stratosphere where it will generate nitric oxide (see Figure 12.8) (another major environ-

$$N_2O \xrightarrow{h\nu} N_2 + O$$

$$N_2O + O \longrightarrow 2\ NO$$

$$N_2O + O \longrightarrow N_2 + O_2$$

$$\underset{R\ 12}{CCl_2F_2} \xrightarrow[(\lambda < 220\ nm)]{h\nu} CF_2 + 2\ Cl$$

Figure 12.8. The generation of nitric oxide and chlorine atoms in the stratosphere.

mental effect of nitrous oxide is its contribution to the anthropogenic greenhouse effect, and this is discussed in the following section). Chlorine atoms (and bromine atoms which are even more potent ozone degrading species) are mainly generated by photolysis of fully halogenated hydrocarbons that are stable enough to be transported to the stratosphere. In the form of chlorofluorocarbons (CFCs) and halons (containing bromine), they have been used in enormous quantities during the last decades as propellants, solvents, refrigerants, in fire-extinguishers and as blowing agents for plastic foams. A major advantage with the CFCs when they were introduced was their very low toxicity, due to their stability, especially compared to the chemicals that they replaced (e.g. ammonia in refrigerators).

Fully halogenated hydrocarbons are in general long-lived (half-life > 100 years in the atmosphere), and many of them are efficiently transported to the stratosphere (see Figure 12.9). As can be seen, there is a strong trend that they are more stable the more fluorine they contain.

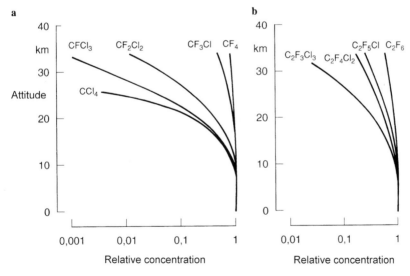

Figure 12.9. The portions of halogenated methanes and ethanes that reach various altitudes of the atmosphere.

While carbon tetrachloride with a half-life of approximately 50 years will not reach 30 km altitude in significant amounts, almost all carbon tetrafluoride emitted (half-life in the atmosphere 50,000 years) will end up in the stratosphere.

The halons are less stable compared to the CFCs but degrade ozone more efficiently because they generate bromine atoms. Halon 1301 (CF_3Br) for example, has a half-life that is 3 times shorter that that of R13 (CF_3Cl), but it degrades stratospheric ozone 10 times more efficiently (1 ton halon 1310 emitted into the atmosphere degrades as much ozone as 10 tons R13). In 1985, the CFCs were responsible for approximately 80 % of the ozone degradation by halogenated hydrocarbons, while halons and chlorinated hydrocarbons were responsible for approximately 10 % each. The compounds that have been developed as direct substitutes for the CFCs and halons are not fully halogenated, and the presence of a hydrogen–carbon bond makes them more prone to oxidation in the atmosphere (and by organisms which makes them more toxic!). In addition, saturated hydrocarbons such as pentane and cyclopentane have also replaced the CFCs as blowing agents for plastic foams.

The use of CFCs and halons have since the awareness of the effects on the stratospheric ozone been restricted, and the production and emission in the industrialized countries have dropped from aproximately 600,000 tons R11 and R12 per year in 1985 to close to zero. However, it will take a long time before the already emitted amounts have been degraded and disappeared form the atmosphere, and we will therefore have to live with the additional UV radiation let through by a lower concentrations of stratospheric ozone for another couple of generations.

12.5 Greenhouse effect

The energy that the earth receives from the sun is to the largest extent dissipated through the radiation of heat from the earth back to space, but the atmosphere reflects some of this heat radiation from the earth. Thanks to this, the temperature at the surface of the earth is approximately, and on average, more than 30 °C higher than it would have been without the atmosphere. The gases in the atmosphere, for example water vapour and carbon dioxide, act similarly to the panes in a greenhouse, letting the radiation from the sun pass through but absorbing the infrared radiation (heat radiation) from the earth and then emitting it in all directions including back to earth, which explains the name of this phenomenon (see also Figure 12.10). Water is the most important greenhouse gas, responsible for over 60 % of the natural greenhouse effect, but its concentration in the atmosphere is not changing significantly due to human activities.

While the natural greenhouse effect is a prerequisite for life on earth in its present form, it is not difficult to imagine that the balances that have been established between the incoming radiation from the sun, the chemical and biological processes that affect the atmosphere, and the heat radiation from earth are sensitive. Changes in for example the sun's activity, or the composition of the atmosphere, can be imagined to have an significant impact on the temperature on earth, leading to changes in for example the weather systems that may have a dramatic impact on life on earth. However, although the greenhouse effect is a reality that we know exists, we know relatively little about how sensitive the systems regulating the greenhouse effect are and exactly

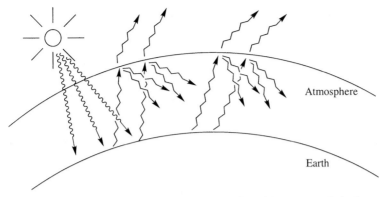

Figure 12.10. Heat radiation from earth is to some extent reflected by compounds in the atmosphere, accounting for the greenhouse effect.

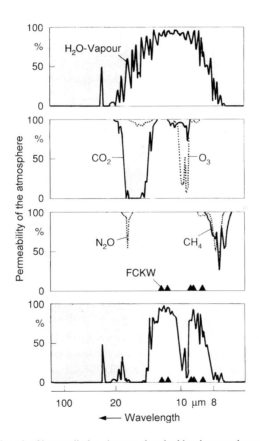

Figure 12.11. The wavelength of heat radiation that are absorbed by the greenhouse gases.

what will be the results of an temperature increase of, for example, 1 °C. The weather and the climate are constantly changing, and if we look a couple of centuries back it is evident that the earth has experienced both cold and warm periods that have no apparent connection with changes in the atmosphere. What has been happening over the last decades is that the emission of greenhouse gases, for example carbon dioxide, methane, nitrous oxide, ozone and the freons, has increased so much that an anthropogenic greenhouse effect is possible and even to be expected. The greenhouse gases differ in several ways, for example in which frequencies of infrared radiation they absorb (see Figure 12.11), and this is of course also the case for their ability to contribute to the greenhouse effect.

Carbon dioxide is the end-product of carbon in the metabolic processes of man and many other organisms, and on average every person expires approximately 700 g carbon dioxide per day. It is not really toxic, although accidents are not uncommon as in some instances it may be present in such high amounts that it supersedes the oxygen necessary for breathing (carbon dioxide is a heavier gas than oxygen), and concentrations exceeding 10 % for a longer time may be dangerous. On the other hand, for plants carbon dioxide is desirable and they will grow stronger and faster in the presence of higher concentrations of it. Carbon dioxide exists as a gas in the atmosphere, as a solute in the hydrosphere, as carbonic acid, hydrogen carbonate and carbonate ions also in the hydrosphere, and as various solid carbonates in the lithosphere and the pedosphere. The hydrosphere contains approximately 50 times more than the atmosphere.

Carbon dioxide is a relatively poor greenhouse gas, but the amounts present in the atmosphere compensate for this weakness and it is responsible for approximately 22 % of the existing and 50 % of the increase of the greenhouse effect presently observed. Almost one thousand billion tons of carbon dioxide are emitted each year, approximately 50 % from biological processes, almost 50 % from the oceans and a few percent from the use of fossil fuels. Most of the anthropogenic carbon dioxide is taken up by the oceans and by organisms, but a significant part (approximately 40 %) stays in the atmosphere and is responsible for the yearly increase of carbon dioxide of about 0.3 % that lately has been registered. Since the 18th century, the concentration of carbon dioxide has increased by almost 30 %, and the increase has accelerated dramatically during the last decades. A rather frightening picture emerges if the concentration of carbon dioxide (and the other major greenhouse gas methane) in the atmosphere is compared with variations in the temperature for the last 160,000 years (data obtained from drill samples obtained from the permanent ice of Antarctica) (see Figure 12.12). There is obviously a strong correlation, although it is impossible to say if it is the concentrations of carbon dioxide and methane that determine the temperature or *vice versa* (it is argued that an increase in temperature will increase the biological processes and thereby increase the amounts of especially methane). The concentration of the two gases have in the last decades increased to levels that are unsurpassed, and if the small increase in temperature that has been observed during the last years (0.3–0.6 °C in 100 years) in fact is due to the increasing concentrations of the greenhouse gases we may be in big trouble.

The concentration of methane in the atmosphere is also increasing, even faster (approximately 1 % per year) than that of carbon dioxide, and even if the total contribution of methane to the greenhouse effect is small today (a few percent) it is contributing approximately 13 % of the present increase. As mentioned in Section 12.1, anaerobic bacteria that produce methane as

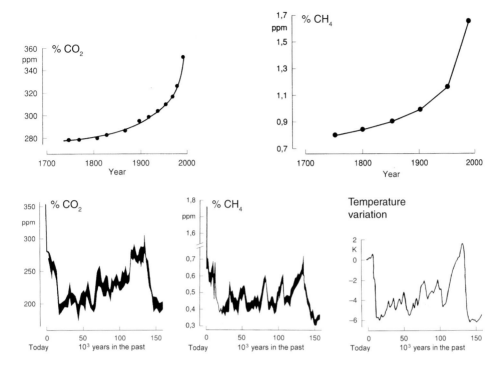

Figure 12.12. The relationship between carbon dioxide and methane concentrations in the atmosphere, and the variation in temperature.

an end-product of their metabolism and degrade organic matter (especially in wetlands, rice paddies and landfills, and in the fore-stomachs of ruminant animals and in termites) are the major source of methane, while for example the use of fossil fuels (especially natural gas) also makes a significant contribution. The degradation of methane in the atmosphere is almost exclusively due to its oxidation by the hydroxyl radical and molecular oxygen in the troposphere (see Figure 12.3). Less than 10 % is transported to the stratosphere where it is degraded photochemically. Methane is only one of many volatile hydrocarbons that are released in large quantities into the atmosphere. All hydrocarbons are potential greenhouse gases and methane is the poorest (nevertheless being approximately 20 times more efficient compared to carbon dioxide), but contrary to all other hydrocarbons methane is relatively stable with a half-life of approximately 4 years.

Ozone (see Section 12.4 for formation and properties), perhaps primarily interesting for its role in photochemical oxidation but also an highly efficient greenhouse gas (approximately 2000 times more efficient than carbon dioxide), contributes to the supplementary greenhouse effect due to the activities of man (in addition to the natural greenhouse effect) with approximately 5 %.

Nitrous oxide is around 200 times more efficient as a greenhouse gas compared to carbon dioxide, it is also involved in the depletion of the ozone layer in the stratosphere and has been discussed in Section 12.4. Its contribution to the supplementary greenhouse effect due to the activities of man is approximately 5 %.

CFCs (e.g. R11 and R12) are extremely potent greenhouse gases, more than 10,000 times more efficient than carbon dioxide, in addition to their effect on the stratospheric ozone. Their present contribution is relatively small, a few percent, but their concentration has increased rapidly during the last three decades of the 20th century (approximately 5 % per year). As discussed in Section 12.4, the use of CFCs has now been restricted, and it is expected that the increase will slowly be exchanged for a decrease.

The global warming due to the greenhouse effect is to some extent balanced by the cooling caused by for example the emission of sulphur dioxide from both natural and anthropogenous sources (*vide supra*). Many other factors, e.g. the increased formation of clouds that reflect the sun's radiation with increased temperature, etc., are also studied and discussed, and it is still not possible to make any conclusions about what will happen.

12.6 Pollution by halogenated hydrocarbons

Halogenated, in particular chlorinated, hydrocarbons are in the consciousness of many generally toxic and environmentally unfriendly. It is true that many of the accidents and scandals involving chemicals during the last decades have been caused by halogenated compounds, but it is the ways that man has used these chemicals that led to the undesired effect, not the chemicals by themselves. Lately we have become aware that many natural products are halogenated, in most cases chlorinated, and chlorine is actually an essential element for life (an adult human contains approximately 100 g chlorine).

12.6.1 Dioxins

Polychlorinated dioxins (PCDDs) and polychlorinated dibenzofurans (PCDFs) are often called dioxins as a group, and the same name is used for the most toxic member 2,3,7,8-tetrachlorodibenzodioxin (TCDD, dioxin). They are formed in very low amounts (typically nanograms per m^3 exhaust) as an impurity when organic material containing chlorine (organic or inorganic) is incompletely burnt in conventional waste dump combustion plants. Due to the chemical stability of PCDDs and PCDFs, the temperature during combustion has to attain 1000 °C in order to degrade them as well, and such temperatures are not normally employed. Another source has been the bleaching of paper pulp in the paper industry when this was carried out using chlorine as an oxidant, but new processes that make use of other oxidants have been developed. A third source, which used to be the most important and that focused attention on the dioxins, is the formation of dioxins as a by-product during chemical reactions with chlorophenols. An example is 2,4,5-trichlorophenoxyacetic acid (2,4,5-T), a chlorophenol derivative that has been produced in large quantities and used as an herbicide (a component of the notorious Agent Orange). 2,4,5-T is formed by the reaction of 2,4,5-trichlorophenol (formed from the reaction between 1,2,4,5-tetrachlorobenzene with sodium hydroxide in an alcoholic solvent at 160 °C) with chloroacetic acid (see Figure 12.13). TCDD is a by-product of this process, as can be seen in Figure 12.13, and Agent Orange contained approximately 10 ppm TCDD. Later it was discovered that the formation of TCDD depends on the temperature and the concentration of the trichloropenoxide anion, and if the reaction was carried out under conditions when these two factors were controlled the amounts of TCDD in the final product was less than 0.1 ppm. The accidental contamination of the surroundings of Seveso in Italy 1976 with TCDD was caused by a chemical reactor in which this reaction was carried out. For some reason it was overheated, TCDD was formed in significant amounts and when the pressure became to high the contents of the reactor, a few hundred kg of 2,4,5-trichlorophenol and a few kg of TCDD, were blown into the air via a pressure relief valve.

The toxicity of TCDD has already been discussed. It is highly toxic to certain mammals (the ground in the contaminated area in Seveso contained enough TCDD to kill 50,000 guinea pigs per square metre!) although humans do not appear to be the most sensitive. A slight increase in the incident of cancers in the population of Seveso has been noted 20 years after the accident. The acute effects of TCDD are believed to be caused by metabolites, although they have not yet been identified, and it appears as if the initial metabolic conversion involves epoxidation of the aromatic ring. However, TCDD is a very poor substrate for the metabolic systems in mammals,

Polychlorinated dibenzodioxins
(PCDDs)

Polychlorinated dibenzofurans
(PCDFs)

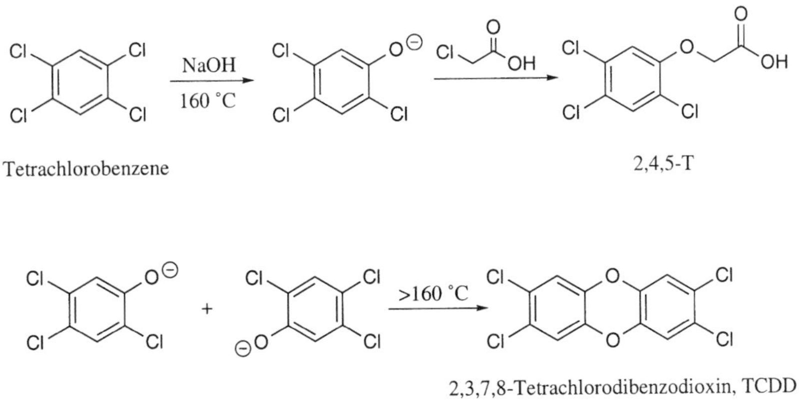

Tetrachlorobenzene

2,4,5-T

2,3,7,8-Tetrachlorodibenzodioxin, TCDD

Figure 12.13. The formation of TCDD from 2,4,5-trichlorophenol.

and the half-life in humans is in the order of 10 years. The risk for bioaccumulation of TCDD over the years must therefore be considered. Other PCDD derivatives containing more than 4 chlorines are less toxic, and the fully chlorinated octachlorodibenzidioxin is approximately 1000 times less potent compared to TCDD. Its carcinogenic activity is believed to be caused by its binding to the Ah receptor (Ah because it also binds other aromatic hydrocarbons), that in a very complicated and not completely understood way is involved in cell division.

12.6.2 PCB

Contrary to the dioxins which were only formed as unwanted by-products, the polychlorinated biphenyls (PCBs) have been produced in large quantities for many years. As the name implies, they are formed by chlorinating biphenyl, a process that cannot be controlled very well. PCBs are therefore always mixtures of many isomers, and the commercial products were simply characterized by the relative amount of chlorine they contained. PCBs are chemically extremely stable, not volatile and will not burn easily, but are still liquid as an oil at room temperature. They were used for many things, as coolants in electrical transformers and capacitors (they are good insulators), as hydraulic oils, as plasticizers for various polymers, to impregnate paper and wood, etc. The PCBs are less toxic compared to the more potent dioxins, although the mechanism for the toxicity of the two classes may well be the same. Instead it is a combination of the amounts of the PCBs produced and emitted to the environment (in total far more than 1,000,000 tons) which together with their chemical stability make them relatively abundant, and their high lipophilicity which makes them subject to bioaccumulation. In addition, they may be contaminated by PCDFs, as PCBs may be oxidized at high temperatures to PCDFs. This possibility also shows how important it is that products that contain PCBs are taken care of properly, and not simply burnt together with other waste.

The toxic effects of the PCBs have been observed in species that are at the top of food chains of exposed species, for example sea birds that prey on fish (that eat other marine organisms, etc). The banning of the PCBs in many countries in the 1970s has resulted in lower concentrations in the species affected, but it will take a long time before the problem has disappeared.

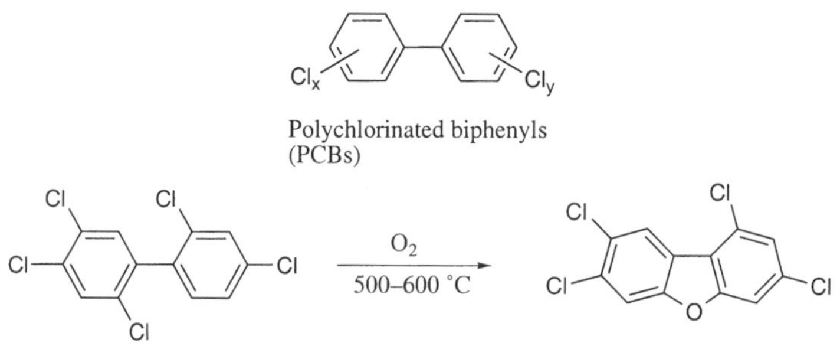

Polychlorinated biphenyls
(PCBs)

Figure 12.14. The general stucture of PCBs, and their transformation to PCDFs.

12.6.3 DDT

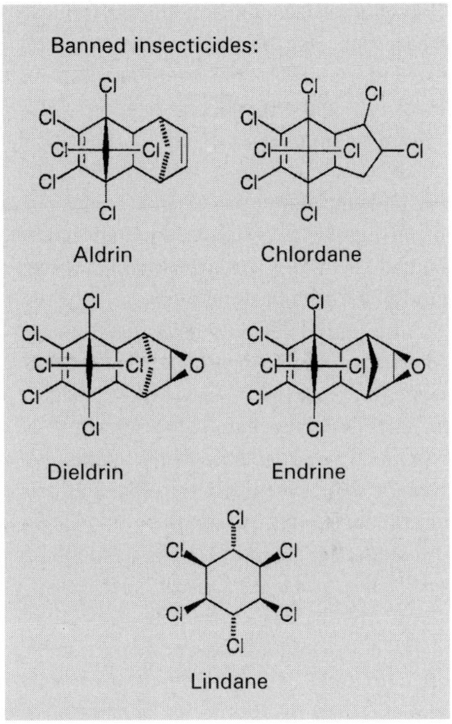

1,1,1-Trichloro-2,2-bis(4-chlorophenyl)-ethane (DDT) has also been mentioned several times, and is taken as a representative for the class of chlorinated pesticides that have been used in large amounts but have been banned and exchanged for more selective and environmental friendly compounds. Other examples that will not be discussed further are aldrin, chlordane, dieldrin and lindane.

Due to its pesistence in the environment, its low volatility and its lipophilicity, it is long-lived and efficiently absorbed by organisms which may accumulate DDT in for example fat tissue. The metabolism in mammals is slow, because DDT is not readily available for metabolic enzymes, and the metabolites are also toxic (e.g. DDE, see Figure 12.15). The effect of DDT on the nervous system has already been discussed, in addition, and perhaps more serious, it has a hormone-like effect (mimicking the female sex hormone oestrogen and affecting the release and conversion of other hormones). This is the mechanism by which DDT lowers the fertility of animals and weakens the shell of birds eggs.

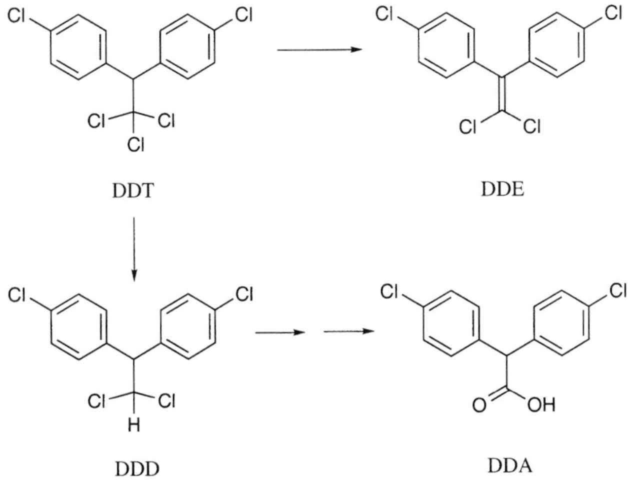

Banned insecticides:

Aldrin

Chlordane

Dieldrin

Endrine

Lindane

DDT

DDE

DDD

DDA

Figure 12.15. The metabolism of DDT in mammals.

12.7 Pollution by metals

Many metals and metalloids are seriously toxic to both man and environment, yet they are produced in large quantities and are commercially very important products. Metals are of course also a natural part of the environment and toxic metals are for example emitted by volcanoes, forest fires and sea spray (*vide supra*). However, these concentrations can be increased as a result of anthropogenic emissions, and it is not only the direct discharge from various activities but also secondary effects due to for example acid pollution (see Section 12.3) that is responsible. Most metals are produced in mines, and waste from mining and refinement of metals are major sources for pollution. Another major source is the combustion of fossil fuels (coal, oil, natural gas) and wood, which contain trace amounts of metals that will be emitted into the atmosphere with the fumes.

Atmosphile elements have their greatest mass transport through the atmosphere, while lithophile elements mainly pass by rivers to the oceans. A characteristic for atmosphilic elements [for example antimony (Sb), arsenic (As), cadmium (Cd), copper (Cu), lead (Pb), mercury (Hg), molybdenum (Mo), selenium (Se), silver (Ag), tin (Sn) and zinc (Zn)] is that they exist in relatively volatile forms as they are, as oxides, or methylated in organometal derivatives, or they are emitted by industrial processes as particles in the atmosphere. In general, atmosphilic metals that easily bind to bioorganic compounds, e.g. to the –NH– or –SH groups of proteins, are the most dangerous for all organisms, in addition to being relatively mobile and readily absorbed by organisms. Such metals may be bioaccumulated in food chains and eventually give toxic effects in certain species.

The effects of metals depend on a number of factors of which some are related to the organism intoxicated, e.g. the ability to metabolize metals, and others on the form that the metal is present in. The oxidation state, the ligands, etc are examples of such factors, and a metal ion is not equally reactive as a "free" ion (free metal ions in water are always hydrated) as a complexed ion. Unicellular organisms, with a large surface to their surroundings, will absorb metals as well as other compounds in a different way than multicellular organisms like man, which have specialized organs for uptake. For obvious reasons, metals in the form of metal ions will not readily pass cell membranes but will mainly be taken up by individual cells by the help of porter molecules (ligands) and carrier proteins. However, as discussed in Chapter 6, for most metals little is known in detail about the mechanisms for uptake.

Resistance in microorganisms may be the result of the complexing (protein complex, sulphide, etc.) of the metal whereby it is rendered immobile, either inside or outside the cell, or by active exclusion from the cell (e.g. by volatilization by methylation of Pb, Hg and Sn.

In the following, a summary of the effects of a selection of the metals that are toxic and cause severe environment effects is given, but it should be stressed that this in no way should be considered to be comprehensive. All metals discussed here are what we call heavy metals, having a density of at least 5 g/cm^3. Often one associates the term "heavy metal" with "toxic metal", but it is important to remember that also other metals (having a density of less than 5 g/cm^3) can be very hazardous.

12.7.1 Arsenic (As)

The earth-crust contains approximately 2 ppm arsenic. It is wellknown by everybody as the classic poison used in detective stories, although its importance has decreased with the development of more potent organic biocides. Arsenic exists in several forms, of which the reduced arsine (AsH$_3$, with the oxidation state -3), arsenic trioxide (As$_2$O$_3$, with the oxidation state +3) and arsenic pentoxide (As$_2$O$_5$, with the oxidation state +5) are important exampels. The consumption of arsenic has decreased during the 20th century, as arsenic compounds used as insecticides and to treat various illnesses in man and animal gradually are replaced by organic compounds, but it is still used for electronic applications and as a component in various alloys.

Arsenic is emitted into the environment from the burning of coal (coal contains approximately 20 g arsenic per ton and brown coal up to 1.5 kg per ton), from the smelting of metals (copper and nickel), and from volcanoes. The natural releases are approximately 8,000 tons per year, which is a small amount compared to the releases caused by human activities. In the atmosphere, arsenic is transported over long distances before it precipitates, and may then circulate many times in various forms through water, soil and air before it eventually is bound in the sediments. Plants do not readily absorbe arsenic compounds, except if arsenic-containing pesticides have been used for their cultivation or if they are grown in a contaminated area, and the consumption of seafood (arsenic is strongly bioaccumulated by algae) is the main source of arsenic for man. In mammals it is absorbed well, in both the intestines and the lungs, it is transported by the blood (bound to hemoglobin in the red blood cells) to the liver and the kidneys. In man, some amounts will be deposited in bones, hair, nails and skin, but approximately 70 % is excreted (mainly with the urine). It is believed to disturb the biochemical reactions either by replacing phosphorus in the phosphate groups of the nucleotides (Section 4.3.6) or by reacting with thiol groups in enzymes, and arsenic is consequently toxic to all organisms. Of concern is also the carcinogenicity and teratogenicity of arsenic, although it is not the most potent of the heavy metals in this respect. Arsine and arsenic trichloride (AsCl$_3$) are considerably more toxic than the arsenic oxides, and is a serious problem in industries where they are used (e.g. for the manufacture of electronic components), but due to its instability it will not survive long enough to be an environmental hazard. As several other heavy metals, arsenic oxides may be reduced and methylated by microorganisms, to organic forms that are more volatile and bioaccumulated more efficiently. A few examples of historically important arsenic compounds are shown in Figure 12.16.

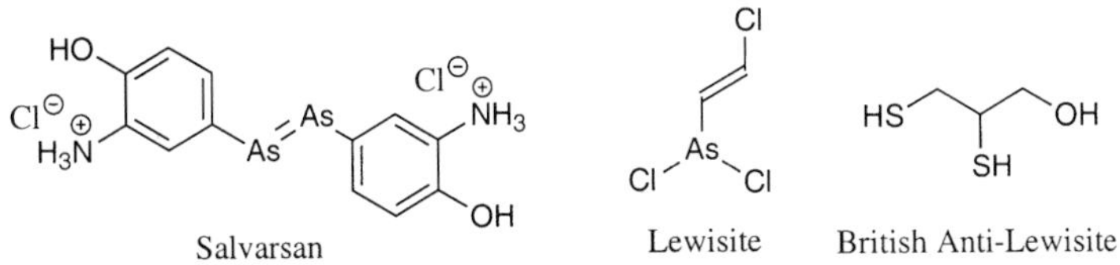

Figure 12.16. Some well-known arsenic compounds.

Salvarsan was developed (by the German chemist P. Ehrlich) in the beginning of the 19th century, and was successfully used to treat syphilis until the arrival of the penicillins in the 1940s. Lewisite was developed as a war gas (by the american chemist W. Lee Lewis), it has similar reactive and irritant properties as mustard gas but shows in addition a long-term toxicity due to the presence of arsenic. British Anti-Lewisite (BAL) was developed as an antidote to lewisite, it chelates (forms a complex with its three heteroatoms) arsenic (and other heavy metals) and renders it less toxic.

12.7.2 Cadmium (Cd)

The amounts of cadmium in the earth's crust is approximately 0.1 ppm, and cadmium minerals (e.g. CdS, CdO and $CdCO_3$) is most commonly found as a minor component in zink minerals. Cadmium compounds have been used in large amounts as a stabilizer in PVC and as pigments, lately these uses have been restricted with the intention to reduce the consumption of cadmium but the exploding use of rechargable batteries (with a nickel-cadmium cell) has instead increase the amounts of the metal in circulation. In addition, cadmium is used for surface treatment of metals (a thin layer of 10 μm on iron will prevent corrosion) and in alloys.

Approximately 8,000 tons of cadmium is emitted into the environment per year, and almost all of that is due to human activities. Acid rain has also increased the amount in circulation, as the solubility and thereby the amounts absorbed by plants increase with decreasing pH value (plants need zinc and cadmium is absorbed by the same system). Cadmium has a very strong affinity for proteins, which will affect its absorbtion in mammals (cadmium in the food is not taken up efficiently in the intestines) and excretion (cadmium has perhaps the longest half-life of all chemicals in man). Many factors will moderate the absorbtion of cadmium as well as any other metal in man, and zinc deficiency is very important for the efficiency that ingested cadmium is absorbed. In man, small amounts are deposited in the kidney bound to metallothionein (the presence of cadmium induces the biosynthesis of this protein), while larger amounts are stored in the liver bound to various proteins. It is toxic, causing lung oedema if an aerosol containing cadmium is inhaled (5 mg cadmium per m^3 for 8 hours may be fatal) or acute toxic effects on the gastrointestinal tract. The chronic effects are severe, for example the kidneys are irreversible damaged when the amounts of cadmium has reached approximately 200 μg per g kidney tissue (in "unexposed" individuals the kidney normally contain 20–40 μg/g), and the kidney dysfunction will eventually affect the metabolism of phosphorus and calcium and promote osteoporosis (the elimination of mineral constituents from bone). For several decades an area close to an abandoned lead-zinc mine in the Jinzu valley of Japan was so contaminated with cadmium that the peole living there got ill (contaminated water was used to irrigate rice fields). The illness was called the itai–itai disease (itai is an expression for pain in Japanese, indicating how it was experienced by the afflicted) and led via damaged kidneys to porous bones that were deformed and easily collapsed. The low pH (approximately 5) of the soil in the Jinzu valley made cadmium stay in solution and facilitated its absorbtion by the rice, similar contaminations with cadmium in areas with soil containing calcium carbonate and having a pH above 7 did not lead to any notable effect. Even if hundreds of people, of which many died, over the years were afflicted in Japan it took a long time to establish the cause (and admit the guilt!) of the itai–itai disease. Cadmium is considered to be a carcinogen, although not very potent. It does not pass placenta very efficiently and does not appear to be teratogenic.

12.7.3 Chromium (Cr)

Chromium is a relatively common metal, and the earth-crust contains approximately 100 ppm. It exists in a number of different oxidation states, but only the oxidation states 0 (metallic chromium), +3 (as chromic oxide, Cr_2O_3), +4 (as chromium dioxide, CrO_2) and +6 (as chromium trioxide, CrO_3) are of practical importance. Most of the chromium produces is of course used for the production of stainless steel, and chromium–plating of various metals is another important application that consumes the metal. Chromic oxide is used as a (green) pigment and as a catalyst, chromium dioxide is magnetic and used in audio, video and data storage tapes, and chromium trioxide is an important oxidizing agent in organic chemistry. Chromium is an essential chemical (Section 4.2.1) that is needed for the biosynthesis of the glucose tolerance factor, and the total absence of chromium in man results in diabetes. Together with cobalt and molybdenum it is the chemical that (so far) we need the lowest amounts of, an adult contains approximately 5 mg chromium which is around 600,000 atoms per cell.

While chromium in oxidation state +3 is very poorly take up by cells, chromate anions (CrO_4^{2-}, oxidation state +6) are readily transported by phosphate–sulphate carrier and chromium toxicity is closely linked to the higher oxidation state. However, inside the cells chromium (+6) is reduced to chromium (+3) and thereby trapped, and it may well be the reduced form that is causing the toxic effects on the molecular level. Chromium in oxidation state +6 is toxic to all organisms, microorganisms, plants as well as fishes. In man it has been shown to be a potent allergen, causing a delayed-type hypersensitivity that results in eczemas, and it is also considered to be a carcinogen. As other metals, chromium ions may bind to proteins and DNA, and the allergenic effect is caused by the transformation of endogenous proteins to forms that no longer are recognized by the immune system, by chromium.

12.7.4 Copper (Cu)

Copper, of which the earth's crust consists of 50 ppm, is found naturally both in its metallic form and as various compounds with the oxidation state +1 or +2. As it is easily available in its metallic form (either directly or from the ores), copper was one of the first metals to be exploited by man. It is hard but yet formable, it is relatively resistant to corrosion, it is an excellent conductor of electricity, and man has consequently found a number of uses for copper. Together with other metals it forms valuable alloys such as brass (with zinc) and bronze (with tin), and it is an essential chemical (Section 4.2.1) of which an adult contains approximately 100 mg. It is part of several enzymes, for example cytochrome oxidase and superoxide dismutase, and severe copper deficiency will consequently lead to, among other things, the blocking of cell respiration and the destruction of cells by superoxide.

In spite of the fact that copper is essential, it is a very toxic metal of which any organism in principle would only be able to tolerate a narrow range of. In mammals evolution has solved this dilemma in an elegant way, by for example making sure that excess copper is eliminated rapidly and that the copper that the cells make use of is tightly bound to specific copper proteins and not available for other reactions. As a result of this, we are less sensitive to intoxication by copper that one might fear. However, excess copper is highly toxic to microorganisms (copper sulphate, $CuSO_4$, and copper oxide, Cu_2O, have been used extensively as pesticides), and humans that are exposed to excessive copper in the diet will initially notice it by the effect that it has on the bacterial flora in the guts.

12.7.5 Lead (Pb)

Lead, 13 ppm of the earth's crust, is not found in nature in its metallic form, and its inorganic compounds are mainly in oxidation state +2 (as for example lead oxide, PbO, and lead sulphide, PbS). The major uses for metallic lead is in lead–acid batteries, the kind used in all normal cars. In addition, metallic lead is used for cable sheatings, in shot, and in alloys (soldering tin for example contains approximately 50 % lead). Inorganic lead compounds are used as pigments and as protective agents in paints and in glassware. Organic lead (which are absorbed very efficiently), especially in the form of tetraethyllead, has been used in enormous quantities (hundreds of thousands of tons, often together with similar amounts of 1,2-dichloroethane and/or 1,2-dibromoethane) as antiknock agents in petrol (see Figure 12.17), but with the entry of catalytic converters in cars, and the consciousness of the hazards with emitting such large amounts of lead into the atmosphere, the use of tetraethyllead has decreased a lot and will (hopefully) soon disappear. Instead, *tert*-butyl methyl ether, which is combusted to water and carbon dioxide, is used as antiknock agent in petrol.

During the combustion process tetraethyllead is heated and dissociates to a lead atom (which is oxidized to lead oxide) and four ethyl radicals. These will quench any radicals formed too early, before the sparking plug have given the ignition spark, and thereby prevent self-ignition (knocking) at inappropriate times during the ignition cycle in a car engine. Lead oxide reacts with dichloroethane (or dibromoethane) to volatile (inorganic) lead compounds that are emitted with the exhausts. In addition, the lead compounds formed in this process act as a lubricant for the valves of the engine.

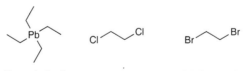

Tetraethyllead   1,2-Dichloroethane   1,2-Dibromoethane

Figure 12.17. Important components in leaded petrol.

The first to experience lead poisoning were the Greek, who realized that drinking acidic beverages from lead containers was harmful and should be avoided. The Romans did not learn from the Greek, of course, they used lead for the water pipes (the word "plumbery" has its roots in the Latin name for lead, plumbum) and were consequently chronically exposed to lead. Although it has not been proven in any way, the hypothesis that this lead poisoning played an important role in the decline of the Roman empire has been brought forward and is not completely inconceivable! In our bodies, lead is able to take the place of calcium in bone tissue, which by itself is not harmful but prevents its excretion. In mammals, lead gives rise to a multitude of toxic effects. Perhaps the most well-known is the effect of lead is on the biosynthesis of haemoglobin, which results in the formation of zinc-porphyrin instead of iron-porphyrin (lead interferes with the enzyme ferrochelatase which inserts iron into the porphyrin). Because the spectroscopic properties of zinc-porphyrin are different from those of iron-porphyrin it is easy to measure and quantify. High exposure to lead will eventually lead to anemia because haemoglobin with zinc instead of iron will not be able to transport oxygen. In addition, considerably lower concentrations of lead appear to be associated with the degeneration of nerve cells of children, resulting in impairments in growth, hearing and mental development. Again, as far as we understand the molecular mechanism of lead is that is will bind to sulphur and nitrogen groups in proteins and other macromolecules. Lead poisonings can also be treated by the administratiuon of a chelator (as BAL, *vide supra*) that will bind to and inactivate the metal. Other organisms are more concerned with traditional "lead poisoning", which not necessarily means getting shot but also includes eating the lead shot that are spread in significant quantities during the hunting season. Especially the waterfowls and the animals that prey on waterfowls are the victims of this effect, and in several countries lead shots have been banned.

12.7.6 Mercury (Hg)

Mercury (0.08 ppm in the crust of the earth) is mainly obtained from mined cinnabar (mercury sulphide, HgS) by rosting the ore at 600 °C and condensing the metallic mercury (which is a liquid that boils at 356 °C). At room temperature mercury has a vapour pressure that corresponds to 14 mg mercury per m^3 air, which is considerably higher that the concentrations tolerated at workplaces (typically 0.05 mg/m^3) and lethal after a couple of hours. The evaporation rate is not high, at 20 °C approximately 6 µg/h/cm^2, but if not properly ventilated rooms where mercury is handled will soon become dangerous to stay in. In contrast to other metals mercury is to some extent soluble in both water and organic solvents, it has a high surface tension and its physical and chemical properties has made it useful for a number of technical applications (in for example thermometers, barometers, relays and fluorescent tubes). The ability of mercury to form alloys with (amalgamate, forming an amalgam) other metals has been used in metallurgy, in chemical processes (e.g. the production of chlorine) and for dental amalgams (composed of mercury, silver, tin and small amounts of copper or zinc). Organic mercury compounds were prepared early, phenyl mercury chloride was for example launched as a fungicide already in 1915 although the use of this and other similar mercury-containing (alkyl-, alkoxy- and arylmercury compounds) pesticides did not peak until the 1960s.

In water organisms mercury is readily absorbed and bioconcentrated in food chains. As fish and shellfish is food for many different organisms, including man, the pollution of waters with mercury may lead to catastrophes (as will be discussed below). In certain areas where the lakes have been severely polluted, there are recommendations that especially pregnant women should not eat the fish. We have already discussed the absorbtion of metallic mercury in man, which is poor for liquid mercury in the intestines but efficient for mercury vapour in the lungs. Mercury salts, especially with mercury in the oxidation state +2, are absorbed in the intestines, and so are of course also organic mercury compounds. The primary target organs for inorganic mercury poisoning are the CNS (chronic exposure) and the kidneys (acute and chronic exposure). Organic mercury will primarily affect the CNS, resulting in anything from hearing defects to tremors and the loss of memory, and in several occasions poisonings of large number of individuals have been reported. Many have been poisoned after consuming bread prepared from wheat that had been treated with a alkylmercury fungicide and that was intended to be used as planting-seed, and the most severe accident took place in Iraq 1971 when approximately 60,000 people were exposed and over 2,000 died. In the 1950s, a chemical company producing vinyl chloride (with mercury oxide as a catalyst) discharged mercury wastes into the Minamata bay in Japan. The mercury were converted to methylmercury by microorganisms in the sediments (*vide infra*), the methylmecury was absorbed by the sea organisms and ended up in the shellfish and fish which were consumed by the local inhabitants. The result was a massive outbreak of methylmercury poisoning, the Minamata-disease, afflicting in total more than 1,000 people and killing approximately 100. The cause of the disease was soon identified and 1958 it became illegal to *sell* fish caught in the Minamata bay, but the inhabitants continued to fish for household use which aggrevated the catastrophe. Interestingly, the chemical plant was not closed until 1968, and then for other reasons, and during the last 10 years the pollution of the Minamata bay continued as if nothing had happened.

Mercury is circulated more efficiently than other metals, and the amounts that are released as a result of human activities will "hang around" for a long time. The major reason for this is that microorganisms, present everywhere and also in the sediments, are able to methylate inorganic mercury to monomethyl- and dimethylmercury as well as convert organic mercury compounds to atomic mercury. For the microorganisms, such conversions can be regareded as an excretion mechanism because the result is that they get rid of mercury, but for other organisms it will mean that mercury is mobilized. Monomethylmercury, CH_3Hg+, will be absorbed by organisms, e.g. algae and fish, while dimethylmercury, $(CH_3)_2Hg$, and atomic mercury are volatile and will be transported to the atmosphere. In this way mercury may be transported long distances before it is brought in its original form or in an oxidized form.

12.7.7 Tin (Sn)

Tin (2–3 ppm in the crust of the earth) is an essential chemical for man, it is a part of the hormone gastrine which is release from the stomach into the blood-stream after feeding. If it is absent from our bodies we will suffer from impaired digestion and growth. Tin compounds have many different uses in modern society, but especially interesting are the organotin compounds that for example are used as biocides in various situations. Tributyltinoxide is used as an anti-fouling agent in boat paints (to prevent algae and other sea-living organisms to attach to the body of the boat), as a preservative in cooling liquids, paint and wood, and as a molluscicide. Other organotin derivatives are used as catalysts and stabilizers for plastics, but it is especially the trialkyltin derivatives that are significantly toxic (trimethyl- and triethyltinoxide are considerably more toxic compared to tributyltinoxide). Tributyltinoxide, which is used in large amounts, is toxic to a number of algae, molluscs and fish. In mammals they interfere with the oxidative phosphorylation in the mitochondria, as well as giving rise to cerebral oedema.

13 The assessment of organic structures

As has been discussed in this book, there are strong relationships between the structure of an organic compound and it's chemical properties, and, even if it not always is apparent how, the chemical properties will determine the possibilities of the compound to be toxic. With a relatively limited chemical knowledge about how functional groups and combinations of functional groups influence the properties and reactivity it is possible to identify compounds that can be suspected to be hazardous just by analyzing their chemical structures with pen and paper. When assessing a structure in this way, it is important not only to imagine how the various functionalities may interact but also to hypothesize how chemical transformations and metabolic conversions might change the structure. It is of course not easy to predict in detail how a compound will be metabolized, and as we have seen compounds may be metabolized by several parallel routes. However, if reasonable transformations/conversions will result in a structure that has functionalities that are associated with toxicity it is wiser to anticipate the worst instead of hoping for the best.

Some additional examples of what kind of reactive functionalities one should look for are shown in Figure 13.1 (many other examples are give throughout this book). Isolated carbon–carbon double bonds and keto groups are not associated with high reactivity, but the combination of the two as an α,β-unsaturated ketone is potentially very reactive and should be treated with respect. When comparing the saturated acetic acid ester with the unsaturated analogue in Figure 13.1, the latter is considerably more toxic because it will generate the toxic compound acrolein after ester hydrolysis (by an esterase) and oxidation of the primary alcohol to an aldehyde (by alcohol dehydrogenase). N-Methyl-N-nitrosourethane is easily hydrolyzed because the nitroso group on the nitrogen is electron withdrawing and makes N-methyl-N-nitrosoamine a good leaving group, and it will react spontaneously with water. N-Methyl-N-nitrosoamine in its enol form will spontaneosly generate diazomethane in its protonated form (see also Figure 8.15), an efficient methylating agent that is known to be carcinogenic. Bis-chloromethyl ether is one of the most potent chemical carcinogens known, which previously was use for the manufacture of various resins. It has an ether oxygen bound to the same carbon as a chlorine atom, and the unshared electrons of the oxygen will push electrons towards the chlorine that becomes more negatively charged. As the leaving group (chloride ion) already has "started to leave", bis-chloromethyl ether is an excellent electrophile and the fact that it is bifunctional enables it to cross-link DNA making it's genotoxic effects much worse. In Chapter 7 we discussed dehydrohalogenation of chlorinated compounds, which were hydroxylated on the α-carbon by cytochrome P-450 and gave a product that immediately eliminated HCl and were transformed to the carbonyl compound. Bis-chloromethyl ether will also be hydrolyzed (to formaldehyde), but the fact that the oxygen is not present as a hydroxyl group but an ether hampers this hydrolysis and makes it more long-lived. A similar functionality obtained from the reaction of glutathione with methylene chloride (a chloromethyl thioether) was discussed in Section 8.3, it is also a good electrophile and may be responsible for the toxicity of methylene chloride.

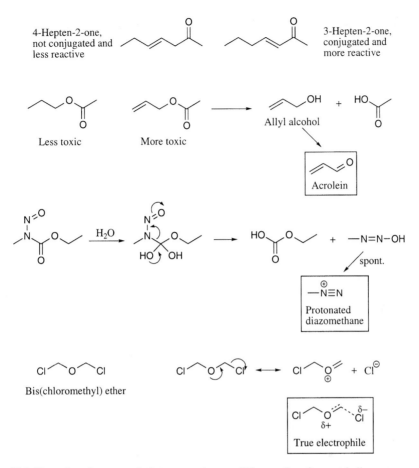

Figure 13.1. Examples of compounds that are reactive or will be reactive after metabolism.

In Figure 13.2 some additional and relatively simple structures are shown, and in the following the structures are assessed for their possibilities to give rise to toxic effects. Before you continue, take the opportunity to practice with the 12 structures and see if you come to similar conclusions.

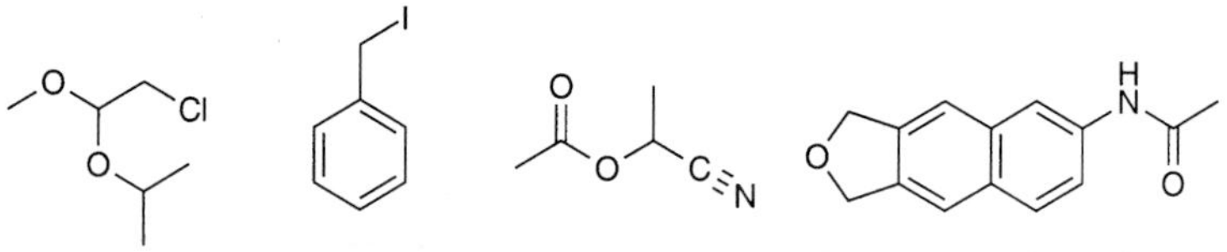

Figure 13.2. Some structures to practice with.

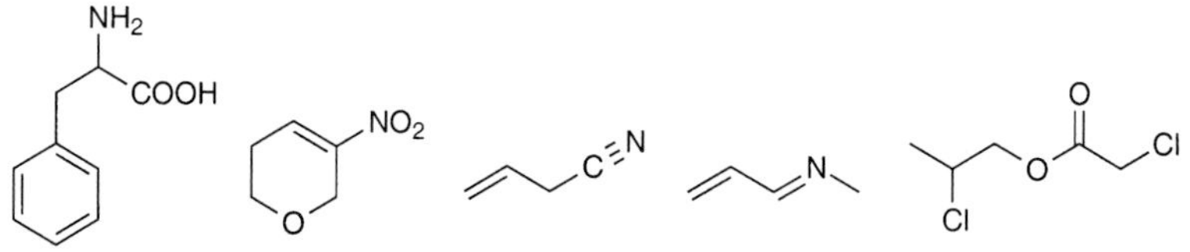

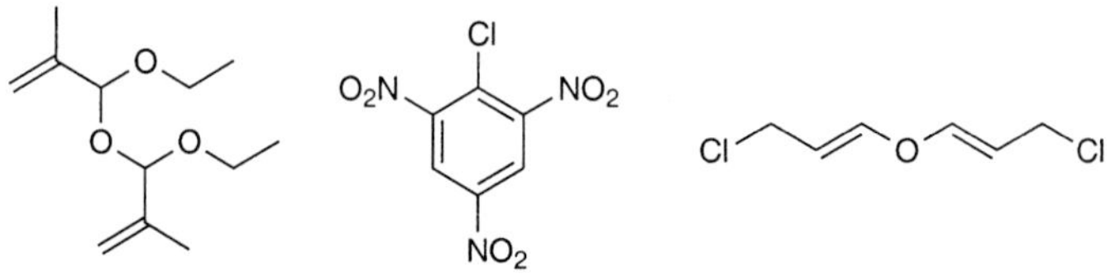

Compound **1** contains a primary chloride functionality, and we can expect it to be electrophilic although steric hindrance on the β-carbon would probably make it a poor substrate for nucleophilic substitution. Both oxygen atoms are bound to the same carbon, making this an acetal group, and acetals may be hydrolyzed spontaneously in contact with water. If compound **1** is hydrolyzed the products are ethanol, 2-propanol and chloroacetaldehyde, and the latter is highly reactive and toxic (discussed in Section 8.3). Although it is difficult to say how fast this spontaneous hydrolysis is, it is reasonable to assume that it is fairly rapid and that compound **1** should be used with care.

Compound 1

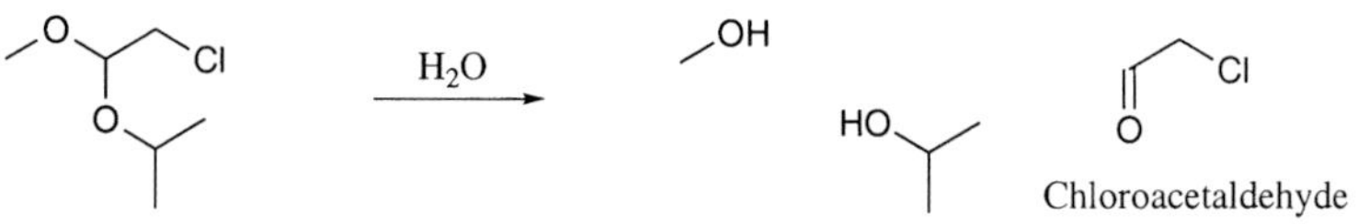

Compound **2** is aromatic, and one can always suspect that it undergoes oxidation by cytochrome P-450 to form different phenols. However, this is not relevant because the compound is simply benzyl iodide which is a good electrophile as it is.

Compound 2

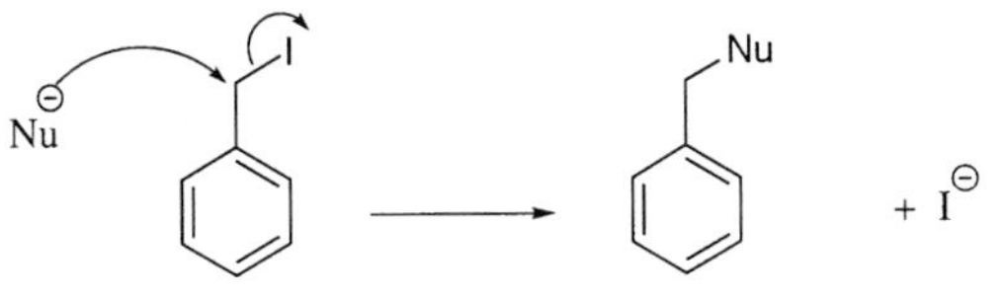

Compound **3** is a nitrile, and in addition an ester. Nitriles may be α-hydroxylated by cyto-chrome P-450 and esters may be hydrolyzed, and the latter reaction is normally much faster. Ester hydrolysis results in the generation of an α-hydroxynitrile, which spontaneously will eliminate hydrogen cyanide and give acetaldehyde. Because the capacity for ester hydrolysis is high, substantial amounts of **3** could in a short time produce toxic amounts of hydrogen cyanide.

Compound 3

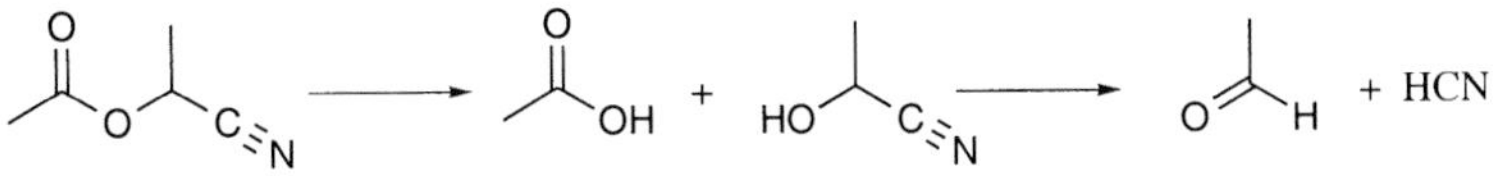

Compound **4** is a naphthalene derivative, with a dihydrofuran ring in one end and a acet-amide substituent in the other. The cyclic ether might be hydroxylated by cytochrome P-450, the saturated carbons are activated as they are both benzylic and have a heteroatom attached, but this will not lead to any metabolites that are associated with toxicity. However, the right-hand end of the molecule is identical to *N*-acetyl-2-naphthylamine, a well-known carcinogen previously discussed in Section 8.2. This is activated metabolically by *N*-hydroxylation (before or after hydrolysis of the acetamide) whereafter the *N*-hydroxyl derivative is conjugated with sulphate to form the final electrophile (see Section 8.2 for alternative routes).

Compound 4

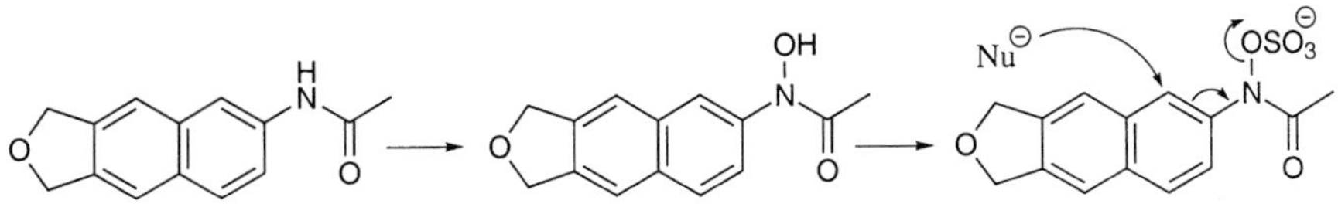

Compound **5** is phenylalanine, an essential amino acid that we need to survive. Although being endogenous or even essential does not mean that a compound is non-toxic we normally do not have to worry about the amino acids, and it is of course important to recognize their structures.

Compound **6** is another example of an activated carbon–carbon double bond, this time by a nitro group which actually is one of the most potent electron-withdrawing functional groups.

Compound 6

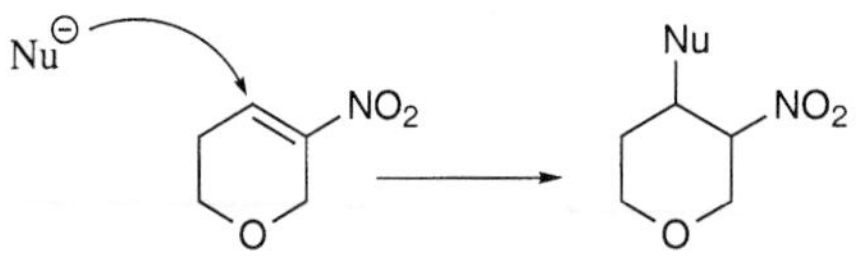

Compound **7** is again a nitrile, and at first glance the carbon–carbon double bond may appear to be conjugated with the electron-withdrawing cyanide group. However, it is not, and compound **7** is not reactive as it is. One can imagine that the carbon–carbon double bond is epoxidized or that the allylic position is hydroxylated, and the fact that the cyanide group can be regarded as a heteroatom (providing additional activation for the allylic position) suggest that the latter oxidation dominates. This would after spontaneous elimination of hydrogen cyanide from the α-hydroxynitrile formed yield reactive acrolein in addition to hydrogen cyanide.

Compound 7

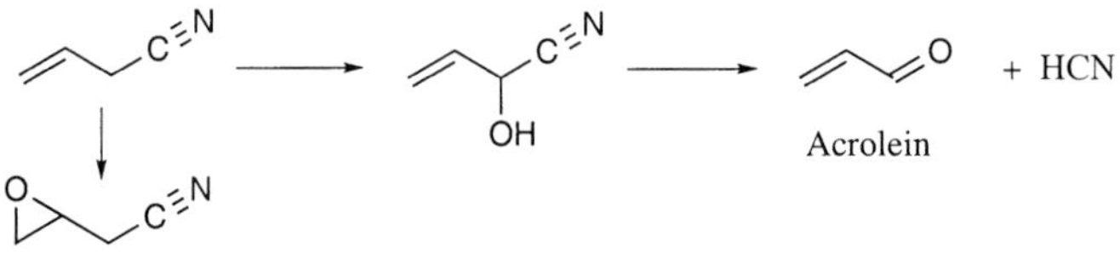

Compound **8** is an imine, and imines should always be considered to be subjected to hydrolysis as this in most cases would be expected to proceed rapidly in contact with water. Hydrolysis of **8** yields acrolein and methyl amine, and the compound is with certainty toxic.

Compound 8

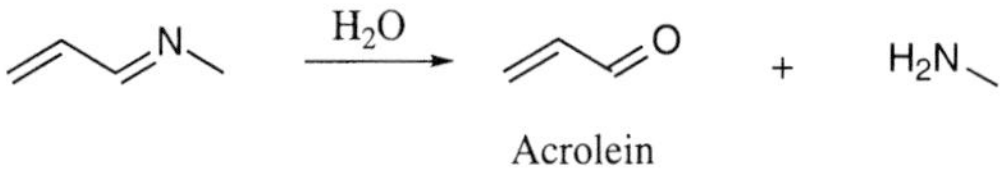

Compound **9** is another ester, and in addition to noting that one of the chlorines probably is a decent leaving group (primary chloride, α to a carbonyl group) we can try to hydrolyze the ester bond. This will produce chloroacetic acid, a compound that inhibits the citric acid cycle, and 2-chloropropanol, and the latter may be oxidized to the more reactive and toxic aldehyde by alcohol dehydrogenase.

Compound 9

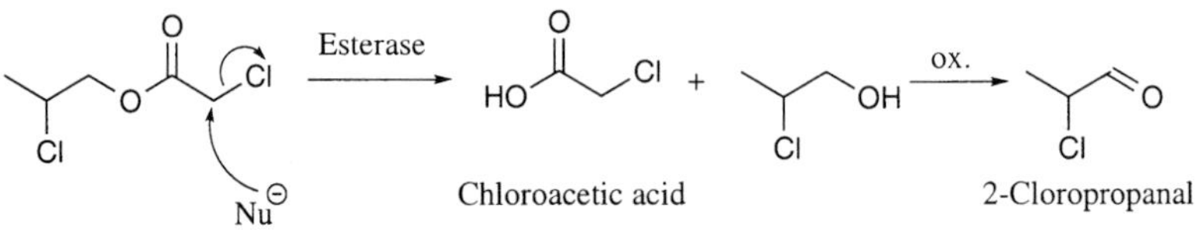

Compound **10** is symmetric, and contains 2 acetal functionalities. If they are hydrolyzed in contact with water the reactive and toxic aldehyde 2-methylpropenal is formed (besides ethanol). In addition, there is also a possibility that the carbon–carbon double bonds are epoxidized.

Compound 10

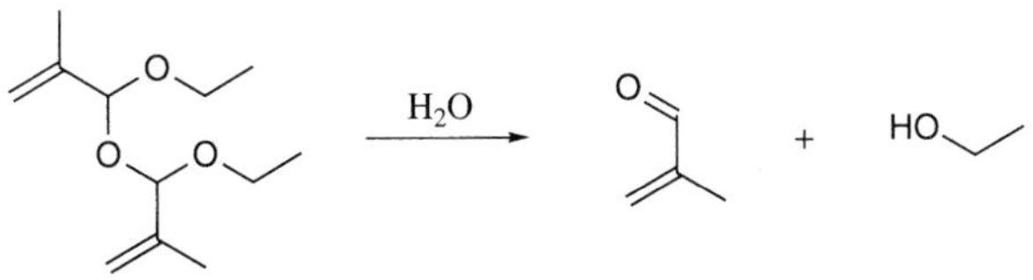

Compound **11** (2,4,6-trinitrochlorobenzene) is something as unusual as an aromatic derivative that is electrophilic. This is due to the exceptional situation with several strongly electron-withdrawing substituents in the correct positions (position 2, 4 and 6). The three nitro groups will reduce the density of electrons on carbons 1, 3 and 5, and a nucleophile can attacks and bind to carbon 1 (to which the chlorine is attached). This breaks the aromaticity of the compound, which is regained by throwing out either the nucleophile or a chloride ion.

Compound 11

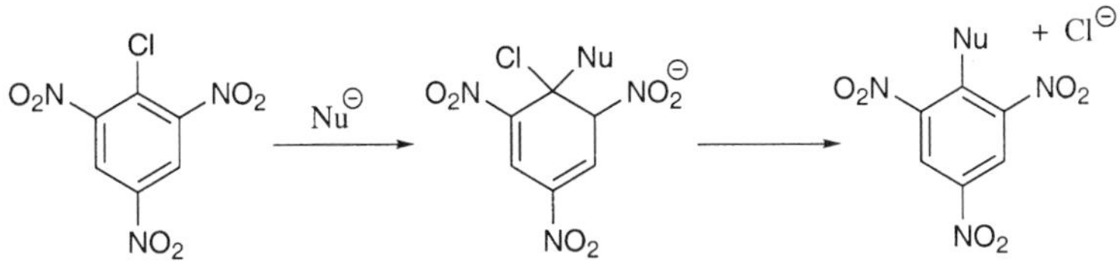

Compound **12** can be seen as an analogue of bis-chloromethyl ether (*vide supra*), the only difference is that there is an extra conjugated carbon–carbon double bond between the electron donating oxygen and the leaving group (the chlorine). Compound **12** should therefore be suspected to be extremely hazardous and treated accordingly.

Compound 12

Glossary

Abiotic	The non-living, physical parts of ecosystems.
Abortion	Expulsion of an embryo or a foetus from the uterus before it is viable.
Acetyl coenzyme A	A metabolic intermediate that transfers acetyl groups to the citric acid cycle and other metabolic pathways.
Acetyl group	Derived from acetic acid by the removal of the hydroxyl group in the carboxylic acid functionality, $CH_3(C=O)-$.
Acetylcholine	A nervous system transmitter that is inactivated by ester hydrolysis by the enzyme acetylcholine esterase.
Acid rain	Rain with a pH below 5.0, formed because the atmosphere is polluted by anthropogenic acids.
Acidosis	The condition when the pH of the blood is lower than normal.
Activation energy	The energy necessary to make a chemical reaction pass the state with highest energy (transition state) and proceed to products.
Active site	The region of an enzyme or another protein to which the substrate (ligand) binds (before it for example is converted), also called the binding site.
Active transport	A protein-assisted transport system that transports molecules across a membrane, also from a region with low concentration to a region with high concentration at the expense of energy in the form of ATP.
Acyl group	A substituent with the general formula $R(C=O)-$, derived from the corresponding carboxylic acid by removal of the hydroxyl group.
Addition reaction	Any reaction in which a molecule or parts of two molecules add to a double or triple bond.
Adipose	Fatty. The adipose tissue is composed of fat-storing cells and holds most of the body fat.
ADP	See ATP.
Aerobic	In the presence of molecular oxygen.
Affinity	The degree of attraction between the binding site of an enzyme or a receptor and a ligand.
Agonist	A ligand that binds to and stimulates a receptor.
Albumin	A protein found in relatively high concentrations in the blood and an important component during the transport of chemicals in the blood.

Alkaloid	A class of basic nitrogen-containing natural products that are produced by plants, of which many have pharmacological activity.
Alkalosis	The condition when the pH of the blood is higher than normal.
Alkane	A saturated hydrocarbon, also called an aliphatic hydrocarbon.
Alkene	A hydrocarbon containing one or more carbon–carbon double bonds.
Alkoxy group	An R–O– substituent, that is derived from an aliphatic alcohol by removal of the hydroxyl group hydrogen.
Alkyl group	A substituent that is derived from an alkane by removal of a hydrogen atom.
Alkyne	A hydrocarbon containing one or more carbon–carbon triple bonds.
Allergenic	The ability of something, e.g. a chemical, to provoke an allergic response by the immune system.
Allylic position	An sp^3 carbon attached to a carbon–carbon double bond.
Alveolus	A thin-walled air-filled bubble in the lungs where the exchange of molecular oxygen and carbon dioxide between the air and the blood takes place.
Amidases	Enzymes that hydrolyze carboxylic acid amides to the corresponding carboxylic acid and amine.
Amino acids	The building blocks of proteins. Human proteins contain 20 different amino acids, although many more are known.
Amphipathic compounds	A compound which molecules have both a polar or ionized part and a nonpolar part, that preferentially take position at surfaces and lower the surface tension. In larger amounts they may form for example lipid bilayers.
Anaerobic	In the absence of molecular oxygen.
Anemia	A decreased concentration of haemoglobin in the blood.
Anion radical	A negatively charged ion that has an unpaired electron and therefore also will react as a radical.
Antagonist	A ligand that binds to the active site of a receptor without stimulating it, but blocking if from agonists.
Anthropogenic	Something that is caused, created or changed by the influence of human activities.
Antibiotic	Literally a compound that is "instead of life", i.e. a toxic compound, but today a denomination for compounds that are used to treat bacterial infections.
Antibody	A protein synthesized by the B cells of the immune system, that specifically binds to the antigen that induced its synthesis.

Antigen	A foreign chemical structure of certain complexity (e.g. with a molecular weight exceeding 5000) that provokes a response by the immune system.
Arene	An aromatic hydrocarbon.
Aromaticity	A highly stabilizing electronic configuration obtained in rings of atoms with p orbitals, when the number of electrons is 2, 6, 10, 14, etc. Note that the number of p orbitals does not have to be the same as the number of electrons.
Artery	A larger, thick-walled vessel that carries blood away from the heart. In the tissues, it is branched into finer vessels, arterioles, that regulate the blood flow from arteries to the capillaries.
Aryl group	A substituent that is derived from an arene by removal of a hydrogen atom, often shown as Ar–.
Atmosphere	The air stratum surrounding the earth, consisting of the troposphere, the stratosphere, the mesosphere and the thermosphere.
ATP	Adenosine triphosphate, the primary repository of energy that can be used in chemical reactions in cells. Energy is released by the hydrolysis of ATP to ADP (adenosine diphosphate) and P_i (inorganic phosphate).
B cells	Lymphocytes capable of becoming antibody-secreting plasma cells.
Base pair	Two nitrogenous bases in opposite strands in double-stranded DNA that pair by hydrogen bonding.
Benign tumour	A mass of tumour cells that divide more rapidly than normal cells, but only grow to a certain size and do not invade surrounding tissues.
Benzylic position	An sp^3 carbon attached to a benzene ring.
Bile	A yellowish fluid containing bile salts, lecithin, cholesterol, haemoglobin derivatives and other end products of the metabolism, and certain metals that are being excreted.
Bimolecular reaction	A reaction in which two molecules are involved in the rate-limiting step, e.g. an S_{N^2} reaction.
Binding site	See active site.
Bioaccumulation	The term used for the mixture of bioconcentration and biomagnification in ecosystems.
Bioconcentration	The extraction by an organism of a chemical from the abiotic environment, resulting in a higher concentration inside the organism than outside.
Biodegradation	Compounds or materials that can be degraded by organisms, that either use it as a food source or co-metabolize it.

Biomacromolecules	Big biomolecules like proteins and DNA.
Biomagnification	The concentration of a chemical in a food chain.
Biomolecules	Compounds that take part in the normal biochemical re-actions that occur in the majority of cells.
Biosphere	The parts of the earth where living organisms are found.
Biotic	The part of ecosystems consisting of organisms or the products of organisms.
Biotransformation	The metabolism of an exogenous compound in an organism.
Blood–brain barrier	A barrier that controls the types of compounds, that enter the brain from the blood and their transport rates.
Botulin toxins	Extremely toxic proteins (seven have been isolated so far) with molecular weights between 200,000 and 400,000, produced by the bacterium *Clostridium botulinum* which may infect food causing botulism. The toxins are heat sensitive and are destroyed if heated to 100 °C.
Bowman's capsule	The sac in the beginning of the kidney nephron that accepts the fluid produced in the glomerulus (primary urine).
Bronchi	The parts of the airways that enter the lungs, and eventually are divided into fine tubes called bronchioles that reach the alveoli.
Capillary	The smallest blood vessel, surrounded by a single layer of endothelial cells permeable for small and nonpolar molecules.
Carboanion	An organic anion in which a carbon atom has an unshared pair of electrons and a negative charge.
Carbocation	An organic cation in which a carbon atom only is surrounded by 6 valence electrons.
Carbonate group	Derived from carbonic acid, produced by the addition of water to carbon dioxide. Organic carbonates are esters of carbonic acid.
Carbonyl group	A functionality containing a carbon and an oxygen connected by a double bond, present in for example ketones, esters and amides.
Carcinogenicity	The potency of something, e.g. a chemical, to increase the number of tumours or to make the tumours appear at an earlier stage compared to those that arise spontaneously.
Catalyst	A chemical or material that accelerates a chemical reaction, by lowering the activation energy, without undergoing any net chemical change itself.
Cell	The basic structural and functional unit into which an organism can be divided and still retains the characteristics that are associated with life.

CFCs	See freons.
Chemophobia	The abnormal fear of chemicals.
Chiral	Compounds are said to be chiral when their mirror image cannot be superimposed on the molecule itself. A chiral compound may exist as one of the two mirror images (enantiomers) or as a mixture of the two.
Chromatid	Half (one arm and one leg) of the "chromosome man" when the genetic material is condensed prior to a cell division. A chromosome consists of two identical chromatids called sister chromatids.
Chromatin	The genetic material, comprised of DNA and proteins (histone and nonhistone proteins).
Chromosome	A threadlike structure containing the genes. Prokaryotic cells contain one chromosome, a DNA molecule; eukaryotic cells contain several chromosomes (in a nucleus) which consist of DNA molecules complexed with proteins.
Chromosome aberration	A change in the number of chromosomes or the structure of a chromosome, also called chromosome abnormality or chromosome mutation.
Chyme	A solution of partly digested food in the stomach and the intestines.
Cilia	Hairlike projections from the surface of specialized epithelial cells that sweep back and forth in a way to propel material along the surface.
Citric acid cycle	A series of enzymatic conversions that convert acetic acid (as acetyl coenzyme A) to carbon dioxide and reduced cofactors, that later are used to produce ATP. Also called Krebs cycle.
CNS	Central nervous system, the brain plus the spinal cord.
Coenzyme	A small organic molecule that is associated with an enzyme and essential for its activity, by serving as a carrier molecule which transfers atoms or small groups during the enzymatic conversion, without being consumed itself.
Coenzyme A	A coenzyme essential for the enzymatic transfer of acetyl and other acyl groups.
Cofactor	A coenzyme or a metal ion that is important for the activity of an enzyme, for example by maintaining the shape of an active site, without being consumed itself.
Colloids	Particles that are so small (<1 μm) and thereby so light that they behave as solutes in a solution, and do not precipitate as heavier particles would do.

Conformation	Any three-dimensional arrangement of the atoms in a molecule that is the result of rotation about single bonds.
Conjugated addition	Nucleophilic addition to the double or triple bond in alkenes or alkynes where the unsaturation is conjugated with an electron withdrawing group (e.g. a carbonyl group), also called Michael addition.
Conjugated unsaturations	Double or triple bonds that are separated by exactly one single bond.
Covalent bond	A chemical bond formed between two atoms that share one or more electron pairs.
Cytochromes	A group of iron-containing proteins that take part in redox reactions by transferring electrons.
Cytoplasm	Cytosol plus cell organelles.
Cytosol	The water solution of a cell around the organelles.
Cytotoxic	Toxic to cells. All chemicals will in principle be cytotoxic if the dose is high enough, but the term is used to describe potent toxicity.
Deamination	The removal of an amino group from a molecule.
Dehydrohalogenation	Removal of a hydrogen and a halogen atom from adjacent positions, a type of β-elimination.
Deletion	A chromosome aberration that is characterized by the loss of a chromosome segment, also called deficiency.
Density	The specific weight, how much a certain volume of a chemical weighs at a specified temperature and air pressure. Normally given in kg/m^3, g/cm^3 or g/ml.
Deoxyhaemoglobin	Haemoglobin that is not combined with molecular oxygen.
Dielectricity constant	Indicates the ability of a compound to be polarized by an electric field that is applied over it.
Differentiation	The process when cells develope and acquire specialized structures and properties that enable them to take care of special functions.
Diffusion	The transport of a chemical from one place in a solution or a gas to another, due to random thermal molecular motion.
Dipole moment	The sum of the individual bond moments in a molecule, given in Debye units (D).
Dispersion forces	See van der Waals forces.
Disulphide group	A persulphide, with a sulphur–sulphur single bond (–S–S–).
Diuretic	A compound that causes an increase of the amount of urine excreted.

DNA	Deoxyribonucleic acid, a polymer of nucleotides in which the sugar is deoxyribose and the bases adenine, cytosine, guanine and thymine. The carrier of genetic information.
Dominant	A gene, or its corresponding trait, is dominant if it is manifest in an organism that has different genes for the same trait (a heterozygote). A mutation is called dominant if the effects immediately are observable.
Dose–response relationships	How the magnitude of a biological effect caused by an exogenous agent, e.g. a chemical, on an individual or a group depends on the dose. Often assumed to be linear, but frequently found to be non-linear.
Down's syndrome	A set of symptoms in man caused by a chromosome aberration that has added an extra copy of chromosome 21 in all cells.
Duplication	A chromosome aberration that is characterized by the presence of an additional copy of a chromosome segment, in the same chromosome or in another.
Ecology	The study of the relationships between various organisms and their interaction with the physical environment.
Ecosystem	An ecological system, a limited area consisting of organisms and the environment they live in, that is more or less self-supporting. A lake, a forest, and a greenhouse may be regarded as ecosystems, although it is important to remember that there are no closed systems on earth.
Electromagnetic radiation	Radiation of various frequency and energy, from gamma rays, X-rays, ultraviolet, visible and infrared light, to radio waves, that consists of waves with both an electric and a magnetic component, and that may travel through matter.
Electron withdrawing groups	Chemical functionalities that are conjugated with carbon–carbon double or triple bonds and by resonance withdraw electrons from the unsaturated carbon furthest away, also called EWGs.
Electrophile	Any molecule or ion that can accept a pair of electrons from a nucleophile and form a new covalent bond.
Electrophilicity	How efficient an electrophile reacts with various nucleophiles, compared to other electrophiles.
Elemental composition	Which atoms and how many of each that constitute a molecule. The elemental composition of water is H_2O, and of ethanol C_2H_5O.
β-Elimination	An elimination of two parts from adjacent atoms in a molecule, which results in an unsaturation. The reversal of an addition to a double or triple bond.

Embryo	The first stage of the development of an organism, for a human being the first 8 weeks of intrauterine life.
Emulsion	A suspension of small (approximately 1 µm in diameter) lipid droplets in a water solution.
Enantiomers	A pair of molecules that are mirror images, they have identical chemical properties except when it comes to their interaction with achiral molecules, e.g. biomolecules such as proteins and DNA.
Endocytosis	A transport mechanism by which cells absorb solid or liquid material. The plasma membrane invaginates and is pinched of to form a small intracellular vesicle which is dissolved.
Endogenous	With an origin from within the body.
Endonuclease	An enzyme that hydrolyzes internal phophodiester bonds in for example DNA.
Endoplasmic reticulum	A cell organelle consisting of a network of membrane tubules, vesicles and sacs interconnected with the nuclear envelope, where for example protein synthesis and metabolism of exogenous compounds take place.
Enterohepatic circulation	The recycling of chemicals that are excreted by the liver with the bile, by their (in most cases after conversion by the intestine bacteria) reabsorption in the intestines and transport with the blood back to the liver.
Epidemiological investigation	The comparison of the frequency of an illness, e.g. the number of cases of cancer, in an group of individuals exposed to for example a chemical at the workplace compared to an non-exposed but otherwise similar control group.
Essential chemicals	Chemicals that are required for normal and optimal body functions but are not produced in adequate amounts by the body itself, and have to be supplied in the food.
Esterases	Enzymes that hydrolyze carboxylic acid esters to the corresponding carboxylic acid and alcohol. Some are able to hydrolyze thioesters, to the carboxylic acid and thiol, and even amides, ureas, phosphoric esters and thiophosphoric esters.
Eukaryote	A cell, or an organism with cells, that has a nucleus.
Eutrophication	The enrichment of organisms with nutrients, or growth-limiting factors, something that may disturb ecological equilibria and thereby affect ecosystems.
Excretion	The transport of chemicals present inside the body to the outside, via for example the lungs (in the expired air), the liver (in the faeces) or the kidneys (in the urine).

Exocytosis	A transport mechanism by which cells can discharge material to the extracellular fluid, by fusing the membrane of intracellular vesicles with the plasma membrane.
Exogenous	With an origin from outside the body.
Exonuclease	An enzyme that hydrolyzes terminal phophodiester bonds in for example DNA.
Extracellular fluid	The water solution that surrounds the cells, comprising the interstitial fluid and the plasma and accounting for 30 % of the water in the body.
Facilitated diffusion	Passive protein-assisted transport, by proteins that recognize certain chemicals and help them to diffuse through a membrane.
FAD/FADH$_2$	Flavin adenine dinucleotide, an agent that takes part in biological oxidations and reductions. During oxidations FAD is reduced to FADH$_2$, and vice versa.
Ferric ion	Fe^{3+}.
Ferrous ion	Fe^{2+}.
Filtration	Transport of solutions driven by a pressure gradient across any type of barrier.
First-pass effect	Nutrients absorbed in the intestines, the organ intended for the uptake of all exogenous chemicals except molecular oxygen, pass the liver where they are modified before they are distributed to other organs with the blood.
Foetus	The later stage of development of an organism, after the embryo, for a human being the period between 8 weeks and birth.
Frameshift mutation	A point mutation caused by the deletion or the insertion of one or several base pairs in DNA, and that changes the reading frame for protein synthesis.
Freons	Chemically extremely stable derivatives of the smallest hydrocarbons such as methane and ethane, in which one or several of the hydrogen atoms have been exchanged for chlorine and/or fluorine. Freons lacking hydrogens are called CFCs.
Fumigant	A airborne biocide in general, but often used for pesticides which will disinfect the soil from various parasites, e.g. nematodes.
Gamete	A germ cell that has matured to a reproductive cell capable of fusing with a similar cell of the opposite sex during fertilization, also called sex cells.
Gastrointestinal tract	The mouth, oesophagus, stomach, and the small and large intestines.

Gene	A sequence of nucleotides in the genome that has a specific function, for example coding for a protein or regulating the transcription of another gene.
Genome	The genetic content of a cell.
Genotoxicity	The potency of something, e.g. a chemical, to damage the genetic material and thereby cause a toxic effect.
Germ cell	A cell line in animals from which the next generation of gametes will be derived.
Glomerular filtration	The transport of the essentially protein-free plasma from the blood to Bowman's capsule in the glomerulus.
Glomerulus	The unit where the blood is filtrated in the nephron, intimately associated with Bowman's capsule.
Glycolipid	A plasma membrane component with an attached carbohydrate, situated so that the carbohydrate is at the extracellular surface.
Glycoprotein	A plasma protein with one or several carbohydrates attached, situated so that the carbohydrate(s) are at the extracellular surface.
Glycoside	A carbohydrate derivative in which the hemiacetal hydroxyl group has been replaced by another group, the so called aglycone, attached by a heteroatom. If the carbohydrate is glucose, glycosides are called glucosides.
Glycoside bond	A covalent bond between a carbohydrate (the acetal carbon) to the heteroatom in alcohols (forming *O*-glycosides), thiols (forming *S*-glycosides) and amines (forming *N*-glycosides).
Golgi apparatus	A cell organelle that processes and secretes newly synthesized proteins in the cell.
Greenhouse effect	The warming of the atmosphere due to the reflection of infrared radiation from the earth by gases such as water vapour, carbon dioxide and CFCs. In a greenhouse, heat retention results both from the corresponding effect (the panes reflect infrared radiation) but also from the conservation of warm air.
GTP	Similar to ATP, but containing the base guanine instead of adenine.
Haem	An iron-containing group that is essential for the function of for example haemoglobin and the cytochromes.
Haemoglobin	A protein composed of four polypeptide chains with a haem group each, located in the red blood cells and responsible for transporting molecular oxygen.

Halons	Chemically stable halogenated derivatives of the smallest hydrocarbons, which may contain all four halogens but always are brominated.
Heat capacity	The amount of energy that it takes to increase the temperature of a certain amount of a chemical with 1 °C.
Heat of evaporation	The amount of energy that it takes to evaporate a certain amount of a liquid chemical to its gaseous state.
Helper T cell	A type of T cell that aids and regulates B cells so that the production of antibodies is enhanced.
Hepatotoxic	Toxic to the liver.
Herbicide	A chemical that is toxic to plants, and may have found a use as a weed-killer.
Histone proteins	Proteins with many basic amino acids that are complexed with DNA in the chromosomes of eukaryotes.
Hormone	An endogenous compound (messenger) that is synthesized and secreted into the blood by specialized cells and generally affects the function of distant cells and tissues..
Humification	The processes when the products of the degradation of dead organisms in the soil react with each other and with products of the microorganisms, to form complex polymers that are of vital importance for the organisms living in the soil.
Hybride orbital	An orbital formed by the combination of two or several atomic orbitals.
Hydrocarbon	A compound that may be saturated or unsaturated but is composed of only hydrogen and carbon atoms.
Hydrogen bond	A relatively strong dipole–dipole attractive force between hydrogens attached to strongly electronegative atoms and other strongly electronegative atoms.
Hydrophilic	Loving water, from Greek.
Hydrophobic	Fearing water, from Greek.
Hydrophobic effect	The tendency of nonpolar molecules and groups to stay together in a separate phase, and not be dissolved by water solutions.
Hydrospere	The parts of the world that contain water, oceans, seas, lakes, rivers, floods, streams and rills.
Hypersensitivity	An acquired sensitivity to an antigen that can result in serious reactions if the same antigen is encountered again.
Inducer	An effector molecule that is responsible for the induction of enzyme synthesis.
Induction	The initiation or increase of enzyme synthesis in response to an environmental stimulus.

Inductive effect	The polarization of the electron density of a covalent bond because one end has more electronegative substituents. The inductive effect will stabilize any negative charge present in the positive end of the dipole, and any positive charge present in the negative end.
Inflammation	A non-specific response of the immune system to an injury or the presence of an antigen, characterized by swelling, pain and increased blood flow in the affected region.
Intermediate	An intermediary product formed between successive reaction steps.
Interstitial fluid	The extracellular fluid surrounding the cells of a tissue, excluding the blood plasma.
Inversion (mutation)	A chromosome aberration that is characterized by the reversal of a chromosome segment.
Inversion (weather)	A weather condition characterized by lower temperature of the air that is close to the ground compared to the air at higher altitudes. The cooler air will remain stagnant, and the inversion over a city will create problems as the concentrations of air pollutants increase.
Irreversible	Proceeding essentially in one direction only.
Isomers	Compounds with the identical molecular formula but with different chemical structures and thereby different properties.
Ketene group	A carbon–carbon double bond in which one carbon forms an additional double bond, with an oxygen, and the other carbon binds two groups (e.g. hydrogen and/or alkyl groups), i.e. $R_1R_2C=C=O$.
Larynx	The part of the respiratory system that connects the pharynx and trachea and contains the vocal cords.
Leukocyte	A white blood cell.
Ligand	A chemical that by non-covalent forces will bind to a molecule, normally a protein.
Ligases	A class of enzymes that catalyze reactions that combine two molecules.
Lipophilic	Loving fat, from Greek.
Lipophobic	Fearing fat, from Greek.
Lipoprotein	A complex between fat and protein which may be rich in the former (low-density plasma proteins) or the latter (high-density plasma proteins).
Lithosphere	Stone and rock material of the earth's crust down to approximately 100 km from the face of the earth.
Lymph nodes	Small organs that are situated at the lymph vessels, where lymphocytes are formed and stored.

Lymphocyte	A cell type which mainly is responsible for the specific defence of the immune system against foreign matter, a type of leukocyte.
Lysosome	Cell organelle surrounded by a single membrane, which contains digestive enzymes that are used to break down materials (e.g. macromolecules and even bacteria) that have been absorbed by endocytosis, as well as damaged components of the cell itself.
Macrophage	A cell type that is part of the immune system and which function is to take care of foreign matter by phagocytosis, a type of leukocyte.
Malignant tumour	A mass of cancer cells, that divide rapidly and invade surrounding tissue, and if not treated normally will kill an organism.
Membrane	A structural barrier that surrounds cells and organelles, composed by phospholipids and other amphipathic compounds as well as proteins, that regulates the flow of many chemicals through the cell and provides a framework for many enzymes and receptors.
Micro-, μ	A part per million, one microlitre (μl) is 0.000001 litre and one micrometre (μm) is 0.000001 m.
Microbes	Minute organisms, for example bacteria, fungi and protozoans.
Microsomal	Associated with the smooth endoplasmic reticulum, especially of the liver cells.
Microtubules	Filaments that provide internal support for the cell, maintain and change the shape of cells, and participate in the movements of organelles and cell components.
Microvilli	Fingerlike projections which increase the absorptive surface of the epithelial cells of the small intestine and the kidney nephron.
Milli-, m	A part per thousand, one millilitre (ml) is 0.001 litre and one millimetre (mm) is 0.001 m.
Mineralized	Completely degraded to inorganic materials, e.g. carbon dioxide, nitrate, and water, leaving no organic materials behind.
Mitochondria	The cell organelle that is responsible for the production of energy in the form of ATP.
Molecular weight	The weight (in gram) of one mol ($6.02 \cdot 10^{23}$) molecules of a compound.
Monomolecular reaction	A reaction in which only one molecule is involved in the rate-limiting step, e.g. an S_{N^1} reaction.

Mutagenicity	The potency of something, e.g. a chemical, to increase the number of mutagenic effects or to make them appear at an earlier stage compared to spontaneous effects.
Mutation	A hereditary alteration in the nucleotide sequence of the DNA of a cell.
NAD/NADH	A coenzyme that facilitates the transfer of hydrogens in various reactions.
Nano-, n	A part per billion, one nanolitre (nl) is 0.000000001 litre and one nanometre (nm) is 0.000000001 m.
Nematicide	An agent that will kill nematodes, small worms (< 1 mm long) that are parasites in many economically important plants.
Nephron	The functional unit of the kidney, composed of the glomerulus, Bowman's capsule, proximal tubule, Henle's loop, distal tubule and collecting duct.
Nephrotoxic	Toxic to the kidneys.
Neuron	A nerve cell.
Neurotoxic	Toxic to the nervous system.
Neurotransmitter	A compound that is released by a nerve cell as a response to stimulation, and that will act on other excitable cells.
N-Nitrosamide group	An amide with a nitroso group connected to the amide nitrogen by a nitrogen–nitrogen bond.
N-Nitrosamine group	An amine with a nitroso group connected to the amine nitrogen by a nitrogen–nitrogen bond.
Noble gas electron configuration	The most stable electron configuration, with the s and p subshells of the valence shell filled with electrons (8 valence electrons).
Nonbonding electrons	See unshared electrons.
Nuclear envelope	A double membrane that surrounds the nucleus of a cell.
Nuclear pores	Openings in the nuclear envelope that facilitate the passing of RNA from the nucleus to the endoplasmic reticulum.
Nucleophile	A molecule or an ion that is able to donate an unshared electron pair to an electrophile in a way that a new covalent bond is formed.
Nucleophilic addition	A reaction between a nucleophile and an electrophile that in some way can accommodate the additional electron pair, for example resulting in the addition of the nucleophile to a double or triple bond in the electrophile.
Nucleophilic substitution	A reaction between a nucleophile and an electrophile that has a leaving group, that is forced to leave the electrophile together with an electron pair.

Nucleophilicity	How efficient a nucleophile reacts with various electrophiles, compared to other nucleophiles.
Nucleoside	A building block of nucleic acids, composed of a sugar (ribose or 2-deoxyribose) bonded to a purine or pyrimidine base by a β-N-glycoside bond.
Nucleotide	An ester between phosphoric acid and one of the sugar hydroxyl groups (position 3 or 5) of a nucleoside.
Nucleus	A membrane-enclosed organelle of eukaryotic cells that contains the chromosomes.
Octane number	The amount (in %) of isooctane (2,2,4-trimethylpentane) in a isooctane/heptane mixture that has the same knock properties as the petrol being evaluated.
Octet rule	The tendency of the atoms in the first rows of the periodic system to react in ways to achieve filled outer shell of eight valence electrons.
Oedema	The accumulation of excess fluid in the interstitial space.
Operator gene	A nucleotide sequence that is recognized by a specific repressor protein, which by binding to the operator inhibits the transcription of the associated genes.
Orbital	A region in space where an electron or a pair of electrons spends most (90–95 %) of its time.
Organ	A collection of tissues that form a structural unit and that collaborate to serve a common function.
Organ system	A collection of organs and tissues that collaborate to serve an overall function.
Organelle	Microscopic cellular components that are separated from the rest of the cytoplasm and perform specialized functions.
Organogenesis	The period during the gestation when the major organs of the body form.
Osmotic pressure	An important driving force for the transport of chemical between solutions of different composition that are separated by for example a membrane.
Ovum	The female gamete.
Oxidative phosphorylation	A process that takes place in the mitochondria and that makes use of the energy released by the oxidation of nutrients to produce ATP from ADP and inorganic phosphate.
Oxyhaemoglobin	Haemoglobin that is combined with molecular oxygen.
Pancreas	A gland connected by a duct to the small intestine which secretes digestive enzymes and bicarbonate into the intestines and also produces hormones (e.g. insulin).
Pedosphere	The top soil layer of the lithosphere, formed by the weathering of rocks and stone and the decay of dead organisms.

Peptide bond	A covalent bond between the amino group of one amino acid and the carboxylic group of another, with the elimination of water, to form an amide.
Peripheral nervous system	PNS, the remaining parts of the nervous system, except the CNS.
Persistent	Long-lived, a compound that is very difficult for various organisms to degrade and that withstands sunlight and other chemicals.
pH	The measure of the acidity of a water solution, at pH 7 a solution is neutral while it is acidic at lower values and basic at higher. The pH value corresponds to the negative logarithm (base 10) of the oxonium ion concentration.
Phagocytosis	A form of endocytosis when larger particles (e.g. cells or parts of cells) are absorbed by a cell by the invagination of the plasma membrane.
Pharynx	The throat, the common passage for both air and food.
Phospholipid	A lipid with a diacylglycerol backbone to which a phosphate group plus a strongly polar or ionic group is attached, an important component of cell membranes.
Phosphorylation	The addition of a phosphate group to an organic molecule.
Pi (π) bond	A covalent bond formed by the overlap of parallel p orbitals.
Pinocytosis	A form of endocytosis when liquid is absorbed by a cell by the invagination of the plasma membrane.
Placenta	The border between foetal and maternal tissues in the uterus, where the exchange of nutrients and excretion products between the two circulations takes place.
Plasma	The blood minus cells.
Plasma cell	Active B cell secreting antibodies.
Plasma membrane	The outer barrier of cells, which separates the cytoplasm from the extracellular fluid.
Polycyclic aromatic hydrocarbons	Hydrocarbons composed of three or more fused aromatic rings formed as a by-product during the combustion of organic material, also called PAHs.
Polymer	A big molecule that consists of many single units (monomers) that have been linked together.
Polymerase	An enzyme that assembles identical or similar subunits into a large molecule or a polymer, e.g. DNA.
Polypeptide	A chain of at least 20 amino acids linked by peptide bonds.
ppb	Parts per billion, for example 1 μl per m^3 or 1 mg per ton.
ppm	Parts per million, for example 1 ml per m^3 or 1 g per ton.

Pre-electrophilic	An organic compound that is converted to a significantly electrophilic compound, by the mammalian metabolism or by some other biological or chemical transformation.
Prokaryote	A cell, or an organism with cells, that lacks a nucleus.
Promoter gene	A nucleotide sequence adjacent to the operator gene at a structural gene, to which the decoding enzymes bind and initiate the transcription if the operator is not blocked by a repressor.
Protease	An enzyme that breaks the peptide bonds between amino acids in peptides.
Protein	A macromolecule consisting of one or several polypeptide chains.
Protein kinase	A class of enzymes that catalyze the addition of a phosphate group to certain amino acids of proteins.
Radical	Any chemical with one or several unpaired electrons.
Reactive oxygen species	Collective term used for the reactive species formed when molecular oxygen is reduced, superoxide, hydrogen peroxide and the hydroxyl radical.
Receptors	Proteins normally situated in the plasma membranes to which ligands bind and exert its biological effect.
Recessive	A gene, or its corresponding trait, is recessive if it only is manifest in organisms that have identical genes for the same trait (a homozygote). A mutation is called recessive if the potential effects are not directly observable.
Repressor	A protein that binds to an operator gene and prevents the transcription of the associated genes by inhibiting the decoding enzymes to bind to the promotor gene.
Resonance hybrid	The description of a compound as a composite of contributing resonance structures.
Resonance stabilization	The lowering of the energy and thereby stabilization of a molecule or an ion due to resonance.
Resonance structure	An electronically alternative but still reasonable structure that can be assumed to make a significant contribution to the resonance hybrid.
Reversible	Proceeding in either direction, permitting the establishment of an equilibrium.
Ribosome	An organelle attached to the endoplasmic reticulum or dispersed in the cytoplasm consisting of RNA and proteins, the function of which is to synthesize proteins with the amino acid sequence specified by a gene.
RNA	Ribonucleic acid, a polynucleotide in which the sugar is ribose and the bases adenine, cytosine, guanine and uracil. The transmitter of information from DNA to the proteins.

Rodenticide	Toxic to rodents.
Scavengers	Antioxidants such as carotene, vitamin C and vitamin E, that react with and inactivate radicals.
Sigma (σ) bond	A single bond, in which the shared electrons are present along the axis joining the atoms.
Solute	The component in a solution that is dissolved by the solvent.
Somatic cells	All body cells except the germ cells and the gametes.
Sorbate	The chemical that is being sorbed during adsorption and absorption.
Sorbent	The organic or inorganic surface (two dimensions) or material (three dimensions) that sorbes a sorbate.
Sorption	The adsoption of a chemical onto a surface or the absorption of it into a material.
sp Hybrid orbital	A hybrid atomic orbital that has been formed by the mixing of one *s* and one *p* orbital.
sp² Hybrid orbital	A hybrid atomic orbital that has been formed by the mixing of one *s* and two *p* orbitals.
sp³ Hybrid orbital	A hybrid atomic orbital that has been formed by the mixing of one *s* and three *p* orbitals.
Sperm	The male gamete.
Structural gene	A gene that codes for a polypeptide.
Sulphate group	From sulphuric acid, organic sulphates are esters of sulphuric acid containing a completely oxidized sulphur atom, $-O-S(=O)_2-O-$.
Sulphide group	As an ether group but with S instead of O, $-S-$.
Sulphonate group	Present in sulphonic acids, in which the sulphur atom is oxidized so that it binds two oxygen atoms with double bonds and one with a single bond, $-S(=O)_2-O-$.
Sulphone group	A sulphide group in which the sulphur atom is oxidized so that it binds two oxygen atoms with double bonds, $-S(=O)_2-$.
Sulphoxide group	A sulphide group in which the sulphur atom is oxidized so that it binds one oxygen atom with a double bond, $-S(=O)-$.
Suppressor T cell	A type of T cell that regulates B cells so that the production of antibodies is reduced.
Surface tension	The molecules at the surface of a liquid have higher energy than those in the bulk, and as all systems try to decrease their energy as much as possible a liquid will minimize its surface. This effect gives rise to the surface tension, which for example permits some insects to walk on water.

Synapse	A space into which a chemical neurotransmitter is released as a response to an action potential in a nerve cell and diffuses through in order to influence the activity of an adjacent nerve cell.
T cells	Lymphocytes capable of becoming the effectors of the specific cell-mediated response of the immune system.
Tautomers	Isomers which differ in the location of a hydrogen atom and a double bond relative to a heteroatom, in equilibrium with each other.
Teratogenicity	The potency of something, e.g. a chemical, to increase the number of adverse effects to the embryo leading to malformations or to make them appear at an earlier stage compared to spontaneous effects.
Thiol group	As a hydroxyl group but with S instead of O, –SH.
Threshold level	The minimal dose of a toxic chemical that will cause a harmful effect.
Trachea	The airway that connects the larynx with the brochi.
Transcription	The transfer of genetic information from DNA (a gene) to an RNA molecule, by enzymes called transcriptases.
Transition state	An unstable and transient species formed during the course of a chemical reaction, it has the highest energy and cannot be isolated.
Translocation	A chromosome aberration that is characterized by the change in position of a chromosome segment.
Triglyceride	An ester of glycerol (1,2,3-propantriol) and three fatty acids.
Tubular reabsorption	The transport of compounds from the kidney tubule of the nephron into the capillaries surrounding it.
Tubular secretion	The transport of compounds from the capillary blood into the tubules of the nephron.
Umbilical cord	The cord between the placenta and the foetus, containing the umbilical arteries and vein.
Unshared electron pair	Valence electrons that are not involved in covalent bonds but that may participate in the reactions of the chemical, also called nonbonding electrons.
Valence electrons	Electrons in the outermost (valence) electron shell of an atom, that may participate in bonds to other atoms.
van der Waals forces	Weak intermolecular attractive forces caused by the induction of weak dipoles in nonpolar bonds by neighbouring molecules.
Vein	A larger, thin-walled vessel that carries blood back to the heart.
Vesicle	A small intracellular body surrounded by a membrane.

Villi	Projections from the surface of the small intestine that increase its overall surface, covered by a single layer of epithelial cells.
Vitamin	Essential compound that must be present in small amounts to maintain normal health and growth.
Wavelength	The distance between two wave peaks in electromagnetic radiation, directly correlated to the energy of the radiation (radiation with shorter wavelength contains more energy and is more hazardous).
Zwitterion	An electrically neutral molecule that has both a positive and a negative charge in different parts, for example an amino acid.

Index

The following indications are used in the index: G, glossary; S, structure; T, table